Methods in Molecular Biology

Series Editor
John M. Walker
School of Life and Medical Sciences
University of Hertfordshire
Hatfield, Hertfordshire, AL10 9AB, UK

For further volumes:
http://www.springer.com/series/7651

Salamanders in Regeneration Research

Methods and Protocols

Edited by

Anoop Kumar

Institute of Structural and Molecular Biology, Division of Biosciences, University College London, London, UK

András Simon

Department of Cell and Molecular Biology, Karolinska Institute, Stockholm, Sweden

Humana Press

Editors
Anoop Kumar
Institute of Structural and Molecular Biology, Division of Biosciences
University College London
London, UK

András Simon
Department of Cell and Molecular Biology
Karolinska Institute
Stockholm, Sweden

ISSN 1064-3745 ISSN 1940-6029 (electronic)
Methods in Molecular Biology
ISBN 978-1-4939-2494-3 ISBN 978-1-4939-2495-0 (eBook)
DOI 10.1007/978-1-4939-2495-0

Library of Congress Control Number: 2015931955

Springer New York Heidelberg Dordrecht London

Printed on acid-free paper

Humana Press is a brand of Springer
Springer Science+Business Media LLC New York is part of Springer Science+Business Media (www.springer.com)

Preface

Many of the most fundamental discoveries in experimental biology, such as the embryonic organizers, neuronal specificity, nerve guidance, and units of DNA transcription, originate from salamander research. Salamanders are the only tetrapods capable of repeatedly regenerating entire limbs as adults, and they also display the widest range of regeneration capacities of other complex tissues and organs. These animals constitute unique models for understanding critical processes underlying morphological and functional restoration of lost or damaged structures in vertebrates. The present volume focuses on this particular aspect of salamander biology, which has gained new momentum during the past 10–15 years, partly due to the general interest in stem cells and regenerative medicine. A combined search on *Google scholar* using the terms "salamander" and "regeneration" shows a steady growth in the number of yearly publications with a 140 % increase between 2001 and 2013, resulting in more than 10,000 published articles during this period.

There are considerable variations among the most commonly studied salamanders in the laboratory in terms of their general physiology, life cycle, regeneration spectrum, and also mechanisms by which replacement structures form. The first part of the book outlines the best practices and conditions for maintaining the most commonly used salamander species in the laboratory. The chapters of the two following parts describe experimental manipulations in vivo and in vitro, respectively. These include methods targeting a wide variety of structures, ranging from the limb to the heart and to the brain. The two final sections deal with genetically modified organisms and tools for mining in the genomic databases. These chapters illustrate the boom of recent technical developments, which provide new platforms for understanding salamander regeneration using the most modern molecular tools. The methods chapters of this book are preceded by an inspiring essay on salamander regeneration from phylogenetic and evolutionary perspectives by Jeremy Brockes, who has greatly contributed to revitalize this research field.

Finally, we thank all the colleagues for their invaluable time and efforts to provide with all the finer details to produce this comprehensive collection of methods chapters. We hope that this collection will be useful to all, who already are devoting our activities to salamander regeneration, as well as for those who are just considering to dwell on to this intriguing problem.

London, UK *Anoop Kumar*
Stockholm, Sweden *András Simon*

Contents

Preface .... *v*
Contributors .... *ix*

PART I SALAMANDERS

1 Variation in Salamanders: An Essay on Genomes, Development, and Evolution 3
Jeremy P. Brockes

2 Maintaining Eastern Newts (*Notophthalmus viridescens*) for Regeneration Research 17
Hans-Georg Simon and Shannon Odelberg

3 Housing and Maintenance of *Ambystoma mexicanum*, the Mexican Axolotl 27
Johanna E. Farkas and James R. Monaghan

4 Husbandry of Spanish Ribbed Newts (*Pleurodeles waltl*) 47
Alberto Joven, Matthew Kirkham, and András Simon

5 Maintaining Plethodontid Salamanders in the Laboratory for Regeneration Studies 71
Claudia Marcela Arenas, Andrea Gómez-Molina, and Jean Paul Delgado

PART II EXPERIMENTAL MANIPULATION IN SALAMANDERS

6 Newt Lens Transdifferentiation: From Lentectomy to Immuno-FISH 81
Nobuyasu Maki

7 Studying Newt Brain Regeneration Following Subtype Specific Neuronal Ablation 91
Matthew Kirkham and Alberto Joven

8 The Accessory Limb Model: An Alternative Experimental System of Limb Regeneration 101
Tetsuya Endo, David M. Gardiner, Aki Makanae, and Akira Satoh

9 High-Efficiency Electroporation of the Spinal Cord in Larval Axolotl 115
Aida Rodrigo Albors and Elly M. Tanaka

10 Pseudotyped Retroviruses for Infecting Axolotl 127
Tzu-Hsing Kuo and Jessica L. Whited

11 Thyroxine-Induced Metamorphosis in the Axolotl (*Ambystoma mexicanum*) 141
Peggy S. Coots and Ashley W. Seifert

12 Generation of Aneurogenic Larvae by Parabiosis of Salamander Embryos 147
Anoop Kumar and Jean Paul Delgado

13 In Vivo Modulation and Quantification of microRNAs During Axolotl Tail Regeneration 159
Jami R. Erickson and Karen Echeverri

PART III SALAMANDER CELLS IN CULTURE

14 Derivation and Long-Term Culture of Cells from Newt Adult Limbs and Limb Blastemas 171
Patrizia Ferretti and Anoop Kumar

15 Culture and Transfection of Axolotl Cells 187
Jean-François Denis, Fadi Sader, Patrizia Ferretti, and Stéphane Roy

16 Isolation and Culture of Neurospheres from the Adult Newt Brain 197
Liyakath Ali Shahul Hameed and András Simon

17 Methods for Axolotl Blood Collection, Intravenous Injection, and Efficient Leukocyte Isolation from Peripheral Blood and the Regenerating Limb 205
Ryan J. Debuque and James W. Godwin

18 Assessing Cardiomyocyte Proliferative Capacity in the Newt Heart and Primary Culture 227
Hans-Georg Simon and Shannon Odelberg

19 Long-Term Organ Cultures of Newt Hearts 241
Tanja Piatkowski and Thomas Braun

20 In Vitro Preparation of Newt Inner Ear Sensory Epithelia as a Model for Repair and Regeneration 253
Ruth R. Taylor

PART IV TRANSGENESIS IN SALAMANDERS

21 Transgenesis in Axolotl (*Ambystoma mexicanum*) 269
Shahryar Khattak and Elly M. Tanaka

22 Generating and Identifying Axolotls with Targeted Mutations Using Cas9 RNA-Guided Nuclease 279
G. Parker Flowers and Craig M. Crews

23 Gene Manipulation for Regenerative Studies Using the Iberian Ribbed Newt, *Pleurodeles waltl* 297
Toshinori Hayashi and Takashi Takeuchi

PART V GENE EXPRESSION

24 Transcriptomics Using Axolotls 309
S. Randal Voss, Antony Athippozhy, and M. Ryan Woodcock

25 Sal-Site: Research Resources for the Mexican Axolotl 321
Nour W. Al Haj Baddar, M. Ryan Woodcock, Shivam Khatri, D. Kevin Kump, and S. Randal Voss

26 Data Mining in Newt-Omics, the Repository for Omics Data from the Newt 337
Mario Looso and Thomas Braun

Index *353*

Contributors

Antony Athippozhy, Ph.D. • *Department of Biology, University of Kentucky, Lexington, KY, USA*
Nour W. Al Haj Baddar • *Department of Biology, University of Kentucky, Lexington, KY, USA*
Liyakath Ali Shahul Hameed • *Department of Cell and Molecular Biology, Karolinska Institute, Stockholm, Sweden*
Thomas Braun, Ph.D. • *Max-Planck-Institute for Heart and Lung Research, Bad Nauheim, Germany*
Jeremy P. Brockes, Ph.D. • *Institute of Structural and Molecular Biology, Division of Biosciences, University College London, London, UK*
Claudia Marcela Arenas • *Grupo Genética, Regeneración y Cáncer, Instituto de Biología. Universidad de Antioquia, Medellín, Colombia*
Peggy S. Coots • *Department of Biology, University of Kentucky, Lexington, KY, USA*
Craig M. Crews, Ph.D. • *Department of Molecular, Cellular and Developmental Biology, Yale University, New Haven, CT, USA*
Ryan J. Debuque • *Australian Regenerative Medicine Institute (ARMI), Monash University, Clayton, VIC, Australia*
Jean Paul Delgado, Ph.D • *Grupo Genética, Regeneración y Cáncer, Instituto de Biología. Universidad de Antioquia, Medellín, Colombia*
Jean-François Denis • *Department of Biochemistry, Faculty of Medicine, Université de Montréal, Montréal, QC, Canada*
Karen Echeverri, Ph. D. • *Department of Genetics, Cell Biology and Development, Stem Cell Institute, University of Minnesota, Minneapolis, MN, USA*
Tetsuya Endo, Ph.D. • *Division of Liberal Arts and Sciences, Aichi Gakuin University, Nissin, Aichi, Japan*
Jami R. Erickson • *Department of Genetics, Cell Biology and Development, Stem Cell Institute, University of Minnesota, Minnesota, USA*
Johanna E. Farkas • *Department of Biology, Northeastern University, Boston, MA, USA*
Patrizia Ferretti, Ph.D. • *UCL Institute of Child Health, University College London, London, UK*
G. Parker Flowers, Ph.D. • *Department of Molecular, Cellular and Developmental Biology, Yale University, New Haven, CT, USA*
David M. Gardiner, Ph.D. • *Department of Developmental and Cell Biology, University of California, Irvine, CA, USA*
James W. Godwin, Ph.D. • *Australian Regenerative Medicine Institute (ARMI), Monash University, Clayton, VIC, Australia*
Andrea Gómez-Molina • *Grupo Genética, Regeneración y Cáncer, Instituto de Biología. Universidad de Antioquia, Medellín, Colombia*
Toshinori Hayashi, Ph.D. • *Department of Biomedical Sciences, School of Life Science, Faculty of Medicine, Tottori University, Tottori, Japan*
Alberto Joven, Ph.D. • *Department of Cell and Molecular Biology, Karolinska Institute, Stockholm, Sweden*

SHIVAM KHATRI • *Paul Laurence Dunbar High School, Lexington, KY, USA*
SHAHRYAR KHATTAK, PH.D. • *Technische Universität Dresden, DFG Center for Regenerative Therapies Dresden, Dresden, Germany*
MATTHEW KIRKHAM, PH.D. • *Department of Cell and Molecular Biology, Karolinska Institute, Stockholm, Sweden*
ANOOP KUMAR, PH.D. • *Institute of Structural and Molecular Biology, Division of Biosciences, University College London, London, UK*
D. KEVIN KUMP • *Department of Biology, University of Kentucky, Lexington, KY, USA*
TZU-HSING KUO • *Brigham Regenerative Medicine Center and the Department of Orthopedic Surgery, Brigham & Women's Hospital, Cambridge, MA, USA; Harvard Medical School, Harvard Stem Cell Institute, Cambridge, MA, USA*
MARIO LOOSO, PH.D. • *Max-Planck-Institute for Heart and Lung Research, Bad Nauheim, Germany*
AKI MAKANAE, PH.D. • *Research Core for Interdisciplinary Sciences (RCIS), Okayama University, kitaku, Okayama, Japan*
NOBUYASU MAKI, PH.D • *Institute of Protein Research, Osaka University, Osaka, Japan*
JAMES R. MONAGHAN, PH.D. • *Department of Biology, Northeastern University, Boston, MA, USA*
SHANNON ODELBERG, PH.D. • *Molecular Medicine Program, Eccles Institute of Human Genetics, Department of Internal Medicine, Cardiology Division, University of Utah, Salt Lake City, UT, USA*
TANJA PIATKOWSKI • *Max-Planck-Institute for Heart and Lung Research, Bad Nauheim, Germany*
AIDA RODRIGO ALBORS, PH.D. • *DFG Center for Regenerative Therapies TU Dresden (CRTD), Technische Universität Dresden, Dresden, Germany*
STÉPHANE ROY, PH.D. • *Department of Stomatology, Université de Montréal, Montréal, QC, Canada*
FADI SADER • *Department of Biochemistry, Faculty of Medicine, Université de Montréal, Montréal, QC, Canada*
AKIRA SATOH, PH.D. • *Research Core for interdisciplinary sciences (RCIS), Okayama University, Okayama, Japan*
ASHLEY W. SEIFERT, PH.D. • *Department of Biology, University of Kentucky, Lexington, KY, USA*
ANDRÁS SIMON, PH.D. • *Department of Cell and Molecular Biology, Karolinska Institute, Stockholm, Sweden*
HANS-GEORG SIMON, PH.D. • *Department of Pediatrics, Lurie Children's Hospital of Chicago Research Center, Northwestern University Feinberg School of Medicine, Chicago, IL, USA*
TAKASHI TAKEUCHI, PH.D. • *Department of Biomedical Sciences, School of Life Science, Faculty of Medicine, Tottori University, Yonago, Tottori, Japan*
ELLY M. TANAKA, PH.D. • *DFG Center for Regenerative Therapies TU Dresden (CRTD), Technische Universität Dresden, Dresden, Germany*
RUTH R. TAYLOR, PH.D. • *UCL Ear Institute, University College London, London, UK*
S. RANDAL VOSS, PH.D. • *Department of Biology, University of Kentucky, Lexington, KY, USA*
JESSICA L. WHITED, PH.D. • *Harvard Medical School, Harvard Stem Cell Institute, Cambridge, MA, USA*
M. RYAN WOODCOCK, PH.D. • *Department of Biology, University of Kentucky, Lexington, KY, USA*

Part I

Salamanders

Chapter 1

Variation in Salamanders: An Essay on Genomes, Development, and Evolution

Jeremy P. Brockes

Abstract

Regeneration is studied in a few model species of salamanders, but the ten families of salamanders show considerable variation, and this has implications for our understanding of salamander biology. The most recent classification of the families identifies the cryptobranchoidea as the basal group which diverged in the early Jurassic. Variation in the sizes of genomes is particularly obvious, and reflects a major contribution from transposable elements which is already present in the basal group. Limb development has been a focus for evodevo studies, in part because of the variable property of pre-axial dominance which distinguishes salamanders from other tetrapods. This is thought to reflect the selective pressures that operate on a free-living aquatic larva, and might also be relevant for the evolution of limb regeneration. Recent fossil evidence suggests that both pre-axial dominance and limb regeneration were present 300 million years ago in larval temnospondyl amphibians that lived in mountain lakes. A satisfying account of regeneration in salamanders may need to address all these different aspects in the future.

Key words Newt, Axolotl, Limb regeneration

1 Introduction

This collection of articles is about regeneration in salamanders and various experimental approaches for working with these animals. It is largely focused on the axolotl and some species of newt, which are the two most widely used laboratory animals. The newt and axolotl fall into two of the ten families of salamanders, and regeneration research is mainly concerned to identify common, ancestral, or unifying aspects of the underlying mechanisms. In a recent study that provided an interesting perspective, the origin of muscle-derived cells in the limb blastema was traced to satellite cells in the axolotl, but to multinucleate myofibers in the newt [1]. Salamanders present considerable variation between and within families in certain aspects of their biology, and this chapter explores some examples. I will also consider the related problems of diversity in salamander genomes and in embryonic and larval development. The origin of salamanders is discussed in relation to the fossil evidence from dif-

Anoop Kumar and András Simon (eds.), *Salamanders in Regeneration Research: Methods and Protocols*, Methods in Molecular Biology, vol. 1290, DOI 10.1007/978-1-4939-2495-0_1, © Springer Science+Business Media New York 2015

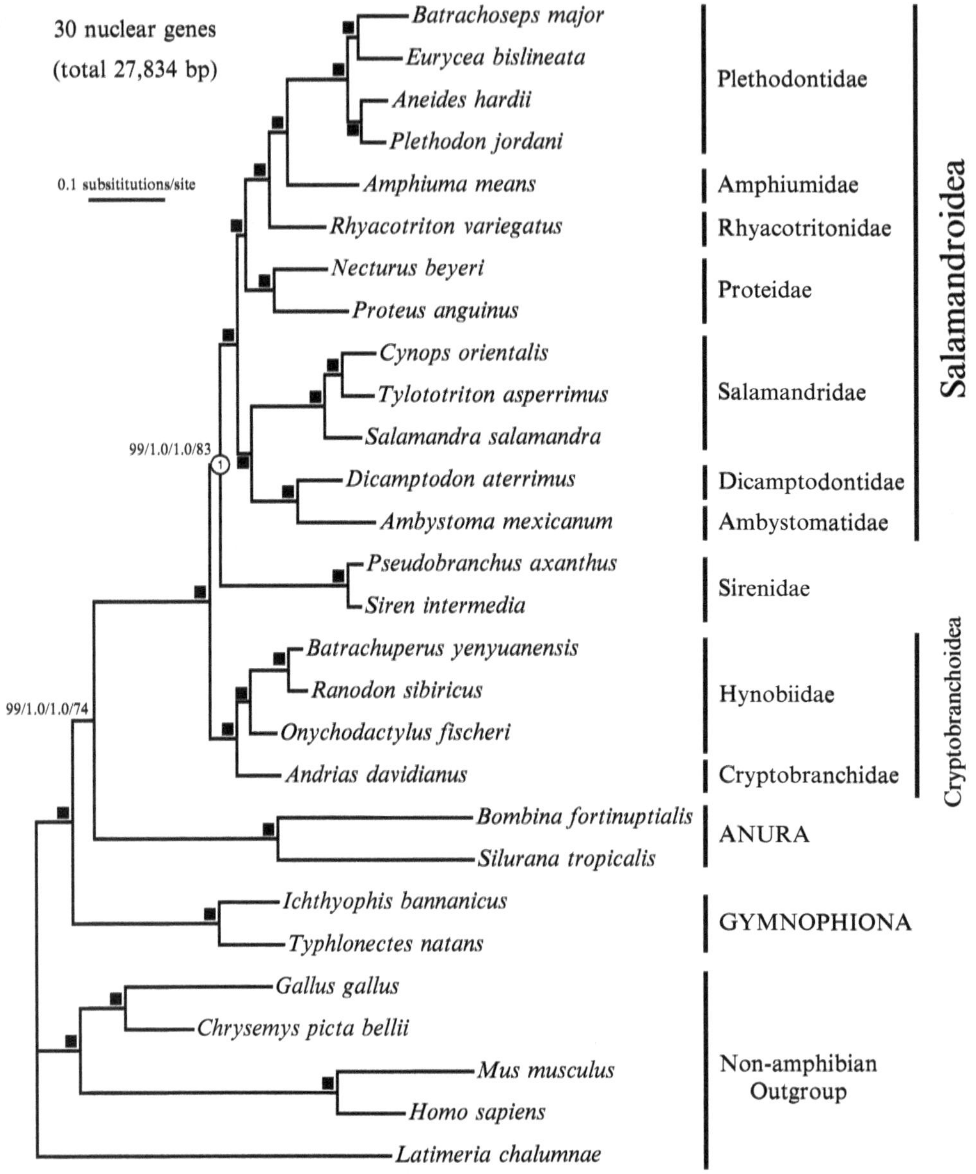

Fig. 1 This tree is taken from Shen et al. [4] and shows the tree for the 10 salamander families based on the analysis of 30 genes in the 19 species shown here. Note that the basal group comprises the hynobiid and cryptobranchid salamanders

ferent Paleozoic amphibians. I suggest that ultimately it will be necessary to understand the evolution of regenerative ability in conjunction with these different aspects of salamander biology.

The ten families have historically been grouped into different trees using a variety of approaches [2, 3], but perhaps the most compelling of these has been the recent molecular analysis by Zhang and colleagues [4]. It employed 30 nuclear protein-coding loci that were identified by PCR from 19 salamander species representing the 10 families. The resulting tree is shown in Fig. 1.

According to this analysis, the first group to diverge from the lineage of extant salamanders were the Cryptobranchoidea, composed of two families, the Hynobiidae and the Cryptobranchidae. The time at which the split occurred can be estimated from molecular dating to be approximately 190–200 million years ago (MYA) at about the beginning of the Jurassic (P. Zhang, personal communication). The two families lack internal fertilization with a spermatophore which is found in all other salamanders, with the possible exception of the Sirenidae [5]. From this and other analyses, hynobiids and cryptobranchids seem to be the most basal group of modern crown-group salamanders. The Cryptobranchidae are giant aquatic salamanders found in the eastern USA, China, and Japan. The Hynobiidae are land-living salamanders from Central and Eastern Asia.

The family Plethodontidae, sometimes referred to as the lungless salamanders, is noteworthy from the viewpoint of variation [6, 7]. There are more species of plethodontids than the sum of all the other families. They include aquatic, terrestrial, and arboreal forms and are almost all found in the New World. The variation in genome size (haploid DNA content) is about sevenfold, so they are an important resource for evaluating the consequence of differences in this property. They also exhibit a range of life histories, including larvae of various durations before metamorphosis, perennibranchiate (retaining gills throughout life) or neotenic species, and notably many direct developing species [7]. In the latter case the eggs may be laid in water or on land, and so the embryo develops outside the reproductive tract. There is no free-living aquatic larva in such species, and most adult features form in the embryo and are present at hatching. There is evidence that direct development arose on several occasions during plethodontid evolution.

2 Variation in the Salamander Genome

Some of the largest vertebrate genomes are found in salamanders, which range from approximately 14–120 Gb [8]. DNA content is positively correlated with nuclear and cell size [9], and large salamander cells and chromosomes have provided a valuable resource for studies on topics such as cytogenetics [10], transcription architecture [11, 12], and microtubule dynamics [13]. Genome size also affects aspects of organismal biology such as the rates of metabolism, differentiation, and growth, as well as the time course of embryonic development and regeneration, and even the biogeography, which the species inhabit [10, 14]. In a noteworthy study nearly 30 years ago, Sessions and Larson analyzed 27 species of plethodontid with a sixfold range in haploid DNA content [15]. These investigators observed an inverse correlation with the rate of limb regeneration, for example, a difference in the time required for complete regeneration varying from 2 to 9 months. These relationships were observed in certain plethodontid lineages and not others. In later studies, large genomes in salamanders have been correlated with changes in the neural circuitry [16] and in the circulatory system [17].

The most informative studies to date on the causes of "genomic gigantism" in salamanders have emphasized the importance of transposable elements (TE). Although at the time of writing no complete salamander genome sequences are available, low-coverage shotgun data has been generated for six plethodontid species covering a threefold range of genome size [8]. Up to 47 % of the genome (for the species *Aneides flavipunctatus*) can be attributed to TEs, but the predominant class is the long terminal repeat (LTR) retrotransposon, which is estimated to be 30 % of the *A. flavipunctatus* genome. These data raise the question of when these sequences entered the genome during salamander evolution. Recent analysis of the basal group genome of the hellbender, a cryptobranchid species, has indicated that the high level of LTR retrotransposons is likely to be characteristic of modern salamanders as a whole and to reflect persistence and diversification of ancestral TE families [18].

It is likely that much remains to be discovered about the role of TEs in salamander biology, but the study by Zhu et al. provides a valuable entry point for some of the issues in relation to limb regeneration [19]. The non-LTR long interspersed nucleotide element-1 (LINE-1) retrotransposon was found to be markedly upregulated in axolotl limb regeneration in both the mesenchymal blastema and the wound epidermis. The genomic content of LINE-1 elements was found to progressively increase by 1.5–2-fold comparing the normal limb with two successive rounds of regeneration, suggesting that productive retrotransposition is induced by limb transection. Presumably this would apply to tissue removal or injury in other regenerative contexts in the salamander. If this occurs, and the data remain fragmentary at present, it needs to be reconciled with the ability of salamanders to undergo multiple rounds of regeneration, for example, in the case of the newt lens [20], without a detectable change in outcome. The recent transcriptome analyses of regenerating newt heart [21] and axolotl limb [22] have found that many different TE and endogenous retrovirus sequences are markedly upregulated in regeneration. What might be the significance of these findings?

A recent analysis of mammalian development suggests that regulation by endogenous retroelements plays a critical role at early stages [23]. In the totipotent cells of the 2-cell (2C) stage mouse embryo and in rare cells that arise in ES cell populations in culture, retroviral GAG protein is expressed as well as many transcripts with junctions to endogenous retroelements. This promoter activity of retroelement LTRs is turned off by the transition to pluripotency after the 2C stage in mouse embryogenesis or by conversion to ES cells in culture. In this and other contexts [24, 25], transposon sequences play a key role in allowing coordinate expression and subsequent repression of many genes. It is possible to speculate that expansion of the genome by assimilation of TEs early in the salamander lineage was critical in the evolution of an extended repertoire of regeneration, by permitting the coordinated regulation of many genes after injury. A related suggestion has been made in relation to the interspersed repetitive sequences present in the newt *Hox C* and *Hox D* gene clusters. These sequences include a high level of newt-specific repeats as well as various kinds of TE. The authors point out that these elements could contribute to the regulation of postembryonic expression in the context of tissue regeneration. The generation of pseudogenes and gene duplications are other activities attributed to TEs, and they may be relevant for the evolution of regeneration [25].

3 Variation in Limb Development

Limb development in salamanders poses an intriguing set of problems. On the one hand there are aspects of limb development that can be recognized in all salamanders, but are not present in other tetrapods (anuran amphibians and amniotes). Although these are not understood, they challenge the generality of some of our ideas about limb development in amniotes, and I will give one current example. On the other hand many salamander species have fully functional limbs for much of the larval period, in contrast to anurans where the limbs emerge shortly before metamorphosis, or amniotes where limb development occurs in the egg [26]. The selective pressures in different larval ecologies lead to marked variations in the timing of limb and digit formation which have been of great interest for understanding the evolution of development [27]. The connections between limb regeneration and these problems of limb development in salamanders have not been apparent in the past, but this may be changing as outlined here.

Figure 2 is modified from a recent review of these issues [27] and illustrates the difference between the property referred to as preaxial dominance in salamanders and its counterpart in other tetrapods. The anterior, or preaxial, digits 2 and 1 are the first to develop, with the radius before the ulna; the same occurs with the corresponding elements in the hind limb. This difference in timing is maintained in relation to subsequent ossification. In other tetrapods the posterior, or postaxial, digits are the first, with the ulna before the radius. Another unique feature in salamanders is the fusion of the carpal bones at the base of digits 1 and 2 to form the *basale commune*. The differences in timing were recognized at least since the early twentieth century and were even used at one point to propose two separate origins of the tetrapod limb [28]. This is no longer considered seriously as an option, and preaxial dominance is

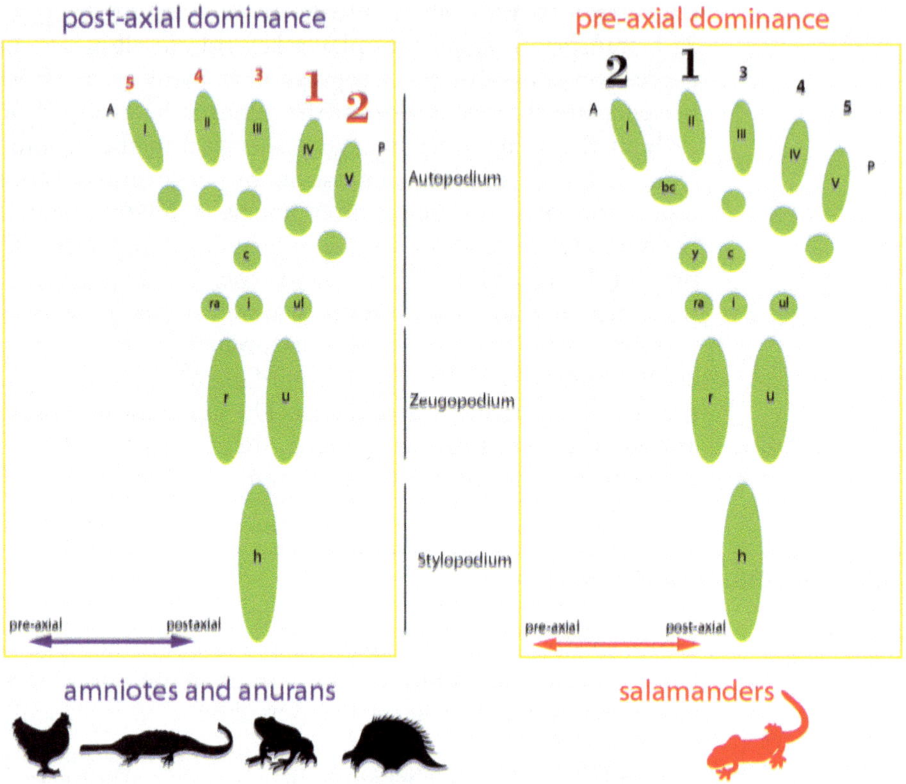

Fig. 2 Diagram illustrating the difference between postaxial and preaxial dominance in outgrowth of the digits. The large 2 and 1 numerals illustrate the first digits to extend in the case of salamanders or the other tetrapods. Modified from a similar diagram in Frobisch and Shubin [27]

regarded as a salamander innovation or apomorphy [27]. The analysis of *Hox A* and *D* complex gene expression in axolotl limb buds has not revealed any differences from other tetrapods that appear to explain the mechanism of preaxial dominance [29].

The significance of preaxial dominance can be appreciated in terms of the selective pressures on aquatic larvae (or embryos for direct developing species) in different locales. Most salamanders have extended larval periods, which can last for more than a year in some species. In species with pond-type larvae, limb development proceeds slowly after hatching while the larvae move around—the limbs are moved as they are developing. These larvae use the preaxial part of the autopod for locomotion and support. Such larvae often have transient balancer organs at hatching to hold the animal in place during the initial phase of limb elongation. The larvae of *Salamandrella keyserlingii* in the Urals have a long fin-like mesenchymal membrane between digits 2 and 1 to assist movement and floating during active feeding [30]. In stream-adapted larvae, for example, the Pacific giant salamander, limb development is advanced at hatching and the forelimb elements are completely differentiated [31]. In plethodontids with direct development, there is no larva, and the limbs develop within the capsule [32]. Nonetheless all of these cases show the extension of the limb axis through digit 2, not digit 4. This is particularly marked for the pond and stream-dwelling larvae, but it has also been observed for the plethodontid case [32]. This is noteworthy because the latter appears in other respects to be more reminiscent of amniote limb development. The limb bud extends in a paddle configuration and the digits form in a concerted fashion, in contrast to the iterative episodes seen in digit development in the larval forms [32]. The variations in limb development illustrate the operation of strong selective pressures, particularly on the free-living aquatic larva, which is considered to be the ancestral life history for salamanders [26, 33]. It is possible to suggest that these selective pressures might operate for regeneration of the limb in a context where predation and density-dependent biting behavior have been observed [34, 35].

Although the *basale commune* is specific to salamanders, it is an example of the reduction of bones in the limb, a widespread feature of tetrapod evolution [36]. In some salamander families—the amphiumids, proteids, and sirenids—the limbs may be reduced so as to have only two digits. For example, the cave-dwelling salamander *Proteus* has two digits in the foot and three in the hand. In these highly reduced salamander limbs, the remaining digits are always the preaxial 1 and 2. It is interesting that in this context the "missing" digits do not form in development, whereas a recent analysis of digit loss in crocodiles and birds has provided examples where digit remnants are clearly observed [37].

The nature of digit development in salamanders is of particular interest at present, when the currently favored model for digit formation in amniotes posits a Turing-type reaction-diffusion network in the limb bud, leading to the concerted formation of the digits [38]. Yet it is well documented that larval salamanders may show extension of digits 2/1, followed by a gap of several days, then formation of digit 3, another gap, and formation of digit 4 [32]. It seems somewhat unlikely that this could reflect the operation of a reaction-diffusion network as proposed for mice. It will be interesting to see if the salamander is just a special case, even a distinct mechanism for generating the digits, or whether the reaction-diffusion network remains the most supported possibility for amniotes.

4 Origin of Salamanders and Salamander Phenotypes in Deep Time

The origin of salamanders in tetrapod evolution poses many difficult problems, yet the current fossil evidence has some fascinating aspects, including some that are of direct relevance here. Fossils of salamanders are known from the Jurassic, but they resemble extant

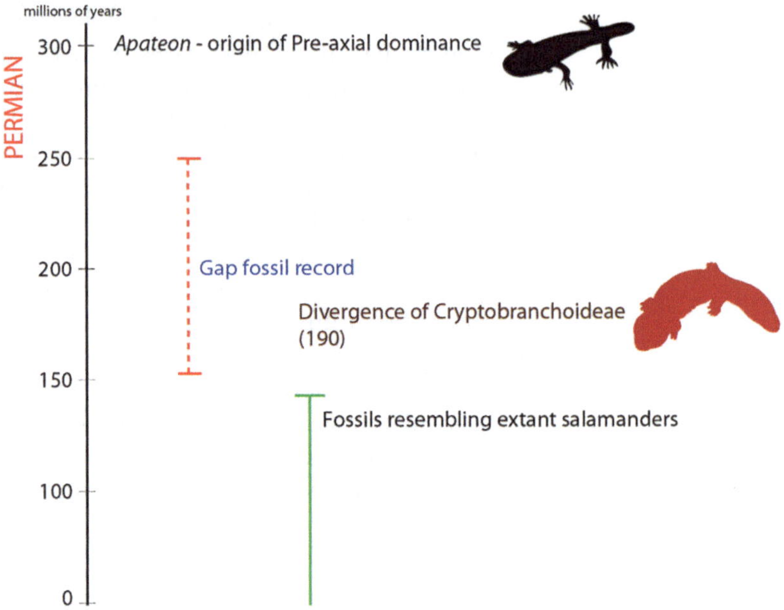

Fig. 3 Schematic outline of salamander evolution and paleontology. Note that although *Apateon* is denoted as being at the origin of preaxial dominance, there may be as yet undiscovered earlier species with this property

salamanders and so do not provide much evidence about origins [5, 39]. One example is the cryptobranchid *Chunerpeton*, a well-preserved fossil from the Middle Jurassic (165 MYA) of China [40]. The life histories of metamorphosis and paedomorphosis (neoteny) have both been identified in Jurassic fossils, but during the preceding Triassic era there is little or no evidence about earlier forms [39]. This takes us back to the Permian era (290–245 MYA) at the end of the Paleozoic (Fig. 3). There is significant uncertainty about how salamanders diverged from anurans and caecilians. The hypotheses vary from proposing a common ancestry for all three groups, to separate origins for each from distinct lineages of Paleozoic amphibians [27, 39, 41, 42]. I will focus on the Temnospondyls, the most anatomically diverse and most specious clade of late Paleozoic amphibians. Although it is possible that preaxial dominance and limb regeneration were found in earlier tetrapods, this group is the oldest for which we have evidence at present.

The Dissorophoidea are a prominent group of Temnospondyl species which include *Apateon*, *Gerobatrachus*, and *Micromelerpeton*. *Apateon* was a prominent neotenic species in the lakes of the Variscan mountain belt in the early Permian (~300 MYA), and some locations in Germany show exceptional preservation of fossils due in part to anoxic conditions at the bottom of the lakes. This is a rare example in paleontology where evidence of a developmental process in deep time, that is preaxial dominance in the limb, is

Fig. 4 Fossil of the early Permian from Rhineland-Pfalz showing a single large *Micromelerpeton* at the right and three smaller examples of *Apateon* at the left (author's collection). Scale bar, 5 cm

directly available by inspection of more than a hundred examples of staged fossil larvae from a single location [43]. In these specimens digit 2 leads in the digital sequence of ossification, while the preaxial zeugopodial elements are advanced relative to the postaxial ones. The presence of this salamander apomorphy in *Apateon* provides strong evidence for a relationship between salamanders and dissorophids [5, 27, 43]. The analysis of cell volumes in Paleozoic tetrapods has just started, and it is interesting that dissorophids appear to have a relatively large volume that is comparable to salamanders. This could be consistent with an expansion in genome size in these animals, but it seems that few samples have been analyzed to date [41].

Micromelerpeton is another dissorophid which is known from well-preserved specimens in the early Permian (~300 MYA) lake deposits. Figure 4 shows a Permian fossil from the Rhineland, with one *Micromelerpeton* and three *Apateon* preserved in association. It was apparently a predator of *Apateon* judging from analysis of gut contents. Several examples of *Micromelerpeton* display limb abnormalities considered characteristic of regeneration in modern salamanders [44]. For example, the fusion of digits along the proximodistal axis is observed, as well as the formation of narrow additional digits. These are distinct from the abnormalities observed in limb development in salamanders. This evidence for limb regeneration in the fossil record is not as direct as that for preaxial dominance [27], but nonetheless it constitutes prima facie evidence that *Micromelerpeton* was capable of limb regeneration.

Gerobatrachus is a single fossil from Texas dating to the early Permian and described originally in 2008 [45]. It is classified as a dissorophid but its precise relationship to urodeles and anurans is a matter of active debate [41]. The leg skeleton possesses a definitive salamander character that is the presence of a *basale commune*. In summary there is strong evidence for preaxial dominance and significant evidence for limb regeneration and the *basale commune* in the possible Paleozoic ancestors of salamanders. It is important to note that this is about 100 MY before the divergence of cryptobranchoids and emergence of crown-group salamanders. The lake system that may have provided the habitat for evolution of salamander-like features in dissorophids, as well as being important for fossil preservation, did not persist beyond the early Permian [41]. The changes that occurred before the appearance of lower or middle Jurassic cryptobranchoids were in part concerned with the challenges of terrestrial life, such as feeding or hearing [39]. They include the development of a tongue, which is apparently not a regenerative structure in modern salamanders, in contrast to the jaws [46].

5 Conclusions

Our ability to deliver meaningful insights into the possible extension of mammalian regeneration will be enhanced by a more detailed understanding of the evolution of regeneration and the factors underlying the extensive repertoire in salamanders. Limb regeneration is a property that distinguishes salamanders from other adult tetrapods, that is, the anuran amphibians as well as amniotes. On the other hand, preaxial dominance is a property of limb development that also distinguishes salamanders from other adult tetrapods. There is a significant consensus among evolutionary biologists that preaxial dominance evolved in the salamander lineage, possibly in response to the selective pressures operating on free-living aquatic larvae [27]. It is plausible, as suggested above, that the same or related pressures may extend to regeneration of the limb. Limb regeneration could be a purely ancestral property for tetrapods that was lost in adult anurans and amniotes, and this remains the most popular hypothesis. Alternatively, it may be that certain salamander-specific novelties were required in addition to an ancestral component in order to confer regeneration on the limb [47, 48]. Both viewpoints would be consistent with the operation of strong selection pressures for limb function in salamander larvae. Recent contributions to the debate on these issues have come from pointing to fin regeneration in *Polypterus* in favor of an ancestral component [49], while a variety of molecular analyses have provided microevolutionary evidence for salamander novelties embedded in the mechanism of salamander regeneration [21, 47, 50]. It is possible that the origins of preaxial dominance and limb regeneration are connected: they share a common target tissue and a similar rationale in terms of selective pressure, and they are both detectable in certain late Paleozoic fossil aquatic dissorophids at a time before the origin of definitive salamanders. The theme of this essay has been that it is interesting to consider connections between these different aspects of salamander biology, and we can expect in the future to include microevolutionary and genomic issues in the mix, as outlined earlier.

Acknowledgments

I thank Peng Zhang for his help in relation to salamander phylogeny and Anoop Kumar for help with the figures.

References

1. Sandoval-Guzman T, Wang H, Khattak S, Schuez M, Roensch K, Nacu E, Tazaki A, Joven A, Tanaka EM, Simon A (2014) Fundamental differences in dedifferentiation and stem cell recruitment during skeletal muscle regeneration in two salamander species. Cell Stem Cell 14:174–187
2. Larson A, Dimmick WW (1993) Phylogenetic relationships of the salamander families: an analysis of congruence among morphological and molecular characters. Herpetol Monogr 7:77–93
3. Duellman WE, Trueb L (1994) Biology of amphibians. Johns Hopkins University Press, Baltimore, MD
4. Shen XX, Liang D, Feng YJ, Chen MY, Zhang P (2013) A versatile and highly efficient toolkit including 102 nuclear markers for vertebrate phylogenomics, tested by resolving the higher level relationships of the caudata. Mol Biol Evol 30:2235–2248
5. Carroll R (2009) The rise of amphibians. The Johns Hopkins University Press, Baltimore, MD
6. Wake DB, Marks SB (1993) Development and evolution of Plethodontid salamanders: a review of prior studies and a prospectus for future research. Herpetologica 49:194–203
7. Wake DB, Hanken J (1996) Direct development in the lungless salamanders: what are the consequences for developmental biology, evolution and phylogenesis? Int J Dev Biol 40:859–869
8. Sun C, Shepard DB, Chong RA, Lopez Arriaza J, Hall K, Castoe TA, Feschotte C, Pollock DD, Mueller RL (2012) LTR retrotransposons contribute to genomic gigantism in plethodontid salamanders. Genome Biol Evol 4:168–183
9. Olmo E, Morescalchi A (1979) Evolution of the genome and cell sizes in salamanders. Experientia 31:804–806
10. Sessions SK (2008) Evolutionary cytogenetics in salamanders. Chromosome Res 16:183–201
11. Miller OLJ, Beatty BR (1969) Visualization of nucleolar genes. Science 164:955–957
12. Kaufmann R, Cremer C, Gall JG (2012) Superresolution imaging of transcription units on newt lampbrush chromosomes. Chromosome Res 20:1009–1015

13. Hayden JH, Bowser SS, Rieder CL (1990) Kinetochores capture astral microtubules during chromosome attachment to the mitotic spindle: direct visualization in live newt lung cells. J Cell Biol 111:1039–1045
14. Litvinchuk SN, Rosanov JM, Borkin LJ (2007) Correlations of geographic distribution and temperature of embryonic development with the nuclear DNA content in the Salamandridae (Urodela, Amphibia). Genome 50:333–342
15. Sessions SK, Larson A (1987) Developmental correlates of genome size in Plethodontid salamanders and their implications for genome evolution. Evolution 41:1239–1251
16. Roth G, Nishikawa KC, Wake DB (1997) Genome size, secondary simplification, and the evolution of the brain in salamanders. Brain Behav Evolut 50:50–59
17. Mueller RL, Gregory TR, Gregory SM, Hsieh A, Boore JL (2008) Genome size, cell size, and the evolution of enucleated erythrocytes in attenuate salamanders. Zoology 111:218–230
18. Sun C, Mueller RL (2014) Hellbender genome sequences shed light on genomic expansion at the base of crown salamanders. Genome Biol Evol 6:1818–1829
19. Zhu W, Kuo D, Nathanson J, Satoh A, Pao GM, Yeo GW, Bryant SV, Voss SR, Gardiner DM, Hunter T (2012) Retrotransposon long interspersed nucleotide element-1 (LINE-1) is activated during salamander limb regeneration. Dev Growth Differ 54:673–685
20. Eguchi G, Eguchi Y, Nakamura K, Yadav MC, Millan JL, Tsonis PA (2011) Regenerative capacity in newts is not altered by repeated regeneration and ageing. Nat Commun 2:384. doi:10.1038/ncomms1389
21. Looso M, Preussner J, Sousounis K, Bruckskotten M, Michel CS, Lignelli E, Reinhardt R, Hoffner S, Kruger M, Tsonis PA, Borchardt T, Braun T (2013) A de novo assembly of the newt transcriptome combined with proteomic validation identifies new protein families expressed during tissue regeneration. Genome Biol 14(2):R16. doi:10.1186/gb-2013-14-2-r16
22. Stewart R, Rascon CA, Tian S, Nie J, Barry C, Chu LF, Ardalani H, Wagner RJ, Probasco MD, Bolin JM, Leng N, Sengupta S, Volkmer M, Habermann B, Tanaka EM, Thomson JA, Dewey CN (2013) Comparative RNA-seq analysis in the unsequenced axolotl: the oncogene burst highlights early gene expression in the blastema. PLoS Comput Biol 9(3):e1002936. doi:10.1371/journal.pcbi.1002936
23. Macfarlan TS, Gifford WD, Driscoll S, Lettieri K, Rowe HM, Bonanomi D, Firth A, Singer O, Trono D, Pfaff SL (2012) Embryonic stem cell potency fluctuates with endogenous retrovirus activity. Nature 487:57–63
24. Macia A, Blanco-Jimenez E, Garcia-Perez JL (2014) Retrotransposons in pluripotent cells: impact and new roles in cellular plasticity. Biochim Biophys Acta doi: 10.1016/j.bbagrm.2014.07.007
25. Muotri AR, Marchetto MC, Coufal NG, Gage FH (2007) The necessary junk: new functions for transposable elements. Hum Mol Genet 16:R159–R167
26. Wells KD (2007) The ecology and behavior of amphibians. University of Chicago Press, Chicago, IL
27. Frobisch NB, Shubin NH (2011) Salamander limb development: integrating genes, morphology, and fossils. Dev Dyn 240:1087–1099
28. Holmgren N (1933) On the origin of the tetrapod limb. Acta Zool 14:187–248
29. Torok MA, Gardiner DM, Shubin NH, Bryant SV (1998) Expression of HoxD genes in developing and regenerating axolotl limbs. Dev Biol 200:225–233
30. Vorobyeva EI, Hinchliffe JR (1996) Developmental pattern and morphology of Salamandrella keyserlingii limbs (Amphibia, Hynobiidae) including some evolutionary aspects. Russ J Herpetol 3:68–81
31. Wake DB, Shubin NH (1998) Limb development in the Pacific giant salamanders, Dicamptodon (Amphibia, Caudata, Dicamptodontidae). Can J Zool 76: 2058–2066
32. Franssen RA, Marks S, Wake D, Shubin N (2005) Limb chondrogenesis of the seepage salamander, Desmognathus aeneus (amphibia: plethodontidae). J Morphol 265:87–101
33. Shubin NH, Wake DB (2003) Morphological variation, development, and evolution of the limb skeleton of salamanders. In: Heatwole H, Davies M (eds) Amphibian biology, vol 5. Surrey Beatty and Sons, Chipping Norton, NSW, pp 1782–1808
34. Semlitsch RD, Reichling SB (1989) Density-dependent injury in larval salamanders. Oecologia 81:100–103
35. Wildy EL, Chivers DP, Kiesecker JM, Blaustein AR (2001) The effects of food level and conspecific density on biting and cannibalism in larval long-toed salamanders, Ambystoma macrodactylum. Oecologia 128:202–209
36. Shubin NH (2002) Origin of evolutionary novelty: examples from limbs. J Morphol 252: 15–28
37. de Bakker MA, Fowler DA, den Oude K, Dondorp EM, Navas MC, Horbanczuk JO,

Sire JY, Szczerbinska D, Richardson MK (2013) Digit loss in archosaur evolution and the interplay between selection and constraints. Nature 500:445–448

38. Raspopovic J, Marcon L, Russo L, Sharpe J (2014) Modeling digits. Digit patterning is controlled by a Bmp-Sox9-Wnt Turing network modulated by morphogen gradients. Science 345:566–570
39. Carroll RL (2007) The palaeozoic ancestry of salamanders, frogs and caecilians. Zool J Linn Soc 150:1–140
40. Gao K-Q, Shubin NH (2003) Earliest known crown-group salamanders. Nature 422: 424–428
41. Schoch RR (2014) Amphibian evolution: the life of early land vertebrates. Wiley-Blackwell, Chichester
42. Hanken J (1986) Developmental evidence for amphibian origins. Evolut Biol 20:389–417
43. Frobisch NB, Carroll RL, Schoch RR (2007) Limb ossification in the Paleozoic branchiosaurid Apateon (Temnospondyli) and the early evolution of preaxial dominance in tetrapod limb development. Evol Dev 9:69–75
44. Frobisch NB, Bickelmann C, Witzmann F (2014) Early evolution of limb regeneration in tetrapods: evidence from a 300-million-year-old amphibian. Proc R Soc B 281:20141550. doi:10.1098/rspb.2014.1550
45. Anderson JS, Reisz RR, Scott D, Frobisch NB, Sumida SS (2008) A stem batrachian from the Early Permian of Texas and the origin of frogs and salamanders. Nature 453:515–518
46. Ghosh S, Thorogood P, Ferretti P (1994) Regenerative capability of upper and lower jaws in the newt. Int J Dev Biol 38:479–490
47. Garza-Garcia AA, Driscoll PC, Brockes JP (2010) Evidence for the local evolution of mechanisms underlying limb regeneration in salamanders. Integr Comp Biol 50:528–535
48. Brockes JP, Gates PB (2014) Mechanisms underlying vertebrate limb regeneration: lessons from the salamander. Biochem Soc Trans 42:625–630
49. Cuervo R, Hernandez-Martinez R, Chimal-Monroy J, Merchant-Larios H, Covarrubias L (2012) Full regeneration of the tribasal Polypterus fin. Proc Natl Acad Sci U S A 109: 3838–3843
50. Looso M, Michel CS, Konzer A, Bruckskotten M, Borchardt T, Kruger M, Braun T (2012) Spiked-in pulsed in vivo labeling identifies a new member of the CCN family in regenerating newt hearts. J Proteome Res 11: 4693–4704

Chapter 2

Maintaining Eastern Newts (*Notophthalmus viridescens*) for Regeneration Research

Hans-Georg Simon and Shannon Odelberg

Abstract

The adult Eastern newt, *Notophthalmus viridescens*, has long served as a model for appendage as well as heart muscle regeneration studies. Newt tissues include all major cell types known in other vertebrates and mammals, including bone, cartilage, tendon, muscle, nerves, dermis, and epidermis. Therefore, these aquatic salamanders make an excellent model for studying the regeneration of complex tissues. Regeneration of adult tissues requires the integration of new tissues with preexisting tissues to form a functioning unit through a process that is not yet well understood. Scale is also an issue, because the regenerating tissues or structures are magnitudes larger than their embryonic counterparts during development, and therefore, it is likely that different physics and mechanics apply. Regardless, regeneration recapitulates to some degree developmental processes. In this chapter, we will describe basic methods for maintaining adult Eastern newts in the laboratory for the study of regeneration. To determine similarities and differences between development and regeneration at the cellular and molecular level, there is also a need for embryonic newt tissue. We therefore also outline a relatively simple way to produce and raise newt embryos in the laboratory.

Key words Eastern newt, Red-spotted newt, *Notophthalmus viridescens*, Embryo, Larva, Red eft, Breeding, Spawning, Regeneration

1 Introduction

1.1 Eastern Newts

Eastern newts are urodele amphibians, commonly called salamanders [1]. A native of eastern North America, Eastern newts belong to the genus *Notophthalmus* of which there are three known species: *N. viridescens* (Eastern newt), *N. meridionalis* (black-spotted newt), and *N. perstriatus* (striped newt). *Notophthalmus* is one of only two genera of newts native to the United States, the other genus being *Taricha*, which inhabits primarily the coastal regions of western North America. Taxonomists currently classify Eastern newts into four subspecies—*N. v. viridescens* (red-spotted newt), *N. v. dorsalis* (broken-striped newt), *N. v. louisianensis* (central newt), and *N. v. piaropicola* (peninsula newt). Red-spotted newts are endemic to the northeastern region of the United States but

Anoop Kumar and András Simon (eds.), *Salamanders in Regeneration Research: Methods and Protocols*, Methods in Molecular Biology, vol. 1290, DOI 10.1007/978-1-4939-2495-0_2, © Springer Science+Business Media New York 2015

range as far north as the Canadian provinces Ontario and Quebec; as far south as Alabama, Georgia, and South Carolina; and as far west as Michigan, Indiana, Kentucky, and Tennessee. Central newts range as far west as Texas and Oklahoma and south to the Gulf of Mexico. Broken-striped newts are mostly restricted to the coastal regions of North and South Carolina, and peninsula newts are found in the Florida peninsula.

1.2 Life Cycle

Eastern newts have two distinct features that set them apart from most other vertebrate species—their remarkable regenerative abilities and their unusual and complex life cycle, which can be divided into four major stages, including the embryo, larval, red eft, and adult stages (Fig. 1). Newts have two fascinating courtship and breeding behaviors. Courtship may involve a stereotypical behavior known as "hula" during which the male undulates his body and tail in an effort to entice the female to nudge his tail. After receiving this stimulus, the male deposits a spermatophore. The female then uses her cloaca to pick up the sperm and stores them in a special cavity known as the spermatheca to be used later for fertilizing her eggs. Alternatively, a more common courtship behavior involves amplexus, in which the male grasps the female's trunk with his large hind limbs and nuzzles her with his snout. Amplexus can go on for several hours before the male dismounts and deposits a spermatophore in front of the female. A female newt fertilizes

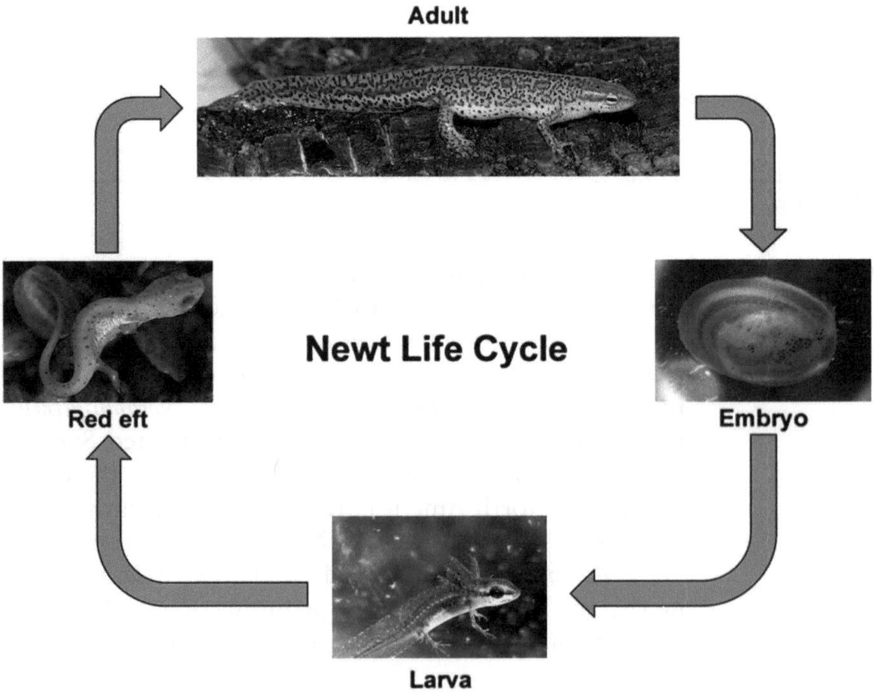

Fig. 1 Newt life cycle

each egg with stored sperm just before depositing the egg on an underwater leaf. She then gently wraps the egg with the leaf to form a protective shield. The fertilized egg develops into an embryo that hatches in about 20–35 days. Over a period of 2–5 months, the gilled larvae develop first forelimbs followed by the hind limbs. When the larvae reach a length of about 35–38 mm, they metamorphose into land-dwelling red efts. These red efts typically grow over a 3-year period before developing into mature adults that are mostly aquatic (Fig. 1). In the wild, newts can live for 12–15 years, which is remarkable longevity for a vertebrate that only weighs 2–3 g. More details concerning the life cycle of the newt have been described elsewhere [2].

1.3 Regenerative Abilities

Although the ancient Greeks knew about regeneration in vertebrate species as evidenced both by Aristotle's notation in *Historia Animalium* that lizards can regenerate their tails and the Greek myth of liver regeneration in Prometheus, the first known scientific study of regeneration in a vertebrate was published by the Italian scientist Lazzaro Spallanzani in 1768 [3]. Spallanzani was able to show that the aquatic salamander (most likely a newt) is able to regenerate its forelimbs and hind limbs, tail, upper and lower jaws, and caudal spinal cord. He also showed that aquatic salamanders could repeatedly regrow a limb even after multiple amputations. Over the past decades, major discoveries have been made using the newt as a model organism for regeneration. In fact, the adult Eastern newt (most often the red-spotted newt) has long served as a primary model for the study of epimorphic regeneration of amputated limbs and tails [4, 5]. Similar to the limb, in a study spanning 16 years, Goro Eguchi and coworkers demonstrated that the regenerative capacity of the newt lens is not altered by repeated regeneration and aging [6]. It has also been shown that during lens regeneration, pigment epithelial cells of the iris can transdifferentiate to lens cells [7]. The plasticity of differentiated cells has been a great focus in regeneration studies. Adult newt cardiomyocytes were shown to reenter the cell cycle [8, 9], and high resolution 3D imaging as well as modern lineage tracing methods have revealed that during limb regeneration, newt multinucleate myofibers dedifferentiate and fragment to form proliferating mononuclear cells that give rise to new skeletal muscle in the regenerated limb [10, 11, 12, 13]. Moreover, the newt not only regenerates the caudal spinal cord following tail amputation but can also regenerate the trunk spinal cord following a complete transection and regain function of initially paralyzed appendages caudal to the injury site [14, 15]. Modern cellular and molecular studies have also shown that parts of the brain can regenerate by activation of quiescent regions of the adult newt brain [16, 17].

As illustrated above, the advent of new methodologies has opened an era in which the newt has become an attractive model to study the cellular and molecular basis of regeneration in a vertebrate species.

This chapter focuses on the maintenance of Eastern newts in a laboratory setting for the purpose of studying their remarkable regenerative abilities.

2 Materials

Prepare all solutions using ultrapure water and analytical grade reagents. Prepare and store all reagents at room temperature (unless indicated otherwise). Diligently follow all waste disposal regulations when disposing waste materials. The housing and maintenance of animals must follow the guidelines of and be approved by the Institutional Animal Care and Use Committee.

2.1 Adult Newts

Notophthalmus viridescens can be purchased from Connecticut Valley Biological Supply Company (Southampton, MA) or Charles D. Sullivan Co (Nashville, TN) (*see* **Notes 1–3**).

2.2 Newt Water

Deionized and dechlorinated water supplemented with 0.0375 % Instant Ocean (Aquarium supply store). We use Instant Ocean salt at a very low concentration, which is not comparable to the concentrations typically used in salt-water aquariums (*see* **Note 4**).

2.3 Aquarium Setup and Maintenance

1. In a dedicated room with controlled temperature of 20–22 °C and day-night light cycle, set up 80 L glass tanks (61 cm × 32 cm × 42 cm) (Aquarium supply store) with lid (Fig. 2a).
2. Cover bottom of aquarium with a 3–4 cm layer of small pea gravel (Aquarium supply store) (Fig. 2a, b).
3. Fill aquarium halfway with approximately 38 L deionized water.
4. In a 4 L plastic beaker, dissolve 15 g Instant Ocean in 2 L of deionized water and add to tank.
5. The conductivity of the water in the aquarium should be approximately 600 μSI. Read the conductivity with a Nester Micro MHO Pen or similar device every month! When conductivity has doubled, dilute water with deionized water.
6. The water pH should be in a range of 7–8. Determine the pH with a calibration-free pH meter every 2 weeks! If pH is getting low, adjust with 1 M NaOH.
7. For water aeration, circulate water using a pump with integrated carbon filter. To maintain water quality, use a filtering system like the H.O.T. Magnum BIO-PRO System (Marineland), which contains in a canister a container with activated carbon (removes toxic chemicals and dissolved organic pollutants), micron cartridge (traps microscopic dirt particles), and Rite-Size sleeve prefilter (screens out free-floating

Fig. 2 Aquarium setup and swimming adult newts. (**a**) 80 L aquariums are supported by metal shelves in temperature- and light cycle-controlled room. (**b**) A layer of pea gravel covers the aquarium bottom with (**c**) aquatic plants rooted into the gravel

dirt and debris) plus a magnetic impeller module that pumps water through the filter. Adjust water level every week.

2.4 Food

1. Beef heart (local butcher). Mince in food processor and keep small portions packaged in plastic bags frozen at –20 °C.
2. *Daphnia* sp., *Tubifex* sp., and California blackworms (Aquatic Foods; *see* **Note 5**). Note that newts prefer live worms over frozen food.

2.5 Breeding, Spawning, and Rearing of Newt Larvae

1. Environmental enrichment for breeding: In addition to the described aquarium setup, root into the gravel a mixed variety of aquatic plants, e.g., *Sagittaria* sp. and *Elodea* sp. (Fig. 2c). Newts also like to spend time out of the water, so you may want to add several rocks that generate a ramp out of the water.
2. Newt eggs and larvae: Keep deposited eggs and rear hatched newt larvae in small clear plastic containers (30 cm × 18 cm × 13 cm). Change water daily.

3 Methods

Carry out all procedures at room temperature unless otherwise specified. The use of animals must follow the guidelines of and be approved by the Institutional Animal Care and Use Committee.

3.1 Animal Housing and Care

1. Keep groups of approximately 30 adult newts in a 20 °C temperature-controlled room in 80 L aquariums with circulating water (Fig. 2a–c).
2. Feed newts live *Tubifex* sp., California blackworms, or frozen beef heart every 3 days.

3.2 Breeding, Spawning, and Rearing of Newt Larvae

We have experienced that females caught in the wild during spring will spontaneously deposit eggs if housed in an aquarium that contains a variety of aquatic plants (Fig. 2c).

1. Mix several sexually mature males and females in a tank and feed well. Noticeable gravid females (females with swollen abdomens) and males with blackened distal hind limb digit pads are the most suitable for mating (*see* **Note 6**). A supply of 100–200 eggs per aquarium with a group of 20 mixed female and male animals can be achieved over a period of 2–4 weeks.
2. Inspect your spawning chamber daily for deposited eggs. Typically, females will attach the eggs to the underside of the plant's leaves, at times rolling a leaf around a single egg or cluster of deposited eggs. Freshly laid eggs are about 1.5 mm in diameter.
3. Remove the egg-laden plant sprigs and transfer them to a clear plastic container. The eggs develop well in room temperature newt water and hatch in 20–35 days to produce free-swimming larvae.

4. Carefully transfer the larvae to a separate clear plastic container filled with newt water and supplemented with *Daphnia* sp., which the larvae will consume.
5. The larvae develop rapidly and once they mature, you can detect prominent gills. These larger larvae require live, moving food, and we found *Tubifex* sp. worms ideally suited.
6. Change the water daily in these small plastic containers. To enhance the animals' environment, add aquatic plants to float in the water.
7. The newt larvae will eventually metamorphose into red efts. This is a physiologically and morphologically significant event, as the animals will resorb their external gills and switch to lung breathing. When you notice signs of this transformation, transfer these more developed animals into a larger chamber that includes an aquatic area with floating *Elodea* sp. plants and a terrestrial component so that the animals can leave the water.
8. Continue feeding the red efts with *Daphnia* sp. and *Tubifex* sp. The mix of land and water with abundant food supply provides an environment the efts will thrive well in. We note that investigators have developed protocols for induced spawning, and we refer to this published work for a more detailed setup description [18, 19].

4 Notes

1. Newts freshly shipped from the vendor should be kept separate for 2 weeks before using for experimentation. Occasionally, a sick newt is among the shipments, and keeping animals "in quarantine" helps to prevent spreading of a potential disease.
2. Newts can be handled by gently picking them up by the tail with gloved hands. Unlike some lizards, which readily discard their tails as a defense mechanism, the tail of the newt cannot be discarded and therefore picking up the newt by the tail is an appropriate method for quickly transferring animals from one location to another. If the newt is to be handled for any length of time beyond a simple transfer, its body should be supported by the palm of the hand.
3. The skin of both Eastern and Western newts often contains a potent neurotoxin known as tetrodotoxin, which binds to and blocks voltage-gated sodium ion channels. The concentration of tetrodotoxin has been shown to vary widely between individuals depending on location, diet, and possibly other unknown factors. Although controversy still remains as to the origin of the tetrodotoxin, several studies suggest that in Eastern newts, the toxin might be of dietary origin and if newts

are fed a toxin-free diet, they can lose their toxicity over a period of several years [20]. Regardless, when handling newts at any stage of their lives, it is best to wear gloves. If a newt is handled without gloves, the hands should be washed immediately afterwards. Needless to say, unlike axolotls, which were regularly eaten in Mexico before they became critically endangered (or possibly extinct in the wild), newts should never be consumed.

4. In the wild, newts live in and by water streams, and therefore, in principle, they can be kept in tap water in the laboratory. However, in most cities, the drinking water is chlorinated to an extent that it could be harmful to the animals. Therefore, if chlorinated tap water is used, it should be sufficiently dechlorinated by allowing it to stand exposed to the air for a sufficient time period for the chlorine to evaporate (usually several days). However, many municipalities are treating their water supply with monochloramine, rather than chlorine. Monochloramine cannot be removed by evaporation. Therefore, we prefer to carefully prepare our own newt water by supplementing the deionized water supply in the laboratory with low amounts of Instant Ocean salt to achieve a conductivity of approximately 600 μSI (equivalent to Evian Water). We have found that the method described in this chapter produces reliable results that allow for the maintenance and care of a healthy population of newts.
5. California blackworms can also be purchased from Eastern Aquatics. Website: currently easternaquatics.com.
6. There is considerable competition among males for the right to mate with females. Often males will try to dislodge a male that is in amplexus with a female. Occasionally, a different male not in courtship deposits his spermatophore between the spermatophore of a male in courtship and the female in an attempt to get the female to collect sperm from his spermatophore rather than the spermatophore from the male in courtship.

Acknowledgments

We would like to acknowledge both current and former researchers in our respective laboratories who helped develop the protocols described in this chapter. These individuals include Paul Khan, Barbara Linkhart, Claudia Guzman, Sarah Calve, Sarah Mercer, Donald Atkinson, Vladimir Vinarsky, Tamara Stevenson, David Kent, and Katherine Zukor. We would also like to acknowledge several of our colleagues who provided us with invaluable advice when we were just beginning our studies on regeneration using this remarkable animal, including Cliff Tabin, Mark Keating, Jeremy Brockes, David Stocum, Roy Tassava, Panagiotis Tsonis, and Anthony Mescher.

References

1. Duellman WE, Trueb L (1986) Biology of amphibians. McGraw-Hill Book Company, New York, NY
2. Petranka JW (1996) Salamanders of the United States and Canada. Smithsonian Institution Press, Washington, DC
3. Spallanzani L (1769) An essay on animal reproductions. Translated from Italian by M. Maty. Printed for T. Becket, and P. A. de Hondt, in the Strand, London. Available through UMI Books on Demand
4. Wallace H (1981) Vertebrate Limb Regeneration. J Wiley and Sons Ltd., Toronto, ON
5. Nye HL, Cameron JA, Chernoff EA, Stocum DL (2003) Regeneration of the urodele limb: a review. Dev Dyn 226:280–294
6. Eguchi G, Eguchi Y, Nakamura K, Yadav MC, Lillan JL, Tsonis PA (2011) Regenerative capacity in newts is not altered by repeated regeneration and ageing. Nat Commun 2:384. doi:10.1038/ncomms1389
7. Eguchi G, Itoh Y (1982) Regeneration of the lens as a phenomenon of cellular transdifferentiation: regulability of the differentiated state of the vertebrate pigment epithelial cell. Trans Ophthalmol Soc U K 3:380–384
8. Oberpriller J, Oberpriller JC (1971) Mitosis in adult newt ventricle. J Cell Biol 49:560–563
9. Bettencourt-Dias M, Mittnacht S, Brockes JP (2003) Heterogeneous proliferative potential in regenerative adult newt cardiomyocytes. J Cell Sci 116:4001–4009
10. Calve S, Odelberg SJ, Simon H-G (2010) A transitional extracellular matrix instructs cell behavior during muscle regeneration. Dev Biol 344: 259-271
11. Calve S and Simon H-G (2011) High resolution 3D imaging: Evidence for cell cycle reentry in regenerating skeletal muscle. Developmental Dynamics 240:1233-1239
12. Lo DC, Allen F, Brockes JP (1993) Reversal of muscle differentiation during urodele limb regeneration. Proc Natl Acad Sci U S A 90: 7230–7234
13. Sandoval-Guzmán T, Wang H, Khattak S, Schuez M, Roensch K, Nacu E, Tazaki A, Joven A, Tanaka EM, Simon A (2014) Fundamental differences in dedifferentiation and stem cell recruitment during skeletal muscle regeneration in two salamander species. Cell Stem Cell 14:174–187
14. Piatt J (1955) Regeneration of the spinal cord in the salamander. J Exp Zool 129:177–207
15. Davis BM, Ayers JL, Koran L, Carlson J, Anderson MC, Simpson SB Jr (1990) Time course of salamander spinal cord regeneration and recovery of swimming: HRP retrograde pathway tracing and kinematic analysis. Exp Neurol 108:198–213
16. Okamoto M, Ohsawa H, Hayashi T, Owaribe K, Tsonis PA (2007) Regeneration of retinotectal projections after optic tectum removal in adult newts. Mol Vis 13:2112–2118
17. Berg DA, Kirkham M, Beljajeva A, Knapp D, Habermann B, Ryge J, Tanaka EM, Simon A (2010) Efficient regeneration by activation of neurogenesis in homeostatically quiescent regions of the adult vertebrate brain. Development 137:4127–4134
18. Khan PA, Liversage RA (1995) Development of Notophthalmus viridescens embryos. Dev Growth Differ 37:529–537
19. Khan PA, Liversage RA (1995) Spawning of Notophthalmus viridescens embryos. Herpetol Rev 26:95–96
20. Yatsu-Yamashita M, Gilhen J, Russell RW, Krysko KL, Melaun C, Kurz A, Kauferstein S, Kordis D, Mebs D (2012) Variability of tetrodotoxin and of its analogues in the red-spotted newt, Notophthalmus viridescens (Amphibia: Urodela: Salamandridae). Toxicon 59:257–264

Chapter 3

Housing and Maintenance of *Ambystoma mexicanum*, the Mexican Axolotl

Johanna E. Farkas and James R. Monaghan

Abstract

The aim of this paper is to assemble a significant amount of information on *Ambystoma mexicanum*, the axolotl salamander, to assist in the basic knowledge needed to raise, breed, and study most aspects of axolotl biology. It is important to understand the basic biology of the axolotl in order to make informed decisions on their proper care and use in experiments. Therefore, we will provide necessary information to the non-herpetologist that will assist in their study of this unique and fascinating animal. We also aim to provide a resource on the general anatomy, behavior, and experimental tips specific to the Mexican axolotl that will be of use to most axolotl laboratories. Axolotls have been actively researched since the 1860s, giving testament to their relatively straightforward maintenance and their versatility as an animal model for development and regeneration. Interest in using the axolotl in laboratory research has grown tremendously over the past decade, so dedicated resources to support the study of this species are needed and encouraged.

Key words Salamander, Limb regeneration, Animal model, Axolotl anatomy

1 Introduction

1.1 Taxonomy, Habitat, and the Laboratory Strain

A. mexicanum, commonly named the axolotl (Fig. 1), are members of the order Urodela (*oura* tail + *delos* evident, also called Caudata) or tailed amphibians, which constitute ten extant families of salamanders found mainly across the temperate regions of the northern hemisphere. Axolotls belong to the family Ambystomatidae, genus *Ambystoma*, which are commonly called mole salamanders and are comprised of approximately 30 species found across North America from southern Mexico to southern Alaska. There are 17 Mexican ambystomatid salamander species that inhabit the mountains of Central Mexico. Five of these species are primarily or obligatorily neotenic, meaning they do not undergo metamorphosis and can breed in the adult larval form [1]. Axolotls are endemic to the Lake Xochimilco area in the Valley of Mexico, which has been reduced over the centuries to a

Anoop Kumar and András Simon (eds.), *Salamanders in Regeneration Research: Methods and Protocols*, Methods in Molecular Biology, vol. 1290, DOI 10.1007/978-1-4939-2495-0_3, © Springer Science+Business Media New York 2015

Fig. 1 Image of female leucistic (*left*), female albino (*middle*), and male wild-type (*right*) axolotls. Notice the large bellies of the two females indicating they are filled with eggs and the large cloaca of the mature male wild-type

~40 km^2 area of artificial canals just outside the city limits of Mexico City [2]. This high-altitude lake system (~2,200 m above sea level) has been inhabited for centuries, most notably by the Aztecs. This is where the name axolotl originates, as the animal was named after the Aztec god Xolotl [3]. The water conditions of Lake Xochimilco between 1978 and 1988 were estimated to be between 16 and 20 °C, pH 7.4–8.0, with a conductivity between 975 and 1,650 microsiemens (μS)/cm [1]. Today, axolotls are critically endangered and are on the brink of extinction due to habitat loss, the introduction of invasive species, and shifts in water quality [4]. It has been estimated that densities were at 6,000 ind./km^2 in 1998, 1,000 ind./km^2 in 2000, and 100 ind./km^2 in 2008 [2], and only a few axolotls were cited in the lake system after months of surveying in 2013 [5]. Unfortunately, the majority of axolotls today are found in aquaria and laboratories around the world.

The modern axolotl strain used in most laboratories is a highly inbred population that most likely arose from a donation of seven wild axolotls (six wild-type and one white mutant) between 1863 and 1866 to the Paris Natural History Museum [6]. In fact, most modern-day laboratory axolotls likely have a direct lineage to these founders, and all white mutants are descendants from this single white animal [7]. A few wild-caught axolotls were introduced into the colony strain in the 1960s including an albino tiger salamander (*Ambystoma tigrinum*) [8], but overall the present-day laboratory strain is likely one of the most long-running inbred strains of any laboratory species. The 150-year history of laboratory breeding seems to have selected against spontaneous metamorphosis (currently <1 % frequency), as it is more prevalent in wild strains than the lab strain [7, 9]. The most extensive colony of laboratory axolotls is maintained at the Ambystoma Genetic Stock Center (AGSC) at the University of Kentucky (www.ambystoma.org), which is a continuation of the Indiana University Axolotl Colony initiated by

Humphrey R.R. in 1957 [10]. Over the past 50 years, the Axolotl Colony has provided the majority of axolotls and housing information to labs worldwide. Valuable information on housing, breeding, and diseases of axolotls can be found at www.ambystoma.org as well as in the archives of the Axolotl Newsletter.

1.2 Laboratory Research Using the Axolotl

Axolotls have classically been used in developmental biological research for practically every organ system due to their large egg size, external development, reliable breeding, acceptance of embryonic and adult tissue grafts, and large clutch sizes [11]. Historically, both newts and axolotls have been the primary types of salamanders used in lab research, but the unique advantage of axolotls is that they can be easily bred in a lab environment in the long term. What truly sets both newts and axolotls apart from other animal models is that they are our closest relatives that display a wide range of striking regenerative capabilities. In fact, axolotls can regenerate and recover from virtually any injury that does not kill them. Regeneration has been observed in parts of the heart, the tail, the jaw, the spinal column, the gills, the brain, and entire limbs (Fig. 2). This regrowth occurs without scarring and with full restoration of function. It can also presumably reoccur an indefinite number of times without any loss of fidelity, although a careful analysis of this point is needed in axolotls. Regeneration can be induced throughout all stages of life, although the process is faster in larvae and less reliable in animals that have been forced to undergo metamorphosis [12, 13].

Other amphibians besides axolotls are studied for their regenerative abilities. Tadpoles of the African clawed frog *Xenopus laevis* can regenerate their tails and spinal columns, while certain species of newts (*Pleurodeles waltl*, *Notophthalmus viridescens*, and *Cynops pyrrhogaster* are the most commonly studied species) can regenerate appendages and organs. Despite these similarities, the axolotl is not closely related to these species. As a member of the order Anura, *X. laevis* is only distantly related to urodele salamanders with a common ancestor approximately 260 million years ago [14], a divergence emphasized by the fact that *X. laevis* loses its regenerative abilities in late larval stages. Furthermore, although newts and axolotls are both urodeles with superficially similar anatomy, they diverged from a common ancestor at least 145 million years ago [14], and recent studies have shown that newt regeneration and axolotl regeneration differ in both mechanism [15–17] and recovery after denervation [18]. Consequently, caution is advised when attempting to compare axolotl protocols and findings with those of other amphibian models.

1.3 Gross Anatomy

Unlike the majority of urodeles, axolotls are obligate paedomorphs which do not undergo metamorphosis unless it is artificially induced via the addition of thyroid hormone to their environment [19]. Consequently, they grow to be large fully aquatic adults and

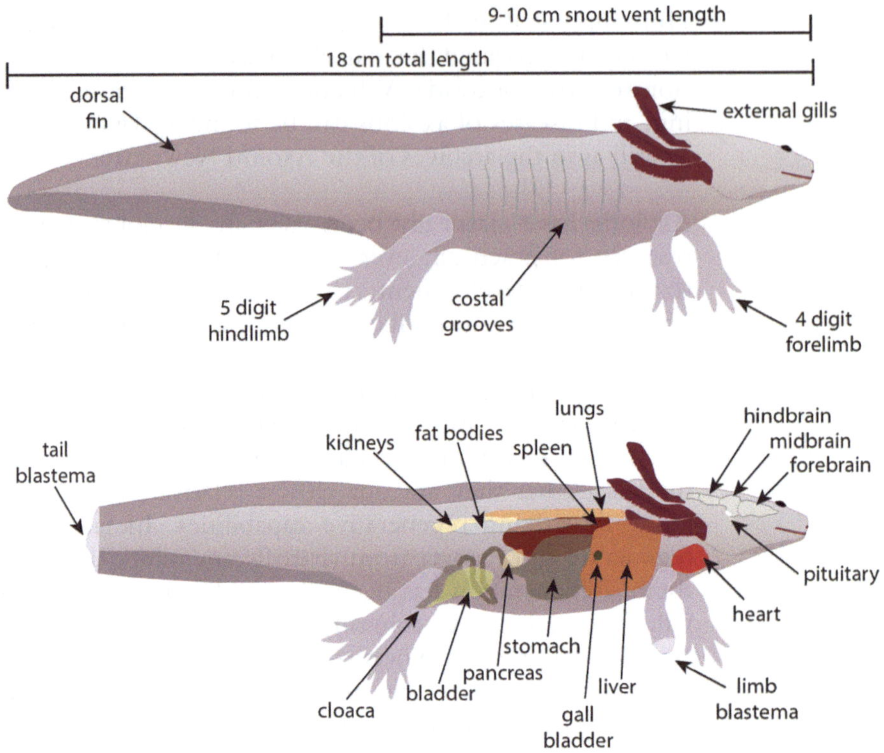

Fig. 2 Cartoons of the exterior (*top*) and interior (*bottom*) of an adult leucistic axolotl. Each organ is drawn to approximate scale and represents the approximate position in the animal. Colors also represent the approximate color of each organ. The regenerating limb and tail represent mid-bud blastema stages. Internal histology of a mid-bud blastema is represented in Fig. 3c

maintain the feathery external gills characteristic of ambystomatid larvae throughout their lives (Figs. 1 and 2). Males reach sexual maturity, indicated by a blackening of the nails, at around 10 months of age, while females tend to mature slightly later from 12 to 18 months. Sexual dimorphism is moderate but visible, as males are slimmer and longer than females and exhibit a cloacal bulge (Fig. 2). Sexually mature females are also more rotund due to their egg supply. While the lifespan of wild axolotls remains unknown, they generally live between 10 and 15 years. Animals can be induced to undergo metamorphosis, which become terrestrial and strongly resemble adult tiger salamanders (*Ambystoma tigrinum*). Axolotls are in fact closely related to tiger salamanders and are capable of crossbreeding with this species in additional to various other ambystomatid salamander species [20]. Studies of *A. tigrinum* have thus provided extensive insight into the anatomy and development of *A. mexicanum* [21].

Although the overall body plan of the axolotl is considered to be among the most primitive of the tetrapods, these salamanders nevertheless share a number of basal characteristics with other vertebrates and have proved useful for the study of many different systems.

Aquatic and lacking a middle-ear structure, axolotls rely on olfaction far more than audition when they are searching for food. The urodele olfactory system is quite complex and allows the animal to discriminate between very similar odorants and chemicals. Because their olfactory epithelium and olfactory bulb are large and easily accessible, tiger salamanders were a favored model for the study of olfaction and neuronal signaling during the 1970s and 1980s [22–24]. Axolotls possess true teeth and a calcified skeleton with cartilaginous joints that are anatomically similar to mammalian joints [25, 26].

Although their eyesight is poor and their vision is largely limited to the detection of movement, the axolotl retina displays the layered structure typically seen in vertebrates. In fact, many early studies of retinal intracellular signaling were conducted in salamanders because of their large and easy-to-access retinal cells [27–29]. Tiger salamanders can detect ultraviolet light, and their UV-sensitive photoreceptors express three different opsins [30, 31]. Furthermore, like fish and other aquatic amphibians, axolotls possess a lateral line system that is used to detect both electrical currents (via ampullary organs located on the head) and water movement (via mechanoreceptive neuromasts that run along the side of the animal). The lateral line develops from neurogenic placodes, and this developmental process may provide insights into the evolution of vertebrate sensory systems [32]. However, the study of the axolotl's development and evolutionary history has been complicated by its massive genome. Consisting of around 32×10^9 base pairs located across 14 haploid chromosomes [33], the axolotl diploid genome is among the largest of all tetrapods. Our understanding of this animal is sure to increase rapidly as genetic technology advances and more techniques are adapted for use with salamanders.

1.4 Structure of the Circulatory and Respiratory System

Axolotls possess a three-chambered heart consisting of two atria and one ventricle. As blood leaves the ventricle, it can pass to either the pulmonary arteries or through an aorta that leads to the rest of the body. Oxygenated blood leaves the lungs and is pumped back into the lone ventricle of the heart, where it mixes with deoxygenated blood that has already circulated through the body. This three-chambered system is consequently less efficient than the four-chambered system seen in mammals and birds, as tissues are nourished by blood that is not saturated in oxygen. The number of aortic arches can vary greatly between and within different urodele species, but axolotls always have four aortic arches [34]. Like all amphibians, axolotls are poikilothermic and their heart rate is strongly influenced by the temperature of their environment. Hematopoiesis arises from the adult liver and spleen [35] and their erythrocytes are very large, nucleated, and strongly autofluorescent. Injury induces rapid vasoconstriction to prevent excessive blood loss, and clotting occurs very quickly even after injury to major arteries. As a result, axolotls are at only minimal risk of bleeding to death during and after surgery. They are a fairly popular model for the study of cardiac development because they can exhibit a recessive mutation, dubbed *c*, that is cardiac lethal within 2 weeks after hatching. The hearts of *c/c* mutants do not beat, and studies have found that these mutant hearts are largely differentiated but lack tropomyosin [36] and organized sarcomeres [37].

Aquatic throughout their lives, axolotls utilize multiple strategies for obtaining oxygen from their environment. They exchange oxygen with their three-paired external gills and can respire through their skin via cutaneous gas exchange. However, axolotls also possess rudimentary lungs and may obtain at least 40–60 % of their oxygen through surface breathing [38]. These lungs are elongated and translucent in appearance and run parallel to the spinal column for virtually the entire length of the body cavity (Fig. 2). Very little research on axolotl respiration has been conducted, but previous studies of tiger salamander larvae have provided likely insights into the mechanism of this process. These larvae inhale using a two-stroke buccal pump system that mixes expired and fresh air in the buccal cavity before pumping it into the lungs. Exhalation is active and very rapid, minimizing the amount of fresh air that is expelled through the mouth and gills. The ventilation efficiency of this system is comparable to that of mammalian ventilation [39]. Axolotls housed in hypoxic tanks make frequent trips to the surface to breathe [40], which can increase the probability that they will swallow air and cause the formation of an air bubble within the body that can disrupt the animal's locomotion and feeding. Because of their numerous options for respiration, axolotls can survive for hours outside of their tanks so long as they are not allowed to desiccate. This ability proves useful for the purpose of surgical procedures, as axolotls will heal very quickly if kept moist and left sedated after surgery.

1.5 The Immune System

Axolotls have a very primitive acquired immune system and are generally described as immunodeficient. They do not induce a humoral response to soluble antigens [41], they produce just two classes of immunoglobulins, and they generate antibodies to antigens extremely slowly if at all [42]. This immunodeficiency is common in urodeles and has worked to the advantage of researchers, as axolotls do not reject tissue from other salamanders and will even readily accept tissues [43] and tissue primordia [44] from other amphibian species such as *X. laevis*. Creative grafting experiments using embryos have done much to elucidate the development of axolotls, while grafting of larval or adult tissue has uncovered some of the mechanisms of regeneration. The recent production of GFP-expressing axolotls has opened up a new avenue of grafting possibilities, and researchers have already begun grafting GFP tissues to white animals in order to reveal the basis of cellular plasticity and positional memory during regeneration [45].

Despite these deficiencies in the acquired immune system, the axolotl is still resistant to bacterial infections. This is likely due to their mucus coat and fairly robust innate immune system, which represent the main line of immune defense for urodeles. Abundant neutrophils and macrophages rapidly engulf and destroy foreign invaders, while antimicrobial peptides provide an additional layer of defense [46]. Past the age of 2 weeks, axolotls are at moderate to minimal risk of bacterial infection even after surgery. However, they are still susceptible to fungal and viral infections and can acquire bacterial infections after chronic stress. Exposure to *Ambystoma tigrinum* virus (ATV) can devastate both wild and laboratory populations, with mortality rates potentially exceeding 90 % [47]. This extreme susceptibility is likely due to a lack of lymphocyte proliferation upon exposure to the virus [48]. Another curiosity of the axolotl immune system is the fact that macrophages seem to be essential for regeneration, as early ablation of macrophages completely inhibits blastema formation and results in excessive collagen deposition after limb amputation [49].

Intriguingly, although they may undergo high levels of cellular proliferation in multiple tissues throughout their lives, axolotls are remarkably resistant to cancer. Repeated studies of various urodeles have failed to induce cancerous growth in regenerating tissues even upon administration of carcinogens [50, 51], although spontaneously occurring tumors have been described in veterinary literature [52, 53]. This curious resistance to malignant growth even during extreme cellular proliferation remains largely unexplored.

1.6 Pigmentation and Structure of the Epidermis

Wild-type axolotls are a mottled gray/brown/olive-green color, and this coloration arises from a combination of three different neural crest-derived pigment cells. Black melanophores are the predominant chromatophores in adult wild-type animals, while yellow xanthophores are more abundant early in development (Fig. 1) [54]. *Tip: shiny iridophores reflect light and are strongly reflective under fluorescent imaging.* White leucistic (white mutant, *d/d*) axolotls are the result of a recessive mutation in an unknown gene. They are oftentimes preferred over wild-type animals for research purposes, as pigment can negatively affect histological staining or fluorescent imaging. Leucistic animals are descendants of a single founder white axolotl that was donated to Auguste Duméril in 1866, hence the genetic designation *d* [55]. Leucistic larvae are lightly pigmented dorsally but lose almost all pigment shortly after hatching due to a lack of pigment cell migration to the flank of the animals [56]. In contrast, albino animals completely lack melanin and have a yellow appearance due to an overabundance of xanthophores (Fig. 1). Although albino eggs are difficult to inject due to their lack of pigment, they are very useful for whole-mount embryonic staining methods. Adult leucistic salamanders can be easily distinguished from albinos by their black eyes.

The adult axolotl epidermis is similar to other larval amphibians and does not contain the distinct cornified layer (stratum corneum) that mammals or metamorphosed amphibians possess [57]. It is highly vascularized in order to facilitate efficient cutaneous gas exchange and is coated with mucus secreted by epidermal Leydig cells dermal mucous glands help retain moisture and withstand microbial threats. Leydig cells are large, club-shaped, and packed with secretory granules (Fig. 3a, b). As they fill with these granules, they rupture and release mucus into the intracellular space where it can then escape via pores onto the epidermal surface [58]. This mucus coat is very sticky and makes for an effective natural adhesive during surgical manipulation of the animals. Epidermal healing and regeneration is completely scar-free and occurs without extensive long-term fibrosis [57].

1.7 Regeneration

Limb regeneration is the classic paradigm of complex tissue regeneration and is currently the primary focus of laboratory research on the axolotl (Fig. 2). Axolotls regenerate limbs by generating a mass of highly proliferative cells called a blastema at the distal tip of a limb stump (Fig. 3c). After amputation, locally derived lineage-restricted progenitor cells or dedifferentiated cells proliferate and recapitulate developmental processes in order to regrow the missing portion of the limb [59]. As long as animals are well cared for, limb regeneration is a robust, dependable assay. A juvenile animal (8.5–10 cm SVL) will reach the differentiation stage of regeneration around approximately 32 days post amputation but will not replace 100 % of the missing limb for at least 100 days post amputation. Three-month-old axolotls (~5 cm SVL) will reach the differentiation stage of regeneration at approximately 22 days post amputation and will replace 100 % of their limbs by approximately 66 days post amputation [12].

Perhaps because axolotls are a neotenic member of an evolutionarily primitive order, their brains are very simple and share several similarities with the mammalian embryonic brain (Fig. 3d). The axolotl brain is relatively flat and elongated with clear delineations between the telencephalon, mesencephalon, and rhombencephalon (Fig. 2). The olfactory bulbs are large, but the optic lobes are small and poorly separated, and the cerebellum is relatively undersized and weakly developed as well [21]. The cerebellum contains the only neurons in the axolotl central nervous system that are not paraventricular—all other neurons remain stationary and do not migrate from the germinal site during development [60] (Fig. 3d). Though less is documented about the terrestrial axolotl brain, it is known that the brain undergoes

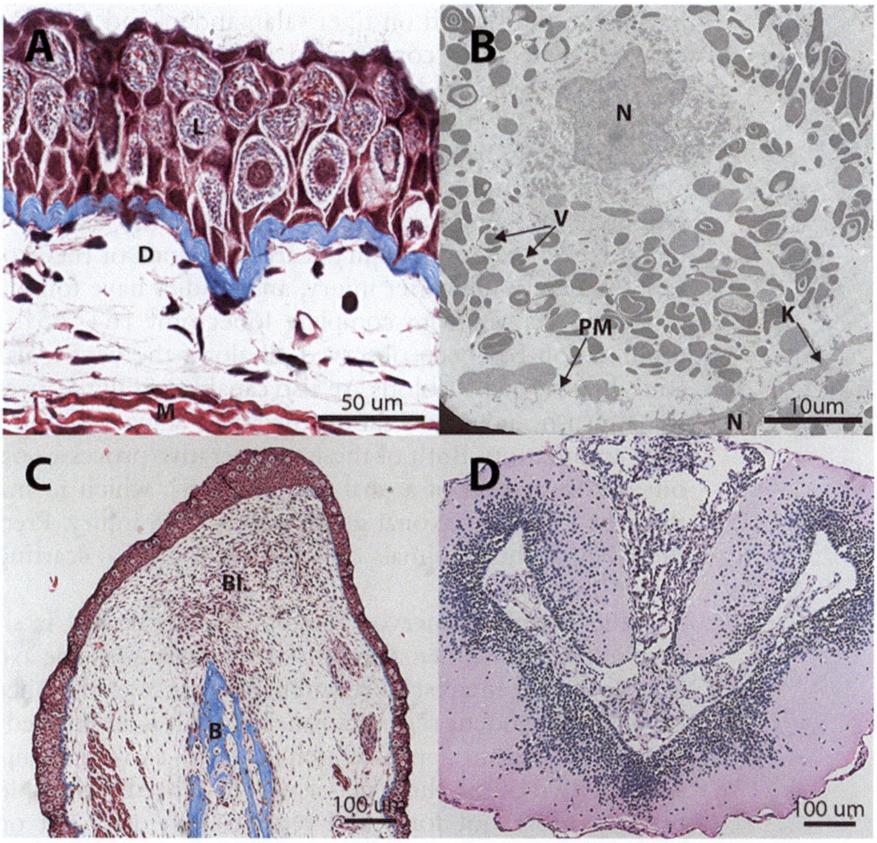

Fig. 3 Histological images of axolotl tissues. (**a**) Masson's trichrome histological staining of a juvenile axolotl epidermis. Notice the large Leydig cells (L) throughout the epidermis, dermal tissue underneath the epidermis (D), and muscle lying underneath the dermis (M). (**b**) A transmission electron micrograph of a Leydig cell in a juvenile axolotl epidermis. Notice the nucleus in the middle (N), which is surrounded by rough endoplasmic reticulum and a large number of dense vesicles (V). PM represents the Leydig cell plasma membrane. K represents a keratinocyte with N representing the nucleus of the keratinocyte. (**c**) Masson's trichrome histological staining of a juvenile mid-bud limb blastema (BL). The bone is indicated with a B. Notice the thickened epidermis on the distal tip of the blastema and mesenchymal cell that makes up the blastema. (**d**) Hematoxylin and eosin histological staining of a juvenile axolotl brain. The cross section is taken through the posterior telencephalon (forebrain; *see* Fig. 2). Notice the ventricular zone that contains highly proliferative neural progenitor cells [66]

structural changes during the metamorphic process, particularly in the optic lobes as they grow to accommodate the animal's new mode of binocular vision [61]. This suggests that the axolotl brain remains plastic and capable of drastic change throughout the animal's lifespan, possibly indicating that axolotls would make a favorable model for the study of neural plasticity. Although seemingly possessed of limited intelligence, laboratory axolotls will learn to associate humans with food and will move to the front of their tank in anticipation of feeding. Few studies on axolotl learning have been conducted, but classical conditioning studies have been performed on tiger salamanders and seem to be most effective when the conditioned stimulus is olfactory in nature [62]. Tiger salamanders have also been trained to complete a T-maze test, a feat which occurs more rapidly and reliably in adult animals than in larvae [63].

Unlike mammals, which possess only limited regenerative potential in the central nervous system, the axolotl CNS can recover from extensive injury. Large portions of the axolotl brain can regenerate fully after injury, and studies have found that they can even recover from complete lobectomy [64, 65]. Neuronal proliferation has been observed all along the ventricular zone of adult brains [66]. Axolotls are also capable of fully recovering from spinal crush and regenerating their tail, spinal column included, after amputation. Both of these regenerative processes occur without the formation of a glial scar [67, 68], which in mammals is thought to inhibit axonal growth after CNS injury. Precisely why the CNS of these animals can heal without glial scarring remains under investigation.

The peripheral nervous system of the axolotl is simple but organized in a manner similar to other tetrapods. The axolotl PNS is of particular interest to researchers because of the phenomenon of nerve-dependent regeneration. If a limb is amputated and then denervated within approximately 7 days—easily accomplished by severing two of the three major nerves at the brachial plexus—the blastema does not form and regeneration does not occur. The cause of this nerve dependency is still under investigation, although it may be due to a loss of nerve-secreted mitogenic factors that are essential for supporting the regenerative process. Axolotl peripheral nerves are myelinated and generally easy to find and sever, particularly in young animals. However, they begin to regrow within 10 days and must be periodically re-severed in studies that last longer than this span. Denervation becomes successively more challenging over time, and consequently it is difficult to maintain a fully denervated state for longer than approximately 20 days.

1.8 Reproductive System Structure

Axolotls reproduce sexually and fertilize internally. Males lay spermatophores which are then taken up and dissolved by the female in the spermatheca, allowing spermatozoa to be stored in the cloaca until it is time for spawning to occur. As eggs leave the oviducts and enter the cloacal chamber, spermatozoa come into contact with the egg and more than one sperm enter the egg cytoplasm [69]. Eggs are usually laid in strings and are protected by a thick coating of sticky jelly, which must be removed prior to embryonic injections or grafting experiments. Though spawnings vary considerably in size, most females will lay between 200 and 1,000 eggs per spawn. Breeding ease and efficiency in the laboratory peaks in the spring and decreases throughout the summer, suggesting that axolotl breeding is seasonal even in animals which have been removed from their natural environment. This seasonality is probably linked to spermatogenesis, which typically begins in early summer and results in the deposition of sperm into the vas deferens during the winter [70]. Some labs have attempted to overcome this seasonal impediment by hormonally inducing ovulation with injections of human chorionic gonadotropin (hCG) [71]. However, this technique does not seem to affect the rate of spermatophore deposition, which seems to be the limiting step in this process.

Despite the fact that axolotls are commonly and reliably bred in the laboratory setting, much remains unknown about the structure and development of their reproductive system. It is known that axolotls share a conserved inductive mechanism of germ cell development with mammals, rather than the oocyte-derived germplasm observed in *Xenopus laevis* and zebrafish [72]. Structural anatomy of the axolotl ovary has yet to be fully described or characterized, though it has been found that adult females constantly undergo oogenesis and thus their ovaries contain oocytes at all stages of maturation and development [73]. The axolotl ovary can take up a considerable portion of the peritoneum and resides next to the fat bodies (Fig. 2). Although breeding fidelity tends to decrease over time, oogenesis never halts completely. Therefore, although few studies of axolotl ovarian development and anatomy have been conducted, the animal remains an intriguing model of ovarian regeneration and long-term adult oogenesis.

Slightly more is known of the male reproductive system. The axolotl testis is comprised of a variable number of lobes, and older animals tend to have increased numbers of these lobes. Lobes are made of lobules, which resemble sacs and are themselves composed of small spermatogonial cysts, each of which arose from a single stem cell [70]. As noted previously, spermatogenesis begins in the early summer after old cysts from the previous reproductive cycle have been broken down and replaced. By August the testis is swollen and heavy, and sperm are released into the vas deferens in the following months as cysts rupture and then degenerate. Not all individuals adhere strictly to this seasonal pattern, and there is even some variation within a single testis as cysts may mature at different rates. Consequently, an axolotl testis is likely to exhibit cysts at many different stages of development.

2 Materials

2.1 Housing

1. A windowless (*see* **Note 1**) or blacked-out room held at a constant temperature of 16–20 °C (*see* **Note 2**).
2. An automatic light timer which controls lighting conditions for a 12L:12D light cycle (*see* **Note 3**).

3. Deionized water or dechlorinated tap water supplemented with a balanced salt solution at a pH between 7.0 and 7.5. Commercial artificial seawater is sufficient for housing all animals. Classically, 40–50 % Holtfreter's solution has been used for adult housing, which has a conductivity of approximately 2,300 µS/cm ($NaCl = 30$ mM, $MgSO_4{\cdot}7H_2O = 0.4$ mM, $CaCl_2 = 0.45$ mM, $KCl = 0.33$ mM) (*see* **Note 4**).
4. Water can be stored in a large recirculating storage tank or in a food service quality 45-gallon wheeled trashcan (*see* **Note 5**).
5. For animal storage, 6QT/5.7 L Sterilite plastic boxes or comparable plastic/glass containers.
6. Animals can also be housed in 5-gallon (18.9 L) aquaria with minimal flow through (*see* **Note 6**).
7. For high animal volume labs, automated flow-through systems can be used to house large quantities of both juvenile (2–3 L tanks) and adult (10 L tanks) animals (*see* **Note 7**). Water temperature should be set to 18 °C, pH 7.0, conductivity to 800–1,500 µS/cm, with 5 % daily water changes. The water flow in each tank should be kept low.

2.2 Raising Axolotls from Embryos to Adulthood

1. Soda-lime glass bowls, 8 in. diameter.
2. Soda-lime glass bowls, 3.5 in. diameter.
3. 20 % Holtfreter's solution.
4. Aeration bubblers.
5. Soft mesh nets.
6. Artemia brine shrimp and brine shrimp hatchery (*see* **Note 8**).
7. California Blackworms (*Lumbriculus variegatus*) (*see* **Note 9**).
8. Soft moist salmon pellets.

2.3 Breeding

1. 28-quart plastic container (58.4 cm L × 41.3 cm W × 15.2 cm H) lined on the outside with foil.
2. Reusable ice packs.
3. Broken terracotta pots or large flat natural rocks.
4. Soft mesh nets.

2.4 Disease Treatment

1. 35 g/L salt solution.
2. 5 mg/mL amikacin solution.

2.5 Shipping

1. At least two water-sealed containers.
2. Packing materials.
3. Thick, sealable plastic bags.
4. Disposable ice packs.

3 Methods

3.1 Housing

1. Three times a week, change water in animal containers manually using a plastic colander or soft mesh net.
2. Scrub containers thoroughly with a brush at each water change (*see* **Note 10**).
3. Feed animals on same schedule as water changes.
4. For aquaria, perform 10 % water changes weekly.
5. Check pH, conductivity, and water level of flow-through systems daily. Check ammonia levels every 2–4 weeks once systems are established.
6. Monitor animal health daily.
7. For further sources on axolotl housing, *see* **Note 11**.

3.2 Raising Axolotls from Embryos to Adults

1. Place embryos in groups of 100 animals in aerated 20 % Holtfreter solution (*see* **Note 12**).
2. The expected death rate is 10 %. Check for and remove dead embryos daily (*see* **Note 13**).
3. Change water three times a week.
4. Embryos will hatch out of their eggs on their own and should be separated from unhatched animals at the next water change.
5. Larvae (*see* **Note 14**) are housed in small groups of 20/container in 6QT plastic containers or 8 in. soda-lime glass bowls.
6. Feed larvae newly hatched artemia brine shrimp daily.
7. Change water three times a week using a soft mesh net.
8. Split groups in half at the first sign of cannibalism and house animals of a similar size together.
9. Transition animals from brine shrimp to aquatic California Blackworms, *Lumbriculus variegatus*, as soon as they are large enough to eat the worms (approx. 3 cm total length) (*see* **Note 15**).
10. At approximately 3 cm SVL, juvenile (*see* **Note 16**) house at a density of 5 per 6QT container or 8 in. soda-lime glass bowl.
11. Animals can be transitioned to sinking soft moist salmon pellets at approximately 7 cm SVL.
12. Housing water is changed and animals are fed every Monday, Wednesday, and Friday.
13. House animals that are being used for limb regeneration experiments alone in a 3.5 in. soda-lime glass bowl (*see* **Note 17**).

14. At approximately 5 cm SVL, animals are transitioned to individual 6QT housing containers or into the automated flow-through systems in 3 L housing containers.
15. Adult (*see* **Note 18**) axolotls are either housed in individual 6QT containers or automated flow-through systems.

3.3 Breeding

1. Place one male and one female together in a 28-quart plastic container covered with aluminum foil. Include two reusable ice packs and a substrate of broken terracotta pots or large flat rocks (*see* **Note 19**).
2. If spermatophores are observed in the tank after the first night, remove the male and add soft mesh nets to the container.
3. The female will lay eggs continuously over the next 12–24 h, usually laying between 200 and 1,000 eggs on the provided substrate.
4. The males can mate once a month and females at maximum every 3 months (*see* **Note 20**).

3.4 Disease Treatment

1. In order to prevent disease, keep animals well fed and in fresh water from the moment eggs are laid to the death of the animal. Chronic stress is the most common cause of illness.
2. Catch signs of poor health early before disease spreads. This usually is apparent by a lack of appetite, loss of coloration in the gills, floating continuously, or loss of the "feathery" appearance of the gills.
3. A *columnaris* bacterial infection manifests as a white fungus-looking growth on the animal's gills or flank.
4. Treat *columnaris*-infected animals with a high salt bath of 35 g/L for 10 min for three consecutive days [74].
5. *Aeromonas* is a common parasite that manifests as red blotchy skin, oftentimes on the animal's leg.
6. Treat *Aeromonas* with three intraperitoneal injections of 5 mg/mL amikacin solution at 5 mg/kg body weight separated 48 h apart.
7. For more information on axolotl diseases, please *see* [75].

3.5 Shipping

1. Always ship axolotls with proper institutional animal care approval (*see* **Note 21**).
2. Shipping should be performed in a water-sealed container such as a cooler or Styrofoam box inside a second cardboard box.
3. Place packing material within the interior container.
4. Place animals and water in a thick plastic bag, seal it, and place the bag within a second bag.
5. Seal bags with heat or securely tape them shut (*see* **Note 22**).

6. Disposable ice packs should also be added to the shipping container.
7. Ship overnight only and remove animals from shipping containers to fresh water as soon as they arrive at their destination.

4 Notes

1. Restricting sources of natural light will increase the chances of successful mating taking place during the typically nonbreeding months.
2. Water temperatures higher than 22 °C are stressful on the animals and can have long-term negative effects on axolotl health.
3. A dimly lit room is preferable to a bright one, as axolotls can be stressed by bright lights.
4. Success is also possible at a lower salinity of 750 μS/cm [76]. Thus, axolotls thrive in a range of salinity conditions.
5. Free-standing water, especially tap water, should be treated with the water conditioners Amquel Plus and Novaqua Plus (Kordon LLC).
6. Aquarium filters should be used, but make sure that the water flow is kept to a minimum, as high flow is irritating to axolotls. Long-term housing in these tanks is reliable without any substrate, but small gravel can be used to act as a host for nitrifying bacteria and supplemented with 20 % crushed coral to facilitate pH balance (Jessica Whited, pers. comm.).
7. The advantage of these systems is that the footprint is relatively small and many of the daily tasks are automated and alarmed over network including water changes, water conductivity, pH, water flow, and temperature. Water-cooling is highly recommended with axolotls considering the heat that is generated by the automated systems. Most modern buildings can leverage process-cooling water, but if this is not available, air-cooled units can be used for chilling as long as an exhaust vent for hot air is built into the animal room. The downside to an automated system is that the water is shared between tanks so disease could spread between animals. Therefore, diligent observation of animal health is critical when using an automated system.
8. Commercial brine shrimp hatcheries work very well for high yields. If the axolotl colony is small, a simple brine shrimp hatchery can be generated using a 500 mL glass bottle with a bubbler set at the bottom placed next to a window. Hatching solution should include 35 g sea salt/L and 1.5 g brine shrimp cysts per 500 mL. Cysts will hatch after approximately 24 h depending upon salinity, heat, light, and batch of artemia so optimization will be needed for each lab.

9. A two-pound shipment of California Blackworms can be kept in the refrigerator in a large plastic container with just enough axolotl housing solution to cover the worms. Worms will survive for weeks if they are washed daily.
10. A small amount of sodium bicarbonate can assist in scrubbing.
11. For advice on establishing a colony, consult the AGSC, read over their guide to axolotl care (www.ambystoma.org), and read the several reviews on raising and caring for axolotls in captivity [11, 55, 76, 77].
12. Embryos left in their jelly coat will hatch in 2–3 weeks. The speed of development can be slowed if embryos are housed in colder temperatures.
13. There is substantial variability in clutch-to-clutch embryo survival so close examination is needed to ensure maximal viability. It is important to remove dead eggs as they will foul the water and negatively affect the development of other embryos. Aerating the water with bubblers will increase animal survival and leads to faster embryo development.
14. Hatched axolotls are considered larvae until their hind limbs have developed to the digit stage.
15. We have found that feeding larval axolotls live blackworms promotes rapid growth.
16. Juveniles are categorized as animals that have completed development, but are not yet sexually mature. Animals range from approximately 3 cm snout tip to vent (SVL) to 10–12 cm SVL. Axolotls do not undergo metamorphosis so it is not possible to determine a true transition from the larval to juvenile stage.
17. This ensures that limbs will be kept intact, as animals housed together may attempt to eat and injure each other.
18. Adults are categorized as animals that are sexually mature. Sexual maturity is highly variable in axolotls ranging from 8 to 15 months. For this reason, we categorize axolotls that are larger than 20 cm snout tip to cloaca length as adults and greater than 10 months of age.
19. Two reusable ice packs are included to lower the temperature of the water to induce male mating behavior. On the first night of mating, males lay spermatophores onto the bottom of the tank, and the female takes up one or more spermatophores into her cloaca. A substrate of broken terracotta pots or large flat natural rocks is added to provide substrate for the males to lay spermatophores.
20. We have found an approximate 30 % success rate in the mating of healthy animals. Reproductive success is variable depending on the season and age of the animals. Matings are generally

more successful during the winter and spring months, with lower success in the summer months.

21. Shipping axolotls is relatively straightforward and safe if the proper measures are taken to safeguard the welfare of the animal.
22. There is enough oxygen in the water for the animals to survive the trip.

Acknowledgments

We would like to thank the Ambystoma Genetic Stock Center for the continued support and advice on axolotl maintenance (R24 OD010435). We would also like to thank Panagiotis Katsaros and William Fowle for the electron microscopy images, Matthew Nguyen and Pankhuri Singhal for the histological staining, and Alex Sweeney and Jason Langshaw for the careful reading of the manuscript.

References

1. Shaffer HB (1989) Natural history, ecology, and evolution of the Mexican "Axolotls". Axolotl Newslett 18:5–11
2. Contreras V, Martínez-Meyer E, Valiente E, Zambrano L (2009) Recent decline and potential distribution in the last remnant area of the microendemic Mexican axolotl (Ambystoma mexicanum). Biol Conserv 142:2881–2885
3. Smith HM (1969) The Mexican axolotl: some misconceptions and problems. Bioscience 19: 593–615
4. Zambrano L, Reidl PM, McKay J, Griffiths R, Shaffer HB, Flores-Villela O, Parra-Olea G, Wake DB (2010) Ambystoma mexicanum. IUCN 2013. IUCN Red List of Threatened Species. Version 2013.2
5. Press A (2014) Axolotl, Endangered 'Water Monster,' Sighted In Mexico after fears of disappearance. 02/24/2014
6. Smith H (1989) Discovery of the axolotl and its history in research. In: Armstrong JB, Malacinski GM (eds) Developmental biology of the axolotl. Oxford University Press, New York, NY, pp 3–12
7. Malacinski GM (1978) The Mexican axolotl, Ambystoma mexicanum: its biology and developmental genetics, and its autonomous cell-lethal genes. Am Zool 18:195–206
8. Humphrey RR (1967) Albino axolotls from an albino tiger salamander through hybridization. J Hered 58:95–101
9. Voss SR, Smith JJ (2005) Evolution of salamander life cycles: a major-effect quantitative trait locus contributes to discrete and continuous variation for metamorphic timing. Genetics 170:275–281
10. Humphrey RR (1979) The axolotl colony at Indiana University. Axolotl Newslett 1:3–8
11. Voss SR, Epperlein HH, Tanaka EM (2009) Ambystoma mexicanum, the axolotl: a versatile amphibian model for regeneration, development, and evolution studies. Cold Spring Harb Protoc 2009:pdb.emo128. doi:10.1101/pdb.emo128
12. Monaghan JR et al (2014) Experimentally induced metamorphosis in axolotls reduces regenerative rate and fidelity. Regeneration 1:2–14
13. R: A language and environment for statistical computing
14. Zhang P, Wake DB (2009) Higher-level salamander relationships and divergence dates inferred from complete mitochondrial genomes. Mol Phylogenet Evol 53:492–508
15. Sandoval-Guzman T, Wang H, Khattak S, Schuez M, Roensch K, Nacu E, Tazaki A, Joven A, Tanaka EM, Simon A (2014) Fundamental differences in dedifferentiation and stem cell recruitment during skeletal muscle regeneration in two salamander species. Cell Stem Cell 14:174–187
16. Blassberg RA, Garza-Garcia A, Janmohamed A, Gates PB, Brockes JP (2011) Functional

convergence of signalling by GPI-anchored and anchorless forms of a salamander protein implicated in limb regeneration. J Cell Sci 124:47–56
17. Kumar A, Nevill G, Brockes JP, Forge A (2010) A comparative study of gland cells implicated in the nerve dependence of salamander limb regeneration. J Anat 217:16–25
18. Liversage RA, McLaughlin DS (1983) Effects of delayed amputation on denervated forelimbs of adult newt. J Embryol Exp Morphol 75:1–10
19. Page RB, Voss SR (2009) Induction of metamorphosis in axolotls (Ambystoma mexicanum). Cold Spring Harb Protoc 2009(8):pdb.prot5268. doi:10.1101/pdb.prot5268
20. Voss SR, Shaffer HB (1996) What insights into the developmental traits of urodeles does the study of interspecific hybrids provide? Int J Dev Biol 40:885–893
21. Francis ETB (1934) The anatomy of the salamander. The Clarendon Press, Oxford, With an historical introduction by Professor Cole FJ
22. Getchell TV (1977) Analysis of intracellular recordings from salamander olfactory epithelium. Brain Res 123:275–286
23. Mackay-Sim A, Shaman P (1984) Topographic coding of odorant quality is maintained at different concentrations in the salamander olfactory epithelium. Brain Res 297:207–216
24. Mackay-Sim A, Shaman P, Moulton DG (1982) Topographic coding of olfactory quality: odorant-specific patterns of epithelial responsivity in the salamander. J Neurophysiol 48:584–596
25. Lee J, Gardiner DM (2012) Regeneration of limb joints in the axolotl (Ambystoma mexicanum). PLoS One 7:e50615. doi:10.1371/journal.pone.0050615
26. Cosden RS, Lattermann C, Romine S, Gao J, Voss SR, MacLeod JN (2011) Intrinsic repair of full-thickness articular cartilage defects in the axolotl salamander. Osteoarthritis Cartilage 19:200–205
27. Grabowski SR, Pak WL (1975) Intracellular recordings of rod responses during dark-adaptation. J Physiol 247:363–391
28. Waloga G, Pak WL (1978) Ionic mechanism for the generation of horizontal cell potentials in isolated axolotl retina. J Gen Physiol 71: 69–92
29. Dvorak D (1984) Off-pathway synaptic transmission in the outer retina of the axolotl is mediated by a kainic acid-preferring receptor. Neurosci Lett 50:7–11
30. Makino CL, Dodd RL (1996) Multiple visual pigments in a photoreceptor of the salamander retina. J Gen Physiol 108:27–34
31. Isayama T, Chen Y, Kono M, Fabre E, Slavsky M, DeGrip WJ, Ma JX, Crouch RK, Makino CL (2014) Coexpression of three opsins in cone photoreceptors of the salamander Ambystoma tigrinum. J Comp Neurol 522:2249–2265
32. Smith SC (1996) Pattern formation in the urodele mechanoreceptive lateral line: what features can be exploited for the study of development and evolution? Int J Dev Biol 40:727–733
33. Straus NA (1971) Comparative DNA renaturation kinetics in amphibians. Proc Natl Acad Sci U S A 68:799–802
34. Putnam JL, Parkerson JB Jr (1985) Anatomy of the heart of the amphibia II. Cryptobranchus alleganiensis. Herpetologica 41:287–298
35. Lopez D, Lin L, Monaghan JR, Cogle CR, Bova FJ, Maden M, Scott EW (2014) Mapping hematopoiesis in a fully regenerative vertebrate: the axolotl. Blood 124:1232. doi:10.1182/blood-2013-09-526970
36. Erginel-Unaltuna N, Dube DK, Robertson DR, Lemanski LF (1995) In vivo protein synthesis in developing hearts of normal and cardiac mutant axolotls (Ambystoma mexicanum). Cell Mol Biol Res 41:181–187
37. Lemanski SF, Kovacs CP, Lemanski LF (1997) Analysis of the three-dimensional distributions of alpha-actinin, ankyrin, and filamin in developing hearts of normal and cardiac mutant axolotls (Ambystoma mexicanum). Anat Embryol (Berl) 195:155–163
38. Whitford WG, Sherman RE (1968) Aerial and aquatic respiration in axolotl and transformed Ambystoma tigrinum. Herpetologica 24:233–237
39. Brainerd EL (1998) Mechanics of lung ventilation in a larval salamander Ambystoma tigrinum. J Exp Biol 201:2891–2901
40. McKenzie DJ, Taylor EW (1996) Cardioventilatory responses to hypoxia and NaCN in the neotenous axolotl. Respir Physiol 106:255–262
41. Ching YC, Wedgwood RJ (1967) Immunologic responses in the axolotl, Siredon mexicanum. J Immunol 99:191–200
42. Charlemagne J (1979) Thymus independent anti-horse erythrocyte antibody response and suppressor T cells in the Mexican axolotl (Amphibia, Urodela, ambystoma mexicanum). Immunology 36:643–648
43. Sessions SK, Gardiner DM, Bryant SV (1989) Compatible limb patterning mechanisms in urodeles and anurans. Dev Biol 131:294–301
44. Harris WA, Cole J (1984) Common mechanisms in vertebrate axonal navigation: retinal transplants between distantly related amphibia. J Neurogenet 1:127–140

45. Sobkow L, Epperlein HH, Herklotz S, Straube WL, Tanaka EM (2006) A germline GFP transgenic axolotl and its use to track cell fate: dual origin of the fin mesenchyme during development and the fate of blood cells during regeneration. Dev Biol 290:386–397
46. Froese JM, Smits JE, Wickstrom ML (2005) Evaluation of two methods for measuring nonspecific immunity in tiger salamanders (Ambystoma tigrinum). J Wildl Dis 41: 209–217
47. Chinchar VG, Hyatt A, Miyazaki T, Williams T (2009) Family Iridoviridae: poor viral relations no longer. Curr Top Microbiol Immunol 328:123–170
48. Cotter JD, Storfer A, Page RB, Beachy CK, Voss SR (2008) Transcriptional response of Mexican axolotls to Ambystoma tigrinum virus (ATV) infection. BMC Genomics 9:493
49. Godwin JW, Pinto AR, Rosenthal NA (2013) Macrophages are required for adult salamander limb regeneration. Proc Natl Acad Sci U S A 110:9415–9420
50. Tsonis PA, Eguchi G (1981) Carcinogens on regeneration. Effects of N-methyl-N′-nitro-N-nitrosoguanidine and 4-nitroquinoline-1-oxide on limb regeneration in adult newts. Differentiation 20:52–60
51. Zilakos NP, Zafiratos CS, Parchment RE (1996) Stage-dependent genetically-based deformities of the regenerating newt limb from 4-nitroquinoline-N-oxide mutagenesis: potential embryonic regulation of cancer. Differentiation 60:67–74
52. Shioda C, Uchida K, Nakayama H (2011) Pathological features of olfactory neuroblastoma in an axolotl (Ambystoma mexicanum). J Vet Med Sci 73:1109–1111
53. Harshbarger JC, Chang SC, DeLanney LE, Rose FL, Green DE (1999) Cutaneous mastocytomas in the neotenic caudate amphibians Ambystoma mexicanum (axolotl) and Ambystoma tigrinum (tiger salamander). J Cancer Res Clin Oncol 125:187–192
54. Frost SK, Epp LG, Robinson SJ (1984) The pigmentary system of developing axolotls. I. A biochemical and structural analysis of chromatophores in wild-type axolotls. J Embryol Exp Morphol 81:105–125
55. Gresens J (2004) An introduction to the Mexican axolotl (Ambystoma mexicanum). Lab Anim 33:41–47
56. Keller RE, Lofberg J, Spieth J (1982) Neural crest cell behavior in white and dark embryos of Ambystoma mexicanum: epidermal inhibition of pigment cell migration in the white axolotl. Dev Biol 89:179–195
57. Seifert AW, Monaghan JR, Voss SR, Maden M (2012) Skin regeneration in adult axolotls: a blueprint for scar-free healing in vertebrates. PLoS One 7:e32875. doi:10.1371/journal.pone.0032875
58. Jarial MS (1989) Fine structure of the epidermal Leydig cells in the axolotl Ambystoma mexicanum in relation to their function. J Anat 167:95–102
59. Simon A, Tanaka EM (2013) Limb regeneration. Wiley Interdiscip Rev Dev Biol 2: 291–300
60. Harris WA (1989) Neurobiology. In: Armstrong JB, Malacinski GM (eds) Developmental biology of the axolotl. Oxford University Press, New York, NY, pp 157–168
61. Stirling RV, Brandle K (1982) Expansion of the visual projection to the tectum of axolotls during metamorphosis. Brain Res 281: 343–345
62. Dorries KM, White J, Kauer JS (1997) Rapid classical conditioning of odor response in a physiological model for olfactory research, the tiger salamander. Chem Senses 22:277–286
63. Schwartz JM, Cogan DC (1977) Position discrimination in the salamander, Ambystoma tigrinum. Dev Psychobiol 10:355–358
64. Kirsche K, Kirsche W (1964) Regenerative processes in the telencephalon of Ambystoma Mexicanum. J Hirnforsch 7:421–436
65. Richter W (1968) Regenerative processes following removal of the caudal sector of the telencephalon including the telencephalo-diencephalic border region in Ambystoma mexicanum. J Hirnforsch 10:515–534
66. Maden M, Manwell LA, Ormerod BK (2013) Proliferation zones in the axolotl brain and regeneration of the telencephalon. Neural Dev 8:1. doi:10.1186/1749-8104-8-1
67. Parish CL, Beljajeva A, Arenas E, Simon A (2007) Midbrain dopaminergic neurogenesis and behavioural recovery in a salamander lesion-induced regeneration model. Development 134:2881–2887
68. Okamoto M, Ohsawa H, Hayashi T, Owaribe K, Tsonis PA (2007) Regeneration of retinotectal projections after optic tectum removal in adult newts. Mol Vis 13:2112–2118
69. Wakimoto BT (1979) DNA synthesis after polyspermic fertilization in the axolotl. J Embryol Exp Morphol 52:39–48
70. Armstrong JB (1989) Spermatogenesis. In: Armstrong JB, Malacinski GM (eds) Developmental biology of the axolotl. Oxford University Press, New York, NY
71. Armstrong JB, Duhan ST (1989) Induced spawnings, artificial insemination and other

genetic manipulations. In: Armstrong JB, Malacinski GM (eds) Developmental biology of the axolotl. Oxford University Press, New York, NY, pp 229–235

72. Johnson AD, Crother B, White ME, Patient R, Bachvarova RF, Drum M, Masi T (2003) Regulative germ cell specification in axolotl embryos: a primitive trait conserved in the mammalian lineage. Philos Trans R Soc Lond B Biol Sci 358:1371–1379
73. Beetschen J-C (1989) Oogenesis. In: Armstrong JB, Malacinski GM (eds) Developmental biology of the axolotl. Oxford University Press, New York, NY
74. Borland S (2000) Practical Axolotl. Axolotl Newslett 28:17–22
75. Duhon ST (1989) Disease in axolotls. In: Armstrong JB, Malacinski GB (eds) Developmental biology of the axolotl. Oxford University Press, New York, NY, pp 264–269
76. Khattak S, Murawala P, Andreas H, Kappert V, Schuez M, Sandoval-Guzman T, Crawford K, Tanaka EM (2014) Optimized axolotl (Ambystoma mexicanum) husbandry, breeding, metamorphosis, transgenesis and tamoxifen-mediated recombination. Nat Protoc 9:529–540
77. Clare JP (2012) Axolotls. www.axolotl.org

Chapter 4

Husbandry of Spanish Ribbed Newts (*Pleurodeles waltl*)

Alberto Joven, Matthew Kirkham, and András Simon

Abstract

Research on urodele amphibians, such as newts, is constantly contributing to our understanding of fundamental biological processes. In the present chapter, we present detailed husbandry protocols for the Spanish ribbed newt (*Pleurodeles waltl*). We describe the main phases of their life cycle, with emphasis on the progressive development of sensory, motor, and integration systems, which lead to the acquisition of specific stereotyped (and conditioned) behaviors. The methods are outlined to manage housing, feeding, handling, captive breeding, health monitoring, and euthanasia in this species under laboratory conditions. With minor changes, these protocols can also be applied to other species of urodele amphibians commonly used in laboratory research.

Key words Salamander, Housing, Environmental enrichment, Reproduction, Breeding, Larvae, Development

1 Introduction

Urodele amphibians became well known by scientists in the past, mostly due to their large embryos, the relatively easy manipulation, and their external and slow development that can be modulated by temperature [1–4]. All these facts together permitted to easily analyze the key stages during ontogenetic development, and as a result, research in urodeles provided significant knowledge regarding the ontogeny of vertebrates [2, 5–9]. In addition, the absence of rejection after transplant [2, 10], their extraordinary regenerative abilities [11–15], the large size of their neurons and neuroendocrine cells [2], and the secondary simplification that characterizes their nervous system [16, 17] have attracted the attention of generations of researchers. Since urodeles have become amenable to modern molecular technologies, such as transgenesis and genome editing, they provide reemerging animal models in particular for studying development and regeneration [2, 15, 18–31]. The Spanish ribbed newt is a relatively large urodele species natural of the Iberian Peninsula and Morocco that can reach 31 cm long,

Anoop Kumar and András Simon (eds.), *Salamanders in Regeneration Research: Methods and Protocols*, Methods in Molecular Biology, vol. 1290, DOI 10.1007/978-1-4939-2495-0_4, © Springer Science+Business Media New York 2015

although most individuals' length ranges between 14 and 18 cm [32–35]. Wild populations develop two types of lifestyle, aquatic and terrestrial, although the latter mostly happens in order to survive either the drying of water masses in which they live or extremely cold environments [33, 35]. In the cases in which the place does not dry up and remain suitable, the Spanish ribbed newts carry out an entirely aquatic lifestyle [32, 36]. Accordingly, in captivity, they can be housed in entirely aquatic systems, provided with environmental enrichments, such as floating surfaces that some individuals eventually use [32]. Wild *Pleurodeles* can be found in shores, shallow and deepest parts of ponds, streams, and small lakes [33], although some authors suggest that they are more difficult to be seen during daytime, when they seem to remain in the deepest parts, hidden either among the aquatic vegetation or under stones [35, 37]. Therefore, for environmental enrichment, it is important for the adults to facilitate them the access to floating surfaces, aquatic plants, and hiding places. The diet during the adult aquatic stage is based on a wide spectrum of invertebrate and vertebrate species such as worms, crustaceans, snails, aquatic insects, small fish, and amphibians, even practicing cannibalism. Similarly, the larvae prey principally on different species of zooplankton, insects, and amphibian larval forms, including conspecifics [33, 35]. Two main conclusions can be obtained from this information: (1) it is recommended to provide more than just one source of food, and (2) the animals must be housed with tank mates of the same size in order to avoid loss of animals due to cannibalistic behaviors.

The breeding season varies in the different wild populations depending upon the environmental conditions, being restricted to the spring time in places where winter is harsh. However, in some temperate areas, the breeding season usually extends from September to April, coinciding the starting with the first rains of autumn [33, 35]. Mating occurs by spermatophore transfer after amplexus, in which the male holds the female with his forelimbs, placing her on his back. After several hours, the male twists around and deposits a spermatophore which is up taken by the female after the male manipulates her body [33, 35, 37–39]. 24–48 h later, the female start to lay a variable number of eggs, which is correlated with her age and size [33, 35, 39]. The embryonic and larval development takes 2–6 months in the wild, as is dependent on environmental factors, such as temperature, and the size of juvenile newts at metamorphosis is very variable [33, 35]. The specific developmental stages can be defined as outlined in Table 1 [40–47]. According to Gallien and Durocher [41], the different stages of the Spanish ribbed newt development are grouped in three principal periods: embryonic life (till hatching), initial larval period (ending at the beginning of feeding),

Table 1
Comparative depiction of developmental periods of commonly used salamander species based on external characters and physiological and behavioral key events

Life history	Am	Pw	Nv	Development of external characters	Development of key physiological and behavioral events
Embryonic stages	1–41	0–34	Tailbud stage, limb field stage bud stage	– Fertilization, first cleavages (*Pw* St. 0–4). – Blastulation (*Pw* St. 5–7). – Gastrulation (*Pw* St. 8–13). – Neurulation, somatogenesis (5 pairs of somites), eye primordium formation (*Pw* St. 14–21) – Somatogenesis (from 5 to 9 pairs of somites): brain flexure and growth, pronephros and tailbud formation, otic placode, and gill primordia appearance (*Pw* St. 22–24) – Tailbud (from 10 to 19 pairs of somites): brain growth and early segmentation, branchial furrowing, body axis straightening, tailbud enlargement, olfactory organ primordium, melanophores in trunk, and balancer appearance (*Pw* St. 25–29) – Limb field (from 20 somites on): brain regionalization, lateral line formation, appearance of melanophores in head region, unbranched gills enlargement, and circulation establishment in gills and tail (*Pw* St. 30–32) – Bud stage I: beginning of brain maturation, forelimb bud primordium formation, eye maturation, operculum and mouth formation, gill branching and elongation, second lateral line bypasses the forelimb on the ventral side (*Pw* St. 33) – Bud stage II: growth and maturation of hindbrain, mouth opening, forelimb bud formation, balancer elongation (*Pw* St. 34) ***Peculiarities*:** – In Notophthalmus viridescens, the embryo is encapsulated in a series of five concentric jelly layers. Due to the limited sac space and the strong layers covering it, the embryo is specially curved as growing proceeds – ***Ambystoma mexicanum*** don't develop balancers	Development of sensory systems – Mechanoreceptive endings in the skin – Skin excitability to noxious stimuli – A visual system firstly based on the pineal eye, which is excited by dimming Development of locomotion in urodeles – First movements (head flexure stage) – C-coil stage – S-wave stage – Swimming Miscellaneous – Primitive reflex mechanism: Muscular response to a mechanical stimulus – Heart beating appearance – Spontaneous muscular movements – Sporadic swimming – Hatching
Initial larval stages	42–44	35–38	Peg stage, e2D stage	– ***Peg stage*:** larva transparent, internal organs including heart, stomach, visceral arches, and auditory capsules are visible; balancer at the maximum point of development, hindbrain and midbrain show more growth and maturation than forebrain areas (*Pw* St. 35–36) – ***e2D stage*:** forelimb bud elongation (final shape: paddle), mouth opening, gill branching and elongation, yolk absorption, hindbrain, and midbrain show more growth and maturation than forebrain areas (*Pw* St. 37–38)	– Use of balancers to remain adhered in all surfaces – Steady behavior – Escape response triggered by visual/ vibrational stimuli – Feeding: stereotyped suction marks the end of this period

(continued)

Table 1
(continued)

Life history	Am	Pw	Nv	Development of external characters	Development of key physiological and behavioral events
Early active larval stages	45–52	39–45	*Forelimb*: 2D, e3D, 3DI, 3DII, 4D *Hind limb*: Limb field, bud I, bud II, bud III	– Formation of three fingers and elbow articulation in forelimbs, hind limb bud primordia appearance, balancer thinning, gill branching and elongation, numerous melanophores distributed in the caudal fin, main subdivisions of the brain already present (*Pw* St. 39–42) – Formation of the forth finger in forelimbs, hind limb bud formation, balancer reduction, differential growth of the forebrain areas (*Pw* St. 43–45)	– Scouting: head movements from side to side – Feeding: acquisition of lunging – Learning: modulation of suction + lunging – Cannibalistic behavior, more common at high density of larvae – Coordinated walking in the bottom with the forelimbs
Late active larval stages	53–57	46–55c	*Hind limb*: eT, 2T, 3T, 4T, 5T	– Balancer disappearance, formation of two fingers in hind limbs (*Pw* St. 46–49) – Formation of the five fingers and articulation in hind limbs, gills at the maximum point of development, growth and maturation of the brain, specifically forebrain basal ganglia (*Pw* St. 50–55a) – Gills reduced to half of their size, fin regression, limbs thickening, skin transformation in trunk region (*Pw* St. 55b) – First molting, skin transformation in the whole animal, flattened head, gills extremely reduced (*Pw* St. 55c)	– Scouting: head movements combined with walking in the active seeking for food – Feeding: acquisition of the use of olfactory cues – Learning: care taker recognition – Cannibalistic behavior modulated by learning – Progressive use of the hind limbs – Pulmonary breathing – Floating
Juvenile, adult	57, 57+	56, 56+	Eft, adult	– Metamorphosis I: complete disappearance of fin and gills, skin change, eyes reorientation (*Pw* St. 56) – Body growing, maturation of the brain (specifically hypothalamus), sexual maturation (*Pw* St. 56+) – Metamorphosis II: skin change, fin development, brain completely developed (*Pw* St. 56+) – ***Ambystoma mexicanum*** don't metamorphose	*Notophthalmus viridescens* and *Pleurodeles waltl* – Terrestrial lifestyle of the eft – Aquatic lifestyle of breeding adults – Terrestrial and aquatic lifestyles of nonbreeding adults depending on environmental cues – Feeding: jaw prehension or biting – Feeding: tongue prehension *Ambystoma mexicanum* – Completely aquatic lifestyle

Am Ambystoma mexicanum, *Nv Notophthalmus viridescens*, *Pw Pleurodeles waltl*, St stage and active larval life (in which limbs and complex behaviors develop) finishing at metamorphosis. In the following paragraph, more details of the different stages are given.

Throughout embryonic development (stages 0–34), in addition to the body plan formation, visual and olfactory primary organ primordia develop, together with the mechanoreceptory and chemoreceptory lateral line [41, 48]. The motor system also appears, and basic stereotyped movement acquisition ends up in swimming [49–51]. Before hatching, at each side of the head a balancer appears (a filament-like appendix used for remaining adhered to different surfaces), followed by the development of three branchial arches. During initial larval life (stages 35–38), behind the branching gills, the forelimb buds arise, and the visual and lateral line systems mature along with the escape response associated to visual and mechanical stimuli, although most of the time the larvae remain steady, absorbing the remaining yolk [26, 27, 52–54]. By the end of this period, larvae start to eat by stereotyped suction mechanism, driven also by visual and vibrational stimuli, but without active food searching. Consequently, in the laboratory, live foods such as *Artemia* nauplii should be provided. At this point, the active larval life begins and lasts about 3 months in *Pleurodeles waltl*, during which many morphological variations occur related to key physiological and behavioral changes [26, 27, 41, 45]. The active larval period can be further divided into two phases: early active larval period (stages 39–45, when forelimbs develop) and late active larval period (stages 46–55c, when hind limbs develop) [26, 27]. In the early active larval period, the most remarkable external changes include the loss of the balancers and forelimb development. In terms of hunting behavior, this period is driven by a refinement of the suction technique, related to scouting the surroundings by first forelimb-driven walking steps together with head movements from side to side, and the acquisition of lunging. During the late active larval period, the hind limbs are formed, a major growth in the body occurs, and the main structures present in the adult brain can be recognized [26, 27, 41, 45]. From this period onwards, the olfactory system start to be important for feeding, as active seeking followed by sniffing results in the capacity of predating on immobile food such as commercial fish pellets or frozen food. In addition, learning processes by classical conditioning can be assessed during this period, as larvae show different behaviors when raised in different conditions.

The present chapter describes husbandry protocols for successfully caring, breeding, and raising *Pleurodeles waltl* in the laboratory in a simplified manner with the aim of making the caretaker job as easy and efficient as possible.

2 Materials

2.1 Tanks, Containers, and Environmental Enrichment Complements

In general, most of the supplies available from Aquarium suppliers are sufficient.

1. Rack system: optional (Fig. 2a).
2. Plastic boxes in different sizes: pay attention to the maximum number of animals that are recommended to be housed together (Fig. 2b).
3. Aquaria: pay attention to the maximum number of animals that are recommended to be housed together (Fig. 2b).
4. Gravel: large size to avoid accidental ingestion.
5. Tiles: can be used either whole for big setups or broken in pieces for smaller tanks/containers.
6. Stones: select the irregular ones that can provide hiding places.
7. Natural or artificial plants.
8. Cork pieces/styrofoam: the borders ought to provide a shore-like structure.

2.2 Aquarium Setup, Maintenance, and Food

1. Filters (e.g., Fluval U2, U3, or U4 internal filters) or similar biofilters.
2. Air pumps with accessories.
3. Syphon (e.g., Marina Easy clean gravel cleaner).
4. Nets and filter-cleaning brushes.
5. *Artemia* hatchery.
6. Tap water conditioner (e.g., Nutrafin Aqua Plus Tap water conditioner).
7. Broad spectrum antibiotic/antimycotic (e.g., General tonic from Tetra Medica).
8. Marine salts (artificial aquarium sea salt mixture).
9. Live *Tubifex* worms.
10. Live Grindal worms.
11. Artemia cysts.
12. Frozen food and pellets.

2.3 Laboratory Supplies and Equipment

1. Surgical no. 5 forceps.
2. Plastic Pasteur pipettes.
3. Surgical scalpels.
4. 24-well plates.
5. Petri dishes (150 mm×25 mm cell culture dish).
6. Stereomicroscope with fiber optic illumination.

2.4 Reagents and Solutions

1. Ultrapure water (distilled water).
2. Human chorionic gonadotropin (e.g., Sigma-Aldrich, cat no. C8554).
3. Cysteine (L-cysteine hydrochloride hydrate).
4. MS-222 (Tricaine methanesulfonate): Dissolve 1 g of MS-222 in 1 L of dechlorinated tap water.
5. 10× Modified Holtfreter's solution (MHS): Dissolve 34.6 g of NaCl, 0.5 g of KCl, 1 g of $CaCl_2$, 2 g of $NaHCO_3$, and 2 g of $MgSO_4$ in 1 L of ultrapure water. Store at 4 °C.
6. 20 % MHS (for embryos): Dilute 20.4 mL of 10× stock MHS in 1 L of distilled H_2O. Add 10 mL of penicillin-streptomycin. Adjust the pH to 7.6 with 1 N NaOH.
7. 40–50 % MHS (for larvae and adults): Dilute 41.6–52.6 mL of 10× stock MHS in 1 L of distilled H_2O. Adjust the pH to 7.6 with 1 N NaOH.
8. 1× MHS (for recovering adults, unfertilized eggs, and sperm solution): Dilute 111.1 mL of 10× stock MHS in 1 L of distilled H_2O. Add 10 mL of penicillin-streptomycin. Adjust pH to 7.6 with 1 N NaOH.
9. Marine salt solution: dissolve 30 g of marine salts in 1 L of distilled H_2O.

3 Methods

3.1 Housing Metamorphic/Adult Newts

Although the metamorphic ribbed newts are air breathers, in nature this species is mostly aquatic and remains all the year in water if the conditions are favorable (*see* **Note 1**). As the laboratory conditions are quite stable, there is no need for the animals to undergo hibernation or aestivation, and newts can be housed in an aquatic-based environment described herein, supplemented with just a floating surface (Fig. 1a).

1. Prepare the location in which the animals will be settled, having in mind ventilation, light/darkness (*see* **Note 2**) conditions, and possible temperature fluctuations (*see* **Note 3**).
2. Select the appropriate container according to the number of animals that are to be housed (Fig. 2).
3. Brush the tank (*see* **Note 4**) under hot tap water, and give a final rinse it in cold tap water.
4. Fill up the tank with chlorine-free water (*see* **Note 5**).
5. Add environmental enrichment objects (*see* **Note 6**) distributed evenly in the tank, providing at least a floating surface, artificial plants, and hiding places (Fig. 1a).
6. Add the filter (*see* **Note 7**), heater, and air pump if desired.

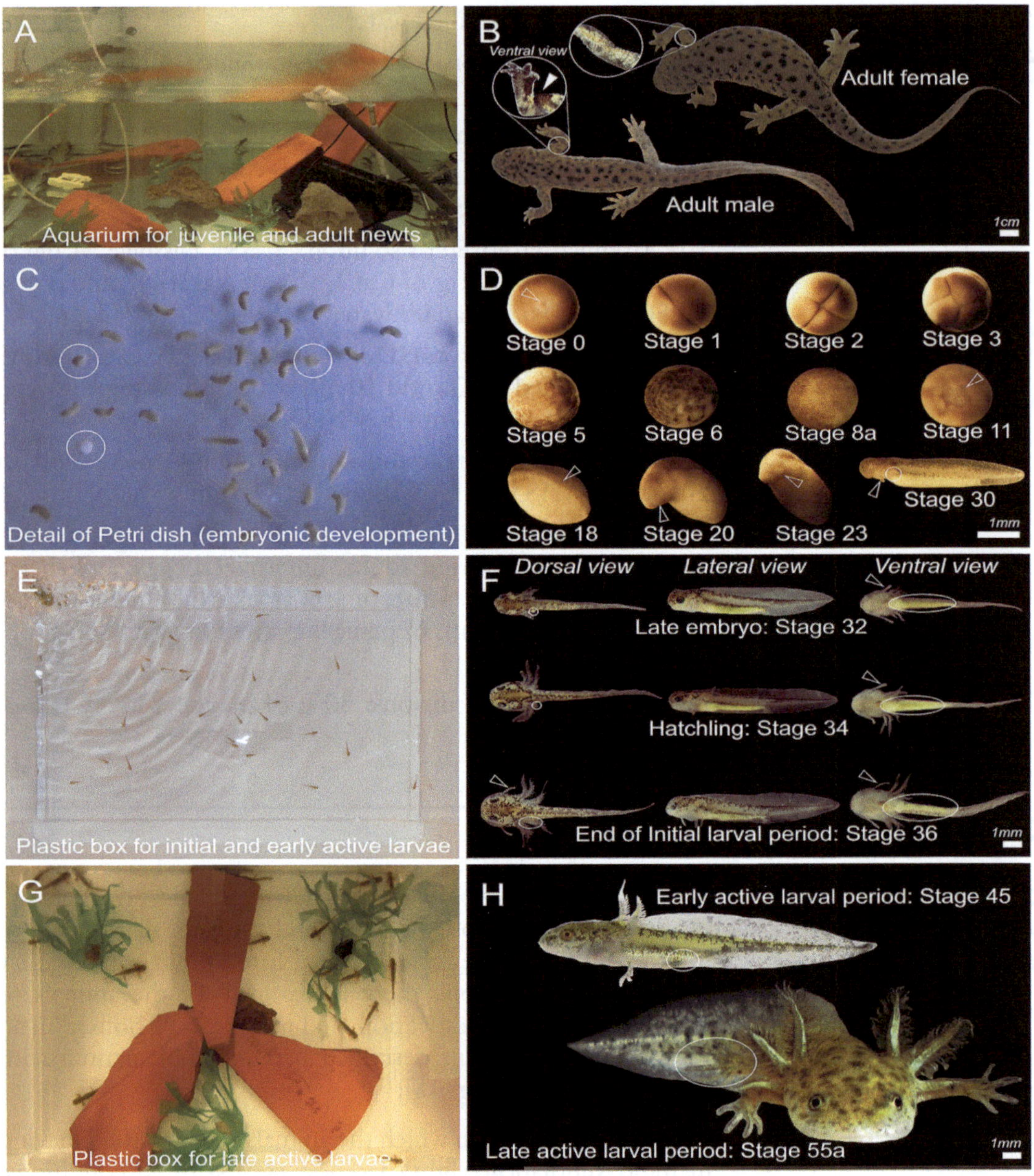

Fig. 1 Setups for Spanish ribbed newts and key features of the different developmental stages. (**a**) Aquarium for juveniles and adults. (**b**) Adult newts. Note the difference in the shape of the body and the specialization of the male forelimbs (*arrowhead*). (**c**) Detail of a Petri dish containing *Pleurodeles* embryos. Note the contaminated eggs that should be removed. (**d**) Representative stages of embryonic development, according to Gallien and Durocher [41]. The *arrowheads* point to the blastopore at stage 11, the closure of the neural tube at stage 18, the eye protrusion at stages 20 and 23, and the balancer at stage 30; the gill primordia are *encircled* at stage 30. (**e**) Box for housing initial and early active larvae. (**f**) Dorsal, lateral, and ventral views of late embryo, hatchling, and initial larva; note the progressive development of the forelimb buds (*small circles* in dorsal view), the development of the digestive system (*big circles* in ventral view), and the maintenance of the balancer (*arrowheads*). (**g**) Box for late active larvae. Observe the even distribution of environmental enrichment objects. (**h**) Early and late active larvae, showing the appearance and development of the hind limbs (*circles*)

A. Modulable rack system for housing newts

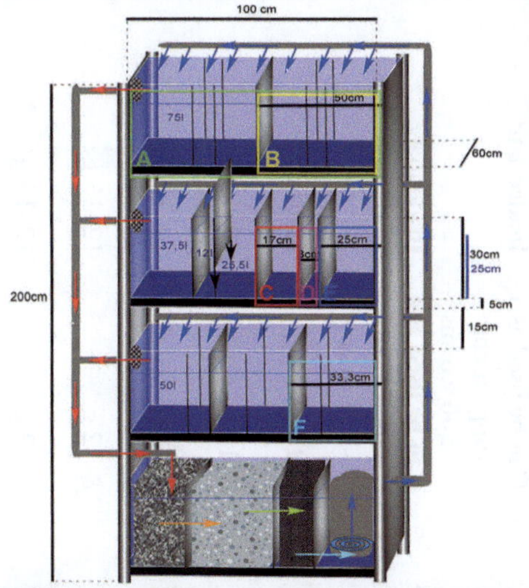

Compartment	Area (cm²)	Volume (liters)	Suitable for...	Ideal number of animals/compartment
A	6000	150	Adults (21-30cm)	7
B	3000	75	Adults (16-20 cm)	7
C	1020	25,5	Big late larvae / Juveniles (5-10 cm)	12 / 6
D	480	12	Late larvae up to 5cm	6
E	1500	37,5	Small adults (11-15 cm)	6
F	2000	50	Small adults (11-15 cm)	7

B. Recommended number of *Pleurodeles waltl* kept in group

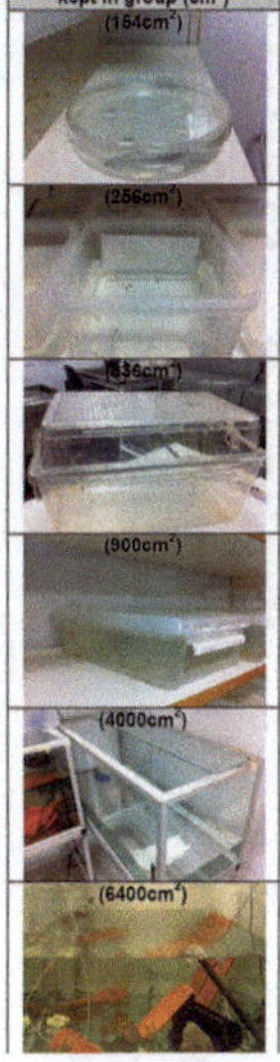

Area \ Animal size	Embryos + Initial larvae (up to 1,2cm)	Early Active Larvae (up to 1,7cm)	Late Active Larvae (up to 2,8cm)	Late Active Larvae (up to 4cm)	Late Active Larvae (up to 5,5cm)	Late Active larvae (up to 6,7 cm)	Juveniles (up to 10 cm)	Adults (11 to 15 cm)	Adults (16 to 20 cm)	Adults (21 to 30 cm)
Minimal area / animal size (cm²)	108	153	252	360	495	603	900	1500	2500	5000
Minimal area / animal if kept in group (cm²)	3	25	42	60	82	100	150	250	400	800
(154cm²)	51	6	3	-	-	-	-	-	-	-
(256cm²)	85	10	6	-	-	-	-	-	-	-
(836cm²)	278	33	19	13	10	-	-	-	-	-
(900cm²)	300	36	21	15	10	9	6	-	-	-
(4000cm²)	-	-	95	66	48	40	26	16	10	-
(6400cm²)	-	-	152	106	78	64	42	25	16	7

Fig. 2 Examples of housing systems for Spanish ribbed newts and advisable maximum number of animals that should be housed together. (**a**) Hypothetical rack system. (**b**) Housing in Petri dishes, plastic boxes and aquaria of different sizes

7. Check that the temperature of the water ranges between 15 and 28 °C.
8. Transfer the newts to the tank (*see* Subheading 3.3 for appropriate handling).
9. Ensure that the lids of the tank allow air exchange with the room and prevent newts from escaping.

3.2 Feeding Metamorphic/Adult Newts

Feed the animals every 2–3 days (*see* **Note 8**).

1. Select the appropriate food types (Table 2), weigh the desired quantity, and mix altogether (*see* **Note 9**).
2. Distribute the food evenly in the tank.
3. If some newts are found in the floating land areas, offer them food (*see* **Note 10**) helped by forceps (*see* **Note 11**).

3.3 Handling and Sexing Post-metamorphic/Adult *Pleurodeles waltl*

1. Use a net to take a newt out of the tank (*see* **Note 12**).
2. Grip the newt by the base of the tail with one hand.
3. Observe the cloaca (*see* **Note 13**).
4. Place it calmly in the palm of your other hand.
5. Observe the forelimbs (*see* **Note 14** and Fig. 1b).

Table 2

Types of food, sources, sizes, adequateness, and suggested quantities

Type	Name (and reference where possible)	Where to find it?	Size (mm)	Adequate for...	Quantity for feeding 200 late active larvae of ~0.2 g/larvae	Quantity for feeding 50 metamorphic juveniles of ~2 g/newt
Free-living food	*Artemia* nauplii from commercial cysts	Pet shop/online	0.4	Early active larvae	Inadequate	Inadequate
	Tubifex worms	Pet shop/aquarist	0.5 × 2 to 1 × 4	Early and late active larvae	Can be used as a supplement: ~500 worms	Inadequate
	Grindal worms (*Enchytraeus* sp.)	Pet shop/aquarist	0.5 × 10 to 1 × 40	Early active larvae	Can be used as a supplement: ~500 worms	Inadequate
Frozen food	Cyclops	Pet shop/online (e.g., zooschatz@t-online.de)	0.5–2	Early and late active larvae	8–12 g	Inadequate
	Bloodworms	Pet shop/online (e.g., zooschatz@t-online.de)	3–10	Late active larvae	8–12 g	12–15 g
	Mysis	Pet shop/online (e.g., zooschatz@t-online.de)	2 × 8	Late active larvae	8–12 g	12–15 g
	Black mosquito larvae	Pet shop/online (e.g., zooschatz@t-online.de)	3–10	Late active larvae	8–12 g	12–15 g
	White mosquito larvae	Pet shop/online (e.g., zooschatz@t-online.de)	3–10	Late active larvae	8–12 g	12–15 g

Pellets	Powder Microvit BASIC (70342)	Pet shop/online (www.tropical.pl)	0.1	Early active larvae	Inadequate	Inadequate
	Tropical energy food S (TENF190)	Pet shop/online (www.aquatic-nature.com)	0.4–0.8	Early and small late active larvae	0.5–1 g	Inadequate
	Cichlid mini granules (T506387-CE)	Pet shop/online (www.tetra.net)	1–2	Late active larvae	0.5–1 g	Inadequate
	TetraMin granules (T505322CE)	Pet shop/online (www.tetra.net)	1–3	Late active larvae and small juveniles	0.5–1 g	1–2 g
	New Life SPECTRUM Community fish formula	Pet shop/online (www.nlsfishfood.com)	1–3	Late active larvae and small juveniles	0.5–1 g	1–2 g
	Tropical Spirulina granulat (60434)	Pet shop/online (www.tetra.net)	1×2	Late active larvae and small juveniles	0.5–1 g	1–2 g
	New Life SPECTRUM Amphibian formula	Pet shop/online (www.nlsfishfood.com)	2	Metamorphic juveniles and adults	Inadequate	1–2 g
	Hikari tropical sinking wafers (21521)	Pet shop/online (www.hikari.info)	6×12	Metamorphic juveniles (previously broken) and adults	Inadequate	1–2 g
	JBL Pond Sterlet (4020000)	Pet shop/online (www.jbl.de)	7–8	Big adults	Inadequate	Inadequate

3.4 Cleaning Post-metamorphic/Adult *Pleurodeles waltl*

For small experimental setups lacking filter, partial cleaning should be done the day after feeding, and a thorough cleaning is recommended at least once every 7 days. In the case of large tanks equipped with filter and air pump, partial cleaning should be done once a week and thorough cleaning once a month.

Partial cleaning

1. Take the tube of the syphon and fill it up with water.
2. Holding both endings of the tube with your thumbs, place one of them inside the tank and direct the other one towards an empty bucket (*see* **Note 15**).
3. Release completely the ending of the tube that is inside the tank, and modulate the vacuum strength with your finger by opening and closing the opposite ending.
4. Use the vacuum to take out of the tank at least one third of the water from the bottom part, absorbing as much detritus as possible (*see* **Note 16**).
5. Refill the tank with chlorine-free tap water (*see* **Note 5**).

Deep cleaning

1. Set up a new tank using the procedure described in Subheading 3.1.
2. Transfer the animals to the new tank using a net.
3. Empty the water of the old tank. Rinse the filter, stones, cork mat, gravel, and plant over a bucket to collect the water.
4. Brush the stones and the cork mat, clean the filter (and all its components), and move the gravel under hot tap water and natural plants under cold tap water.
5. Brush the old tank under hot tap water.
6. Rinse everything in cold tap water.

3.5 Promoting Natural Breeding

Animals kept in good conditions will eventually produce natural spawns if both sexes are housed together, so regular ocular inspections shall be done to detect eggs (*see* **Note 17**). Remove the eggs from the breeding tank to avoid predation from the adults which subsequently provide the best conditions for the offspring to develop (*see* Subheading 3.7). Natural breeding can be promoted by mimicking the conditions in which the Spanish ribbed newts reproduce in their natural environment:

1. Decrease the water level gradually during 15–30 days, increase the temperature to 30 °C, and feed the newts only once a week during that period (*see* **Note 18**).
2. Increase the level of water in three consecutive days, decrease the temperature, and start feeding daily (*see* **Note 19**).

3. Observe the behavior of the newts (*see* **Note 20**) and look for eggs in the tank.
4. Collect all the eggs from the breeding tank using a wide-mouthed plastic pipette (*see* **Note 21**), and place them in a clean container (*see* **Note 22**).

3.6 Artificial Fertilization

For in vitro fertilization, males and females are kept separately, as the female newts are able to keep the sperm for several months after mating [55]. This ensures that fertilization occurs exactly when you want to.

1. Select adult newts that display breeding morphology (*see* **Note 23**).
2. Under anesthesia (*see* **Note 24**), give the female a dose of 50–200 IU of human chorionic gonadotropin (*see* **Note 25**) by injecting it subcutaneously in the lower jaw (*see* **Note 26**).
3. After 8–18 h, place the female in 100 % MHS (*see* **Note 27**); she will start to lay eggs that must be collected every 15 min to perform the in vitro fertilization (*see* **Note 28**).
4. Under anesthesia, squeeze the male abdomen by pressing gently to collect the sperm in a 2 mL tube containing 100 % MHS supplemented with antibiotics (*see* **Note 29**).
5. Place the collected eggs in a Petri dish and remove the excess of MHS. Add some sperm solution over the eggs, and then mix it by collecting the eggs with a widemouthed plastic Pasteur pipette (*see* **Note 30**).
6. Wait 15 min (*see* **Note 31**), and add 20 % MHS (supplemented with antibiotics) to the fertilized eggs to allow water absorption by the jelly (*see* **Note 32**).

3.7 Egg and Embryo Maintenance

A daily careful inspection of the eggs is essential to prevent microbial growth. Remove any unfertilized or contaminated eggs (*see* Fig. 1c), and change the medium (chlorine-free tap water or MHS) every 2–4 days.

1. Fill up the new housing place with chlorine-free tap water or 20 % MHS.
2. If necessary, release the egg from the jelly (*see* **Note 33**) either mechanically using no. 5 forceps or chemically by rinsing them in a solution of 2 % cysteine in 20 % MHS at pH 8.0 for 2–5 min. In the latter case, several washes in freshwater/20 % MHS must be done afterwards.
3. Follow the correct development of the embryos (Fig. 1d), and separate those that show abnormalities.
4. After hatching, when larvae start swimming, monitor daily the development of the embryos for free-living food to be supplemented at appropriate timing (*see* **Note 34** and Fig. 1f).

3.8 Housing Newt Larvae

1. Select an appropriate container (*see* **Note 35**) according to the number of animals that are to be housed and their developmental stage and size, in order to provide them with the required area (*see* **Note 36** and Figs. 1d–h and 2b).
2. Brush the tank under hot tap water, and give a final rinse it in cold tap water.
3. Fill up the tank (*see* **Note 37**) with chlorine-free water (*see* **Note 5**).
4. Fix the heater and air pump (*see* **Note 38**).
5. Add environmental enrichment objects in the case of late active larvae (*see* **Note 39** and Fig. 1g).
6. Check that the temperature of the water ranges between 15 and 28 °C.
7. Transfer the larvae to the tank (*see* Subheading 3.10 for appropriate handling of larvae).

3.9 Feeding Newt Larvae: General Procedures

1. Maintain the alternative sources of *Artemia* (*see* **Note 40**) regarding free-living food: *Tubifex* (*see* **Note 41**) and *Grindal* worms (*see* **Note 42**).
2. Select the appropriate food (Table 2) according to the developmental stage (*see* **Note 43**).
3. Chop the food in small pieces if necessary (*see* **Note 44**).
4. Mix the desired quantity in 500 mL of chlorine-free water.
5. Use a wash bottle (*see* **Note 45**) to evenly distribute the food among all the larvae (*see* **Note 46**).
6. Clean the wash bottle under tap water, and let it dry completely.

3.10 Handling Newt Embryos and Small Larvae

The big larvae can be easily collected using a net, but embryos and small larvae are more fragile and can be easily killed when trying to pick them up with a net. Use Pasteur pipette as described below.

1. Cut the tip of a plastic Pasteur pipette to have an opening larger than the size of the larvae to be transferred.
2. Calmly approach the larva with the widemouthed Pasteur pipette.
3. Capture the larva by a quick vacuum movement.
4. Use your finger tip to close the opening of the pipette (*see* **Note 47**).
5. Release the larva into its new location.

3.11 Cleaning Boxes and Tanks for Larvae

Clean the nonfeeding larvae setups (Petri dishes) at least once a week. The feeding larval containers require cleaning every second day (*see* **Note 48**).

3.11.1 Petri Dishes

1. Collect the eggs/embryos/initial larvae with a tip-cut pipette, and place them directly into a new one.
2. Clean the old Petri dish brushing it with hot tap water, give a final rinse in cold tap water, and let it dry completely. If contamination detected, discard the Petri dish.

3.11.2 Plastic Boxes

1. Place a large net over an empty plastic box or a bucket.
2. Prepare a small container with freshwater.
3. Take out the stones and artificial plants, and brush them under hot tap water.
4. Transfer the entire volume from the container with the newt larvae through the net.
5. Gently move the net to discard detritus and transfer the larvae into the small container with freshwater (*see* **Note 49**).
6. Remove any remaining debris using a pipette from the temporary tank containing the larvae.
7. Check in the empty container that no larvae escaped in the filtering process and discard the water.
8. Brush the plastic tank under hot tap water, and give a final rinse in cold tap water.
9. Fill the tank with chlorine-free water.
10. Replace the stones and artificial plants and gently place the larvae into the clean tank.

3.12 Health Inspection and Treatment of Diseases

Periodically observe the animals for signs of abnormalities in the skin (i.e., bleeding, ulceration, pigmentation abnormalities, fungus), behavior (i.e., slow moving, floating, or upside-down animals), and body constitution (either bloating or extremely skinny animals) (*see* **Note 50**).

1. If a sick animal is detected, isolate it from the main tank.
2. Place the animal in a temporary plastic container filled with marine salt solution for 1–5 min. Skip this step if the affected newt is a larva.
3. Transfer the newt into a tank with an easy access to an artificially created land area and containing 40–50 % MHS supplemented with antibiotic/antimycotic (follow manufacturer's instructions) (*see* **Note 51**). The depth should be around 10 cm.
4. Change the water of the tank every second day.
5. Daily follow the state of the sick newt and check the other newts in the tank in which the sick animal was found.
6. If more animals are affected in the same tank, empty it, brush it with marine salt solution, autoclave the stones/tiles, submerge the filter for 1 h in salt solution after cleaning it, replace

the plants with new ones, and treat all the tank mates as described in **steps 2** and **3**. Refill the tank with chlorine-free water and add antibiotic/antimycotic solution.

7. Every 2 or 3 days, change one half of the volume of the tank.
8. Stop the treatment when the animals look healthy, 2 days after the signs disappear.

3.13 Euthanasia

In those cases where no signs of improvement are observed after treatment, the animal should be euthanized.

1. Place the animal in 0.1 % Tricaine so the depth of the water is about 1–3 cm.
2. Wait a minimum of 30 min to ensure that the newt is deeply anesthetized.
3. Using sharp scissors, cut through the neck of the salamander to remove the head.
4. Quickly crush the skull with the flat side of the scissors or a blade.
5. Freeze the rests of the animal inside a paper towel and discard it according to the local regulations of the laboratory.

4 Notes

1. During the periods of extreme temperatures, *Pleurodeles waltl* may undergo hibernation and aestivation. During these processes, the skin becomes thicker and less permeable; Animals decrease food intake and slow down the metabolic rate. Nevertheless, Spanish ribbed newts have been found all along the year having an active lifestyle in the water [32–36].
2. Natural populations experience slight variations along the year in the light-dark (LD) cycle which approximately ranges from (LD 15:9) in summer to the opposite in winter (LD 9:15). They can be kept in a (LD 12:12) in laboratory conditions, although placing the newts under the influence of natural light (i.e., close to a window) or alternatively adapting the timer of the animal facility trying to mimic the natural LD cycle (±15 min of light/week) may benefit the adult newts and induce the natural breeding at the corresponding time of the year.
3. The Spanish ribbed newt has a high temperature tolerance. In nature, this species survives in regions in which absolute temperatures range from bellow −10 °C to over 40 °C. In the laboratory, the stable temperature prevents most of the newts undergoing hibernation and aestivation. Nevertheless, some animals may decide to spend several days/weeks in the floating surfaces, eating very little, moving slowly, and showing

skin changes typical of those two processes. Unless an animal experiences an alarming loss of weight (such as two thirds of their initial body mass), there is no reason to worry. They will sooner or later go back to the water and eventually breed.

4. You must not use any soap or detergent. Usually, brushing the different components under hot tap water is sufficient, but if deep cleaning is desired, a good option is to use marine salt solution described in Subheading 2.4, **item 9**. Disinfection by autoclaving is advisable for stones, rocks, and tiles. In some cases, 10 % bleach can be used for cleaning or disinfecting filters, but it is very important to rinse the tanks and filters several times with tap water, let it dry for 24 h, and rinse it again several times before use.
5. Keep tap water in a separate container for 24 h to remove chlorine, and to equilibrate with the room temperature. Alternatively, anti-chlorine product can be added to temperate tap water (15–25 °C) according to the manufacturer's instruction.
6. Newts should have the opportunity of choosing between climbing on floating devices (such as cork mat or Styrofoam) or remaining in the water. Under the water, they appreciate hiding places. Entire or broken tiles are probably the best option for this purpose, as they are easy to clean and offer a good hiding area. Live or artificial plants available from aquarium suppliers are a good option to enhance the atmosphere in an aquarium. For making an artificial plant, place some gravel in a light-colored plastic bag (better green color) and make a tight knot close to the base; then, cut the remaining plastic in stripes of 1–2 cm thickness. The use of gravel or stones is optional and it makes cleaning more laborious. If included, the size of the gravel should be large enough so that the newts do not swallow them by mistake.
7. The filter should direct the water flow towards the surface or one of the corners of the tank in order to avoid strong flow.
8. Newts do not require feeding every day; two to three feedings per week is sufficient. The frequency of feeding can be modulated according to the season to mimic natural conditions. Animals can be offered food once a week for mimicking winter time (LD 9:15) and every day for mimicking summer time (LD 15:9).
9. Alternatively, choose one type of food for each feeding day.
10. To avoid rejection of the food in land, the dry pellets may be placed in water for some min before offering them to the newts and the frozen food shall be defrost.
11. In land, newts will rely mostly on their visual system, and thus they recognize the food only if it moves. For that reason, one should offer different types of food with a forceps, moving it

back and forth as if it would be a living prey (or alternatively living preys such as earthworms can be offered with the forceps). After a first recognition, some newts will approach the prey with snout and catch them by jaw prehension, while others may grab them directly by tongue protrusion [38, 57].

12. Regarding body proportions, males present larger tails, while females tend to show a thicker body [32, 33, 35, 54].
13. Males' cloaca shows papillary appearance, whereas females' cloaca presents inward borders. Males' cloaca swells up during the breeding season [32, 33, 35, 54].
14. Males' forelimbs are proportionally longer and stronger than females' ones, and during breeding season the male forelimbs develop black rugosities (nuptial pads) that are used to grip the females during amplexus [32, 33, 35, 54].
15. For the vacuum to work, the level of the water in the tank must be higher up than the bucket.
16. Avoid vacuuming close to the newts. If an animal gets stuck into the tube, simply close the outflow ending of the tube and the animal will be able to swim out of the tube.
17. Eggs can be found attached by the jelly coat either in the substrate, stones, or artificial plants. The eggs have an upper dark animal pole and a lower opaque vegetal pole, so they are usually detectable after softly moving those objects in where they may be present.
18. For the best results, coinciding with the decrease of water level and feeding frequency, gradually lengthen the light cycle to mimic the summer period (LD 15:9). Return slowly towards the autumn-like light cycle (LD 12:12) starting 1 week before increasing water level and food intake.
19. Not only increasing the quantity but also the variety of food may have a positive effect on the induction of natural breeding.
20. The behavior of interest is the amplexus, as it precedes the spawn. It can be recognized easily (*see* Subheading 1).
21. The plastic pipette should be cut in order to get a 3–4 mm tip diameter. The jelly coat surrounding the egg is highly sticky; therefore, once the eggs are drawn into the pipette, close the opening with a finger.
22. Petri dishes can be used to facilitate oxygen exchange in water, or alternatively an air pump can be supplied if the eggs are placed in a large plastic container.
23. Breeding ribbed newts have swollen cloacae and flattened tails.
24. To anesthetize a newt, place it in a plastic box containing 2 cm depth Tricaine solution (*see* Subheading 2.4, **item 4**) and close the lid. Wait a minimum of 10 min before seeing if it is

anesthetized by placing the animal on its back. If it doesn't move within 5–10 s, it is anesthetized.

25. Eggs can be collected from the same animal every 2–4 weeks, although it is better to wait 2–3 months to avoid female's stress.
26. The hormone can alternatively be injected intramuscularly in the dorsal musculature, although the results are less satisfactory.
27. 1× MHS is used rather than water, as it prevents the jelly to absorb water, which in turn inhibits fertilization.
28. Eggs remain fertile for about 30 min. The female can be gently squeezed in order to collect the eggs ready to be laid in the oviduct every hour, although the best results are obtained by simply collecting the eggs every 15 min, avoiding the stress produced by squeezing.
29. Sperm can be collected from the same animal every 2–3 weeks.
30. This step improves the fertilization rate, as the sperm motility is activated by the jelly coat proteins, and the plastic pipette mimics the oviduct in which eggs are fertilized in the case of natural spawns.
31. Do not let the eggs completely dry at any time.
32. Eggs can be kept in chlorine-free tap water; however, the best survival rates are obtained when artificial freshwater is used, especially when supplemented with antibiotics/antimycotics (See Subheading 2.4, **item 6**).
33. Egg jelly is a natural protection of the amphibian eggs but sometimes sticks to undesirable material, such as detritus. In this case, eggs shall be released from the jelly coat.
34. A good knowledge regarding newts' life cycle is crucial for free-living food to be supplemented at appropriate timing, preventing the loss of animals due to cannibalistic behaviors (Table 1). A sign indicating that larvae are ready to start hunting is the disappearance of the whitish/yellowish color in the belly.
35. For eggs and embryos, Petri dishes are a good option, as the oxygen requirement of eggs and early embryos is low and it facilitates daily monitoring to separate unfertilized/dead eggs from the healthy ones. For larvae, plastic tanks with external air pumps are recommended, as they are more adaptable, and it will allow to separate the animals based on size.
36. The surface area of the tank is more important than the volume, as newt larvae mostly lay in the bottom. The abrupt increase in areal requirements from sedentary to early active larvae as shown in Fig. 2b is due to the beginning of feeding. This is to minimize cannibalistic behavior.
37. It is better to keep small larvae (initial and early active) in a container with 5–10 cm depth of water, so they can find the

free-living food easily, they do not accidentally desiccate, and the waste derived from non-eaten food and detritus will not strongly decrease oxygen content. The water depth can be increased for late larvae around 3–4 cm until 15 cm, as they start to use all levels of the water column, albeit 10 cm is enough. This will also help them to get air from the water surface, after they form the lungs and combine oxygen intake from water and air sources (stage 55a).

38. Both the heater and the air pump are recommendable in larval setups, although not essential. The former can be used to increase the growth rate of the larvae, and the latter improves the oxygen concentration of the water, preventing the growth of undesirable microbiota and the consequent death of larvae, especially when the water temperature is high.
39. The environmental enrichment of larvae of stages 45–55a is based on subaquatic objects (i.e., hiding objects and natural/artificial plants). When signs of metamorphosis are seen, such as absorption of the gills and a major change in pigmentation, a floating surface should be provided. In addition, a lid is necessary at this point to prevent young metamorphic newts escaping out of the tank.
40. *Artemia* nauplii provide adequate free-living food for the early active larvae [55, 56]. Commercial *Artemia* cysts will hatch in 1–3 days, and hatched nauplii can be used as food as long as they remain alive, which is only possible if oxygen and food is given. It's good to start a new culture every second day. Filter the *Artemia* to discard the salt water of the culture and distribute the nauplii evenly among the newt larvae using a wash bottle.
41. Keep the *Tubifex* sp. in a plastic tank filled up with water and provided with oxygenation through an air pump. Small doses of fish pellets can be provided as food, but it is not essential. Check the transparency of the water in the *Tubifex* culture; if it is cloudy, change it by chlorine-free tap water (it must be done every 1–3 days). Discard the dead worms. For feeding the larvae, collect some living worms and chop them in small pieces using the lid of a Petri dish and a pair of scalpels.
42. Keep the *Grindal* worms (*Enchytraeus* sp.) in several small plastic boxes filled up with moist soil, avoiding flood and in dark conditions. Pre-soaked dry dog/cat food or fish pellets can be used as food source. Worms can be easily collected with forceps no. 5, a couple of hours after providing a food pellet to the culture (they will surround the pellet). If the worm culture grows too fast (i.e., the food is consumed within 24 h), split the culture into two.

43. During the initial larval period, hatchlings live on their remaining stalk; thus, no feeding is necessary. Once the bellies start to lose their whitish-yellow appearance of the stalk, the initial larval period is about to start, and the first meal (free-living food) should be added. In the early active larval period, larvae start to prey over small live food, which is recognized usually by a combination of visual and vibrational cues. They do not seek for food at these stages, but prey only over the small creatures that pass by their heads. Cannibalistic behaviors become frequent; thus, it is recommendable to keep them at low densities and "ad libitum" conditions. At the beginning of the late active larval period, larvae start to use odor cues to discriminate also steady food, and learning processes modulate food intake. As a consequence, cannibalistic individuals should be isolated if detected. Once cannibalism is solved by housing larvae at appropriate densities, newts can learn that dry pellets are a source of food. The pellet size after hydration should be slightly smaller than the mouth of the animal.
44. For small and medium active larvae (stages 45–54), cut the frozen pellets as small as you can before the tablet gets defrost (also the *Tubifex*). Mix it in freshwater and filter it.
45. Cut the tip of the wash bottle to increase the flow and to avoid that food gets stuck inside the outflow tube.
46. The food tends to sink; thus, shake the wash bottle before adding the food into every tank.
47. Closing the opening of the pipette will prevent the larvae to escape from the pipette and end up on the floor or another dry surface. If they escape, try to save the larvae using a net or a wet piece of paper. Avoid catching them with the fingers or forceps.
48. If the tank is equipped with aeration, partial cleaning can be done every second day (*see* Subheading 3.4), followed by total cleaning after the next 2 days. Do not vacuum the larvae, and always filter the discarded water through a net before disposal to ensure that no larvae are being lost in the process.
49. Try to avoid touching the new clean water with the net; instead, let the larvae gently drop down at a short distance of the water surface. This procedure will prevent transfer of any debris from the net to the water.
50. The Caudata Culture website (www.caudata.org) provides information on the common symptoms and diseases in urodeles [58].
51. You may need to contact the university veterinarian for the treatment.

Acknowledgments

This work could not have been possible without the opportunities of working with various salamander species along the years, under different group leaders: Paschalina Kiriakopoulou-Sklavounou (AUTH, Greece), Agustin González (UCM, Spain), and Elly M. Tanaka (MPI-CBG, Germany). Also, many thanks to all the people ever involved in the caring of the different colonies and for interesting conversations that resulted in improvements of the protocols, with special thanks to Heino Andreas, Shahul Hameed, Heng Wang, Laure Belnoue, Tiago Pinheiro, Ivanna Mayorenko, Jorge Perlado, Patricia Fernández, Joana Branco, Laura Domínguez, Sandra Bandín, Ruth Morona, and Nerea Moreno. This work was supported by a postdoctoral fellowship to A.J. from the Wenner-Gren-Foundation and by the Swedish Research Council to A.S.

References

1. Duprat AM, Gualandris L, Kan P, Pituello F, Saint-Jeannet JP, Boudannaoui S (1986) Archives d'anatomie microscopique et de morphologie expérimentale. Arch Anat Microsc Morphol Exp 75:211–227
2. Eagleson GW (1996) Developmental neurobiology of the anterior areas in amphibians: urodele perspectives. Int J Dev Biol 40:735–743
3. Nieuwkoop PD (1996) What are the key advantages and disadvantages of urodele species compared to anurans as a model system for experimental analysis of early development? Int J Dev Biol 40:617–619
4. Gualandris-Parisot L, Husson D, Foulquier F, Kan P, Davet J, Aimar C, Dournon C, Duprat AM (2001) *Pleurodeles waltl*, amphibian, Urodele, is a suitable biological model for embryological and physiological space experiments on a vertebrate. Adv Space Res 28: 569–578
5. Spemann H (1921) Die Erzeugung tierischer Chimären durch heteroplastische embryonale Transplantation zwischen Triton cristatus und taeniatus. Wilhelm Roux Arch Entwickl Mech Org 120:374–706
6. Spemann H, Mangold H (1924) Über Induktion von Embryonanlage durch implantation artfremder Organisatoren. Wilhelm Roux Arch Entwickl Mech Org 100:599–638
7. Holtfreter J (1933) Der Einfluss von Wirtsalter und verschiedenen Orgenbezirken auf die Differenzierung von angelagertem Gastrulaektoderm. Wilhelm Roux Arch Entwickl Mech Org 127:610–775
8. Duprat AM, Gualandris L, Rouge P (1982) Neural induction and the structure of the target cell surface. J Embryol Exp Morphol 70: 171–187
9. Boucaut JC, Bernard B, Aubery M, Bourrillon R, Houillon C (1979) Concanavalin A binding to amphibian embryo and effect on morphogenesis. J Embryol Exp Morphol 51:63–72
10. Nacu E, Knapp D, Tanaka EM, Epperlein HH (2009) Axolotl (Ambystoma mexicanum) embryonic transplantation methods. Cold Spring Harb Protoc 2009(8):pdb.prot5265. doi:10.1101/pdb.prot5265
11. Ferretti P (1996) Re-examining jaw regeneration in urodeles: what have we learnt? Int J Dev Biol 40:807–811
12. Mitashov VI (1996) Mechanisms of retina regeneration in urodeles. Int J Dev Biol 40: 833–844
13. Parish CL, Beljajeva A, Arenas E, Simon A (2007) Midbrain dopaminergic neurogenesis and behavioural recovery in a salamander lesion-induced regeneration model. Development 134:2881–2887
14. Chernoff EA, Stocum DL, Nye HL, Cameron JA (2003) Urodele spinal cord regeneration and related processes. Dev Dyn 226:295–307
15. Nacu E, Tanaka EM (2011) Limb regeneration: a new development? Annu Rev Cell Dev Biol 27:409–440
16. Roth G, Dicke U, Nishikawa K (1992) How do ontogeny, morphology and physiology of sensory systems constrain and direct the evolution of amphibians? Am Nat 139:S105–S124

17. Roth G, Nishikawa KC, Naujoks-Manteuffel C, Schmidt A, Wake DB (1993) Paedomorphosis and simplification in the nervous system of salamanders. Brain Behav Evol 42:137–170
18. Gonzalez A, Smeets WJ (1991) Comparative analysis of dopamine and tyrosine hydroxylase immunoreactivities in the brain of two amphibians, the anuran Rana ridibunda and the urodele Pleurodeles waltlii. J Comp Neurol 303: 457–477
19. Gonzalez A, Marin O, Smeets WJ (1995) Development of catecholamine systems in the central nervous system of the newt Pleurodeles waltlii as revealed by tyrosine hydroxylase immunohistochemistry. J Comp Neurol 360: 33–48
20. Moreno N, Lopez JM, Sanchez-Camacho C, Gonzalez A (2002) Development of NADPH-diaphorase/nitric oxide synthase in the brain of the urodele amphibian *Pleurodeles waltl*. J Chem Neuroanat 23:105–121
21. Lopez JM, Moreno N, Gonzalez A (2003) Ontogeny of choline acetyltransferase (ChAT) immunoreactivity in the brain of the urodele amphibian *Pleurodeles waltl*. Brain Res Dev Brain Res 140:29–43
22. Brockes JP, Kumar A (2005) Appendage regeneration in adult vertebrates and implications for regenerative medicine. Science 310: 1919–1923
23. Voss SR, Epperlein HH, Tanaka EM (2009) Ambystoma mexicanum, the axolotl: a versatile amphibian model for regeneration, development, and evolution studies. Cold Spring Harb Protoc 2009(8):pdb.emo128. doi:10.1101/pdb.emo128
24. Berg DA, Kirkham M, Beljajeva A, Knapp D, Habermann B, Ryge J, Tanaka EM, Simon A (2010) Efficient regeneration by activation of neurogenesis in homeostatically quiescent regions of the adult vertebrate brain. Development 137:4127–4134
25. Berg DA, Kirkham M, Wang H, Frisen J, Simon A (2011) Dopamine controls neurogenesis in the adult salamander midbrain in homeostasis and during regeneration of dopamine neurons. Cell Stem Cell 8:426–433
26. Joven A, Morona R, Moreno N, González A (2013) Regional distribution of calretinin and calbindin-D28k expression in the brain of the urodele amphibian *Pleurodeles waltl* during embryonic and larval development. Brain Struct Funct 218:969–1003. doi:10.1007/s00429-012-0442-1
27. Joven A, Morona R, González A, Moreno N (2013) Spatiotemporal patterns of Pax3, Pax6 and Pax7 expression in the developing brain of a urodele amphibian, *Pleurodeles waltl*. J Comp Neurol 521:40. doi:10.1002/cne.23385
28. Maden M, Manwell LA, Ormerod BK (2013) Proliferation zones in the axolotl brain and regeneration of the telencephalon. Neural Dev 8:1. doi:10.1186/1749-8104-8-1
29. Sandoval-Guzmán T, Wang H, Khattak S, Schuez M, Roensch K, Nacu E, Tazaki A, Joven A, Tanaka EM, Simon A (2014) Fundamental differences in dedifferentiation and stem cell recruitment during skeletal muscle regeneration in two salamander species. Cell Stem Cell 14: 13. doi:10.1016/j.stem.2013.11.007
30. Kirkham M, Berg DA, Simon A (2011) Microglia activation during neuroregeneration in the adult vertebrate brain. Neurosci Lett 497:11–16
31. Kirkham M, Hameed LS, Berg DA, Wang H, Simon A (2014) Progenitor cell dynamics in the newt telencephalon during homeostasis and neuronal regeneration. Stem Cell Reports 2:12. doi:10.1016/j.stemcr.2014.01.018
32. García-Paris M (1985) Los anfibios de España. Neografis, S.L., Madrid
33. Barbadillo-Escriva LJ (1987) La guía de Incafo de los anfibios y reptiles de la península ibérica, baleares y canarias. Incafo, S.A., Madrid
34. Díaz-Paniagua C, Gómez C, Portheault A, DeVries W (2005) Los anfibios de Doñana. Organismo Autónomo de Parques Nacionales, Ministerio de Medio Ambiente, Madrid
35. Salvador A (2014) Gallipato—*Pleurodeles waltl*. Museo Nacional de Ciencias Naturales. www.vertebradosibericos.org/anfibios/pdf/plewal.pdf
36. Hódar JA, Ruiz I, Camacho I (1993) Régimen alimenticio estival del gallipato *Pleurodeles waltl* (Michaheles, 1830) en una localidad del sureste peninsular. Rev Esp Herpetol 7:5
37. Martínez-del-Marmol G, Jiménez-Robles O (2013) *Pleurodeles waltl* Michahelles, 1830 in Morocco and Western Sahara. www.morocco-herps.com/en/ficha/Pleurodeles_waltl/
38. Griffiths R (1995) Newts and salamanders of Europe. T & AD Poyser natural history, London
39. Schleich HH, Kästle W, Kabisch K (1996) Amphibians and reptiles of North Africa. Koeltz Sci. Books, Koenigstein
40. Glaesner L (1925) Normentafel zur Entwicklung des gemeinen Wassermolchs (Molge vulgaris). Normentafeln zur Entwicklungsgeschichte der Wirbeltiere
41. Gallien L, Durocher M (1957) Table chronologique de développment chez Pleurodeles waltlii Michahelles. Bull Biol 2:1–19

42. Armstrong JB, Malacinski GM (1989) Developmental biology of the axolotl. Oxford University Press, New York
43. Bordzilovskaya NP, Dettlaff TA, Duhon ST, Malacinski GM (1989) Developmental-stage series of axolotl embryos. In: Armstrong JB, Malacinski GM (eds) Developmental biology of the axolotl. Oxford University Press, New York, pp 201–219
44. Khan PA, Liversage RA (1995) Development of Notophthalmus viridescens embryos. Dev Growth Differ 37:529–537
45. Shi DL, Boucaut JL (1995) The chronological development of the urodele amphibian *Pleurodeles waltl* (Michah). Int J Dev Biol 39:427–441
46. Nye HL, Cameron JA, Chernoff EA, Stocum DL (2003) Extending the table of stages of normal development of the axolotl: limb development. Dev Dyn 226:555–560
47. Wong CJ, Liversage RA (2005) Limb developmental stages of the newt Notophthalmus viridescens. Int J Dev Biol 49:375–389
48. Roberts A, Soffe SR, Clarke JD, Dale N (1983) Initiation and control of swimming in amphibian embryos. Symp Soc Exp Biol 37:261–284
49. Coghill GE (1929) Anatomy and the problem of behaviour. Cambridge University Press, London
50. Soffe SR, Clarke JDW, Roberts A (1983) Swimming and other centrally generated motor patterns in newt embryos. J Comp Physiol 152: 535–544
51. Bekoff A (1985) Development of locomotion in vertebrates, a comparative perspective. In: Gollin ES (ed) The comparative development of adaptive skills: evolutionary implications. Lawrence Erlbaum Associates, Hillsdale, NJ, pp 57–94
52. Diesener G, Reichholf J (1986) Lurche und Kriechtiere (Reptiles y anfibios). Mosaik Verlag GmbH, München
53. Nöllert A, Nöllert C (1992) Die Amphibien Europas. Bestimmung-Gefährdung-Schutz (Los anfibios de Europa. Identificación-Amenazas-Protección). Franckh-Kosmos Verlags-GmbH, Sttutgart
54. Salvador A, García-París M (2001) Anfibios españoles. Identificación, historia natural y distribución. Canseco editores, S.L., Talavera de la Reina
55. Hayashi T, Yokotani N, Tane S, Matsumoto A, Myouga A, Okamoto M, Takeuchi T (2013) Molecular genetic system for regenerative studies using newts. Dev Growth Differ 55: 229–236
56. Khattak S, Murawala P, Andreas H, Kappert V, Schuez M, Sandoval-Guzmán T, Crawford K, Tanaka EM (2014) Optimized axolotl (*Ambystoma mexicanum*) husbandry, breeding, metamorphosis, transgenesis and tamoxifen-mediated recombination. Nat Protoc 9: 529–540
57. Deban SM, O'Reilly JC, Nishikawa KC (2001) The motor control of feeding in amphibians. Am Zool 41:18
58. Macke J (2007) Caudata Illness Part 3: Illness Photographs. www.caudata.org/cc/articles/illness3.shtml

Chapter 5

Maintaining Plethodontid Salamanders in the Laboratory for Regeneration Studies

Claudia Marcela Arenas, Andrea Gómez-Molina, and Jean Paul Delgado

Abstract

Limb regeneration studies have been extensively carried out in species of Ambystomatidae and Salamandridae families. So far limited research has been conducted in species belonging to the Plethodontidae family, where some of the species differs from other salamander families due to their direct development, thus absence of a larval life. Here, we describe a protocol to maintain the plethodontid salamanders of genus *Bolitoglossa* species under laboratory conditions to perform regeneration studies.

Key words Plethodontidae, *Bolitoglossa* sp., Husbandry, Limb regeneration

1 Introduction

Ambystomatidae and Salamandridae family species are the most common salamanders that have been used as models in the development and regenerative biology research [1, 2]. There are about 668 species of salamanders that are distributed in ten families [3]. In most of the species, the biological growth is termed indirect development, where a larval stage occurs during development and an ontogenic transition is attained by a process of metamorphosis. However, in the family Plethodontidae, some species have direct development [4]. Besides this, they also show some biological adaptive differences such as anucleate red blood cells, projectile tongue, tail autotomy, and the absence of lungs [5, 6].

South American salamanders are represented only by the family Plethodontidae [7], mainly by the genus *Bolitoglossa*, and about 28 species have been recognized [8]. *Bolitoglossa* species undergo direct development; they are terrestrial animals distributed in diverse habitats and elevations, from high-elevation grasslands to lowland rainforest [9].

Anoop Kumar and András Simon (eds.), *Salamanders in Regeneration Research: Methods and Protocols*, Methods in Molecular Biology, vol. 1290, DOI 10.1007/978-1-4939-2495-0_5, © Springer Science+Business Media New York 2015

In this protocol, we describe the maintenance under captivity of two wild species of salamanders, *Bolitoglossa ramosi* and *Bolitoglossa vallecula*, and use them as a non-model salamanders for regeneration studies.

2 Materials

2.1 Animal Housing and Maintenance

1. The salamanders, *B. ramosi* and *B. vallecula* (Fig. 1a, b), are sympatric species; they are found from 1,200 to 3,000 m.a.s.l elevation, they are nocturnal, and they prefer microhabitats of herbaceous and shrubby vegetation up to 1 m high along the margins of streams and in forest interiors [9]. The average length and weight of the animals are 7 cm and 1.21 g, respectively (*see* **Note 1**).
2. The animals are kept it in plastic tight containers with dimensions of 23×16×9.5 cm. Place a damp tissue paper sprayed with filtered water to keep the humidity.
3. Environmental enrichment objects such as hollow pipes, sticks, and artificial plants (Fig. 2) (*see* **Note 2**).
4. Tap water filtered through a carbon activated filter (*see* **Note 3**).
5. The animals are feed with flies and/or larval *Drosophila melanogaster* three times a week (*see* **Note 4**).

2.2 Reagents

1. Holtfreter's salt stock solution (10×): Dissolve 34.6 g NaCl, 1 g $CaCl_2$, 2 g $MgSO_4$ $7H_2O$, and 0.5 g KCl in 1 L sterile filtered tap water. Dissolve by continuously stirring with heat until the solution is clear.
2. Holtfreter's salt working solution (40 %): Dilute 20 mL of Holtfreter's stock solution to 480 mL with sterile filtered tap water.

Fig. 1 Two species of South American Plethodontidae family members. (**a**) *Bolitoglossa ramosi*. (**b**) *Bolitoglossa vallecula*

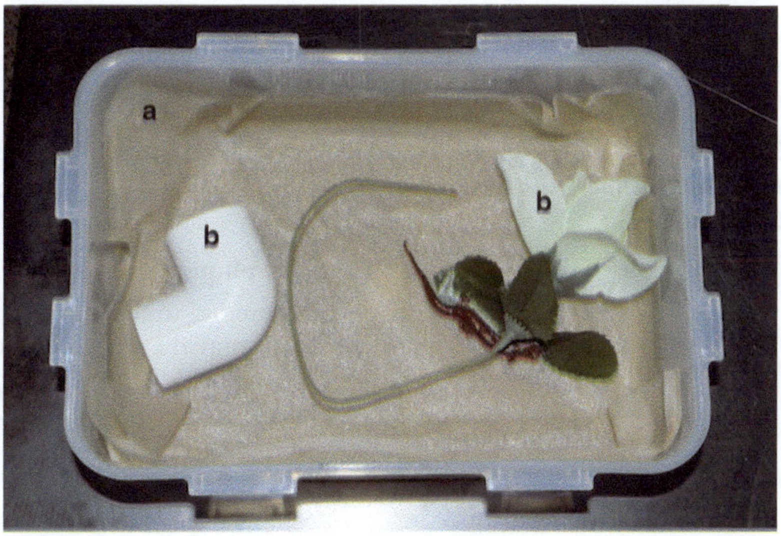

Fig. 2 Plastic container to maintain the salamanders in captivity (**a**) damp paper tissues to keep the humidity. (**b**) Environment enrichment with objects where the animal can climb, hide, or explore

3. 1 M Tris–HCl: Dissolve 0.788 g Tris–HCl in 5 mL of sterile filtered tap water. Adjust the pH to 7.4 with 1 N HCl.
4. Phenol red 0.5 %: Dissolve 0.5 g phenol red in 100 mL of sterile filtered tap water.
5. Anesthetic working solution (1 %): Mix 0.5 mL of Tris–HCl (1 M pH 7.4) and add 250 μL of phenol red (0.5 %). Make up the volume up to 500 mL with Holtfreter's salt working solution (**item 2**), and add 5 g of MS-222. Adjust the pH to 7.4 with 1 N NaOH (*see* **Note 5**) [10].

2.3 Microsurgery Tools

1. Microforceps Dumont #5 (0.10 × 0.06 mm tips) or similar.
2. Microscissors (15 mm blades).
3. Offset Bone Nipper (cutting edge 7 mm).

3 Methods

3.1 Collection of Salamanders

1. The salamanders are collected by the nighttime visual encounter method [11], which is based in the visual registration of the animals in a limited time and area.
2. Once the animal is visualized, collect it carefully with both hands, avoiding taking it by the tail that would lead to caudal autotomy.
3. The specimens are kept in plastic bags and sprayed with filtered tap water, and local vegetation is placed inside the bags to prevent desiccation.

4. Keep the plastic bags inside a polystyrene ice box with an ice pack to maintain low temperature and humidity. Avoid direct contact of the animal with the ice pack.

3.2 Housing and Environmental Conditions

1. The animals are placed inside the plastic container (*see* **Note 6**).
2. Maintain housing temperature at 19–21 °C throughout the period.
3. Use a 12L:12D photoperiodic cycle for the animals.
4. To keep the humidity inside the container, place two layers of tissue paper and spray them with filtered tap water (*see* **Note 7**).
5. Place environmental enrichment objects inside the container to provide the animals with place to hide and to climb.
6. Replace the damp tissues and wash the container with filtered tap water twice a week (*see* **Note 8**).

3.3 Feeding

1. Place the flask with flies inside the freezer 20–30 min (*see* **Note 9**).
2. Place about 20 flies and or larvae per each animal inside the container.
3. Feed the animals three times a week.

3.4 Behavior

Plethodontids move slower and are more docile than newts or axolotls. When handled for long period of time, they can become stressed; therefore, handle them only when necessary (*see* **Note 10**).

In the wild, they can be found in solitary, but in captivity, they can be kept with other plethodontids in the same container.

3.5 Common Diseases and Health Monitoring

Salamanders are susceptible to different microbial diseases. One of the most important microbial diseases that have been reported to contribute to the decline of the amphibian world populations is the fungal infection chytridiomycosis generated by *Batrachochytrium dendrobatidis* [12]; nevertheless, there are few studies reporting infection in plethodontids including Colombia [13, 14]. In our experience, few animals have died in captivity possibly due to fungal infections (*see* **Note 11**).

3.6 Surgical Manipulation

Surgical manipulation is performed under anesthesia; this protocol works for limb amputation of adult salamanders.

1. In a 10 cm Petri dish, add 20 mL of 1 % MS-222 prepared in Subheading 2.2, **item 5**.
2. Place the animal inside the dish and cover with the lid. About 4–5 min is required for the animal to become motionless (*see* **Note 12**).
3. Check the animal's response to stimulus (*see* **Note 13**).

4. Place the animal under the stereomicroscope to perform the surgical procedure.
5. Using microscissors and forceps, amputate the forelimb at the distal or proximal plane (*see* **Notes 14** and **15**).
6. When the procedure is completed, place the animal inside a Petri dish containing excess of sterile filtered tap water for 1 min.
7. Remove the animal from the water.
8. Add 500 μL of 0.5 % sulfamerazine at the injury site of the limb with a syringe or Pasteur pipette.
9. Spray the animal with sterile filtered tap water about every 10 min.
10. When the animal shows signs of awakening, return the animal to the plastic container.
11. Watch the behavior of the animal during the first 2 h.

4 Notes

1. In the Andean Community Nations, the capture of wild animals is tightly regulated by the Department of Environment of the respective country. For experimental studies requiring wild caught salamanders, a specific license is required to collect the animals. It is important to know the local regulations regarding the collection of wild animals in each country and, if required, obtain the respective permits.
2. The plastic containers can be a kitchenware type with rubber seal. It is important to place items that enrich the environment so that the animal can hide from the light. Besides this, sticks or artificial plants can help to increase the area where the animal can climb or explore (Fig. 3).
3. The quality of tap water in our city has a pH of 6.4, which appears to be easily tolerated by the animals; only filtration is needed it to remove the chlorine. A standard kitchen carbon filter is sufficient for this purpose. Another option is to allow chlorine evaporation overnight.
4. When you collect wild animals, it is important to standardize the amount of food that is needed to keep the initial weight of the animal. To do this, measure the weight of the animals once a week. In our experience, 50 flies and/or larvae a week are enough to keep the weight of these species. An important aspect to take into account is that the animals need to hunt for flies when they are alive, the animals do not eat dead flies.
5. It is critical to adjust the pH of the anesthetic before use; use a buffered solution (Holtfreter's salt solution) and adjust the pH

Fig. 3 Limb regeneration *Bolitoglossa ramosi*. Note the regeneration blastema on the left limb

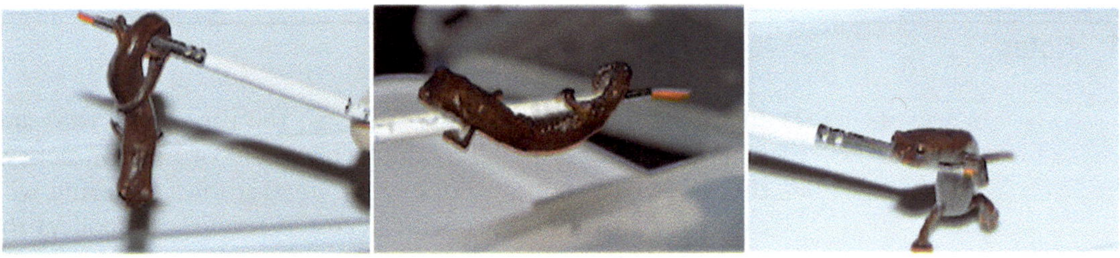

Fig. 4 Handling of plethodontid salamander using a brush stick

to 7.4. When a low pH is used, animals could experience bleeding from the skin.

6. Handle the salamanders with gloves to avoid transmission of any diseases between animals. In order to avoid stress and induction of tail autotomy, handle the animals with a pencil, brush, or stick. Because they like to hang on sticks by using the prehensile tail, reach the tip of the pen or stick to the animal and allow to climb on it (Fig. 4).
7. The quality of tissue paper is important because some brands available from the market are acidic and, in contact with water, will decrease the pH of water. In our experience, the recycled tissue paper works fine.
8. It is important to keep the plastic containers with low microbial density to prevent diseases in the animals. However, amphibians are reported to have antimicrobial peptides that could help them to prevent diseases [15].
9. Flies can be obtained from genetic laboratory facilities. Otherwise, insects can be purchased from local pet suppliers.

Low temperature decreases the fly metabolism, so it is easier to feed the animal when the flies are inactive. It takes about 20–30 min to decrease the metabolism of the flies. Regarding feeding habits, when animals have been for a while in captivity, it is feasible to hand-feed them with a brush.

10. To photograph the animals under the stereo microscope, it is not necessary to anesthetize them, especially for collecting images from the dorsal side. It is better to anesthetize the animals for procedures that imply handling the animal for long periods of time or ventral picture acquisitions.
11. When signs of infection or diseases are detected (abnormal posture, lethargy, loss of righting reflex, reddening of the skin, extended hind limbs) [12], isolate the animal from others and place in quarantine. Use separate gloves and sticks to handle sick animals.
12. Keep the animals for the shortest possible time submerged in anesthetic. When the animal is in contact with the anesthetic in the first minute, they become hyperactive; soon after, their movement of the limbs ceases. The induction and recovery time varies on each animal; in our experience, the recovery time is about 25–50 min.
13. Check the response of the animals to stimulus by touching the limb with tweezers.
14. Once the procedure is completed, it is possible to collect a sample of blood from the amputation plane to carry out experiments by flow cytometry. Use a 10 μL micropipette to take the sample by capillary drawing in about 5–10 μL. It is important to permeabilize the cell with Triton X-100 before flow cytometry experiments.
15. Once the limb is removed, push the muscle tissue attached to the bone rearward with microforceps # 5 and cut the protruding bone with bone nipper. Then, cut the protruding muscle tissue with microscissors to ensure that surface of the amputated limb is flat.

Acknowledgments

Our work is funded by a program grant COLCIENCIAS 569 and Programa-Sostenibilidad 2013–2014 of the University of Antioquia. Claudia Arenas is funded by a fellowship from COLCIENCIAS 567. We thank Juan Manuel Daza for the taxonomic identification of the animals and Melisa Hincapie, Carlos Muñoz, Jose Fang, and Santiago Cuartas for their help during the field work and maintenance of the animals. The animals used to establish this protocol were collected under the environmental license "Permiso Marco de Recolección de Especímenes de Especies Silvestres de la Diversidad Biológica con Fines de Investigación" (Resolución #0524-27/05/2014). This work has the approval of the "Comité de Bioética para Experimentación con Animales" from the University of Antioquia.

References

1. Brockes JP, Kumar A (2008) Comparative aspects of animal regeneration. Annu Rev Cell Dev Biol 24:525–549
2. Stocum DL, Cameron JA (2011) Looking proximally and distally: 100 years of limb regeneration and beyond. Dev Dyn 240:943–968
3. AmphibiaWeb: Information on amphibian biology and conservation (2014). http://amphibiaweb.org/
4. Wake DB (2009) What salamanders have taught us about evolution. Annu Rev Ecol Evol Syst 40:333–352
5. Mueller RL, Gregory TR, Gregory SM, Hsieh A, Boore JL (2008) Genome size, cell size, and the evolution of enucleated erythrocytes in attenuate salamanders. Zoology (Jena) 111: 218–230
6. Mueller RL, Macey JR, Jaekel M, Wake DB, Boore JL (2004) Morphological homoplasy, life history evolution, and historical biogeography of plethodontid salamanders inferred from complete mitochondrial genomes. Proc Natl Acad Sci U S A 101:13820–13825
7. Elmer KR, Bonett RM, Wake DB, Lougheed SC (2013) Early Miocene origin and cryptic diversification of South American salamanders. BMC Evol Biol 13:59
8. Parra-Olea G, García-París M, Wake DB (2004) Molecular diversification of salamanders of the tropical American genus Bolitoglossa (Caudata: Plethodontidae) and its evolutionary and biogeographical implications. Biol J Linn Soc 81:325–346
9. Silva-González N, Páez VP, Bock BC (2011) Morphological variation in *Bolitoglossa vallecula* (Amphibia:Caudata:Plethodontidae) in the cordillera central of Colombia. Actu Biol 33:251–260
10. Lee J, Aguilar C, Gardiner D (2013) Gain-of-function assays in the axolotl (Ambystoma mexicanum) to identify signaling pathways that induce and regulate limb regeneration. Methods Mol Biol 1037:401–417
11. Grover M (2006) Comparative effectiveness of nighttime visual encounter surveys and cover object searches in detecting salamanders. Herpetol Conserv Biol 1:93–99
12. Densmore CL, Green DE (2007) Diseases of amphibians. ILAR J 48:235–254
13. Vazquez VM, Rothermel BB, Pessier AP (2009) Experimental infection of North American plethodontid salamanders with the fungus Batrachochytrium dendrobatidis. Dis Aquat Organ 84:1–7
14. Velasquez EBE, Castro HF, Bolivar GW, Herrera MI (2008) Infection by the chytrid fungus Batrachochytrium dendrobatidis in anurans from Cordillera Occidental in Colombia. Herpetotropicos 4:65–70
15. Meng P, Yang S, Shen C, Jiang K, Rong M et al (2013) The first salamander defensin antimicrobial peptide. PLoS One 8:83044

Part II

Experimental Manipulation in Salamanders

Chapter 6

Newt Lens Transdifferentiation: From Lentectomy to Immuno-FISH

Nobuyasu Maki

Abstract

Newt lens regeneration is achieved by a unique cellular regulation of transdifferentiation where the dorsal iris pigmented epithelial cells (PECs) dedifferentiate and redifferentiate into lens cells. Recent studies have shown that nuclear architecture of PECs is dynamically changed and unique epigenetic regulation in somatic nucleus is crucial in the lens transdifferentiation. Immuno-FISH, detection of protein and gene loci in nucleus, is one of the effective tools to analyze nuclear architecture of PECs. In this chapter a whole process from lentectomy to immuno-FISH is described.

Key words Lens regeneration, Transdifferentiation, Dedifferentiation, Epigenetics, Reprogramming, Salamander, Immuno-FISH

1 Introduction

It is a generalized conception that differentiation is an irreversible event and once cells are terminally differentiated, they lose their multipotency and proliferation ability so that the differentiated cells never change into other types of cells in vivo. An apparent counterexample to this concept is transdifferentiation of iris pigmented epithelial cells (PECs) during newt lens regeneration (Fig. 1) [1]. After lens removal, the PECs lose their original characteristics by shedding their pigment granules (about day 4). At almost the same time, an initial cell cycle reentry of PECs occurs. PECs continue depigmentation and proliferation and finally change to transparent cells, called as dedifferentiated cells, by day 8 and form a lens vesicle by day 10–12. After day 14, lens differentiation occurs from dorsal iris. Even though the dedifferentiation and proliferation take place in both dorsal and ventral irises, lens differentiation occurs only in dorsal iris. Lens transdifferentiation has been demonstrated by clonal culture experiment using newt PECs [1].

Nuclear architecture of PECs dynamically changes during transdifferentiation [1–3]. The original PECs have a small and

Anoop Kumar and András Simon (eds.), *Salamanders in Regeneration Research: Methods and Protocols*, Methods in Molecular Biology, vol. 1290, DOI 10.1007/978-1-4939-2495-0_6, © Springer Science+Business Media New York 2015

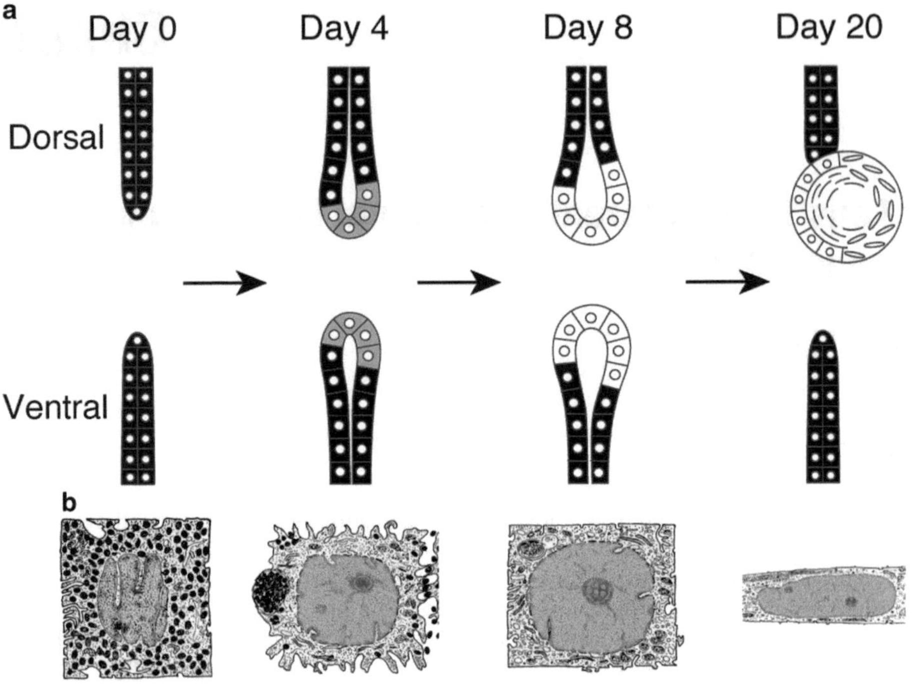

Fig. 1 Illustration of lens regeneration process and changes in nuclear architecture. (**a**) About 4 days after lentectomy, PECs begin to shed their pigments and reenter cell cycle. The transparent dedifferentiated cells are found by day 8. After day 14, the dorsal dedifferentiated cells redifferentiate into lens cells. However, the ventral PECs never differentiate into lens even though these cells show depigmentation and cell cycle reentry. (**b**) Structural change in the nucleus during lens transdifferentiation. The nucleus of original PECs is small and shrunken in morphology. During dedifferentiation, the PEC nucleus swells and its nucleoli become large. *Blue*, nucleus. *Red*, nucleolus. Illustration in (**b**) is reproduced from [9] with permission (Color figure online)

shrunken nucleus measuring about 10 μm in a diameter. During the dedifferentiation of PECs, the nucleus swells and the dedifferentiated cells measure about 20 μm in a diameter. In parallel with the nuclear swelling, the nucleoli of PECs have also enlarged during dedifferentiation. These dynamic changes in nuclear architecture suggest that nuclear regulation plays a key role in newt lens transdifferentiation. Indeed, gene expression analysis has been suggested that newt transdifferentiation shares the mechanisms in somatic nuclear reprogramming by nuclear transfer into oocyte (SCNT) [4] or by forced expression of stem cell factors (as known for iPS) [5]. Oocyte-type linker histone, by which somatic-type linker histone H1 is replaced after SCNT [6], and three of iPS factors, Sox2, Klf4, and c-Myc, are expressed in the transdifferentiation [7]. Therefore, studying the regulation of nuclear architecture and epigenetics during lens regeneration will provide significant advances not only in understanding newt cellular plasticity but also in regenerative biology and medicine.

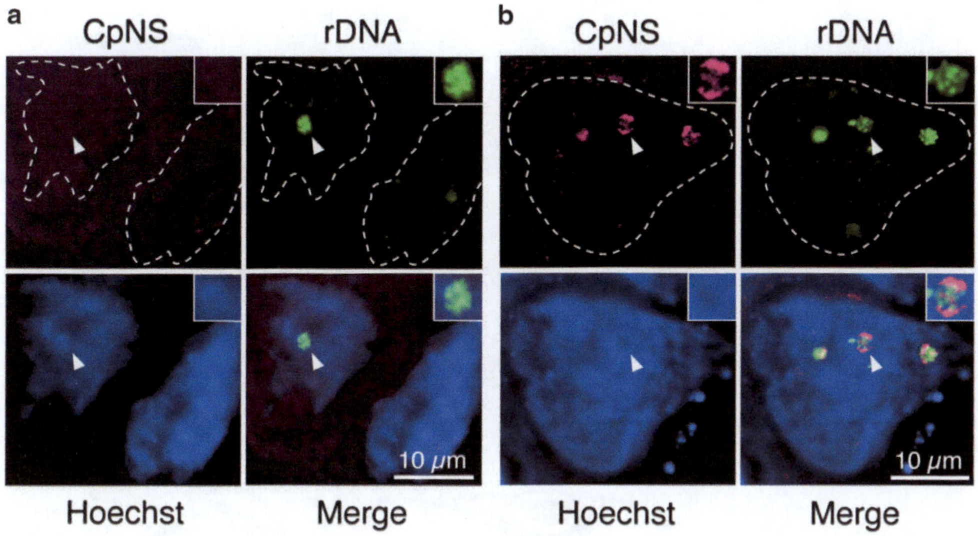

Fig. 2 Immuno-FISH using NS antibody and 18S rDNA probe against PECs before (**a**) and 6 days after (**b**) lentectomy. NS is accumulated in nucleoli of dedifferentiating PECs. Nuclei were counterstained with Hoechst 33258. High-power view of the nucleolus indicated by *arrowheads* is shown in the *inset*. Outline of nucleus is indicated by *dashed line*. Reproduced with permission from Developmental Dynamics [2]

The newt genome DNA contains many types of repetitive sequences, and these sequences localize specific chromatin regions such as centromere and rDNA [8]. To understand changes in nuclear architecture during lens transdifferentiation, we established the immuno-FISH method using those repetitive sequence probes against the whole nucleus in tissue sections (Fig. 2) [2]. In this chapter the entire process from sample preparation, including lentectomy, to immuno-FISH staining, detecting the loci of nucleostemin (NS) protein and the 18S rDNA, is described. The basic techniques and tips for lens regeneration experiments are also described in detail.

2 Materials

2.1 Animal Maintenance

1. Newts: Adult newts (*Cynops pyrrhogaster*) were collected in the northern part of Okayama Prefecture in Japan. All animal care and use protocols were in compliance with the Animal Experiment Handbook at Osaka University.
2. Food: Frozen bloodworm.
3. Plastic container for animal maintenance (35 × 25 × 12 cm).

2.2 Lentectomy

1. Special forceps for lentectomy (Fig. 3a, *see* **Note 1**).
2. Surgical blade: Feather #11 (*see* **Note 2**).
3. Blade holder: Feather #3.

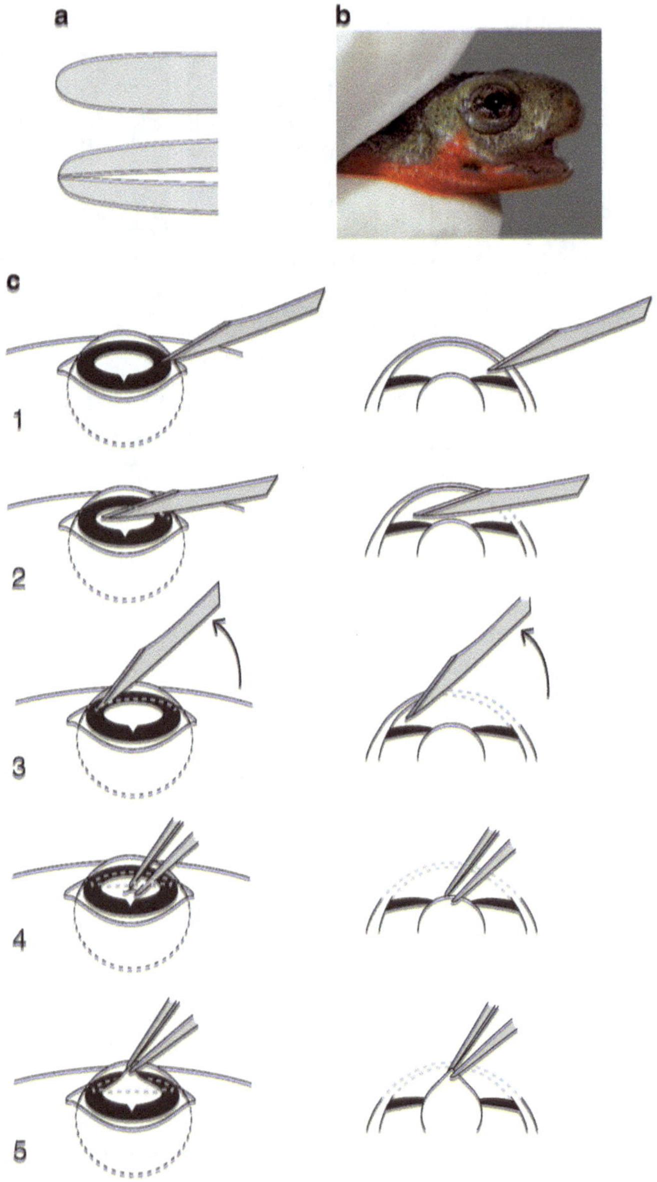

Fig. 3 Process of lentectomy. (**a**) The special forceps for lentectomy. The tip of the forceps is rounded and smoothen with a whetstone. *Upper*, side view. *Lower*, top view. (**b**) Newt held between the thumb and forefinger. (**c**) Illustration of the process of lentectomy. Incision of the cornea (*1–3*). Holding the lens with forceps (*4*). Removing of lens (*5*)

4. Plastic container for anesthesia (12 × 16 × 5 cm).
5. Anesthetic solution: Prepare 0.25 % ethyl 3-aminobenzoate methanesulfonate in 70 % PBS.
6. Stereomicroscope with a double arm fiber illuminator system.

2.3 Sampling and Tissue Sectioning

1. Anesthetic solution: *see* Subheading 2.2.
2. Fixation solution: methanol–acetic acid (3:1) solution.
3. 100, 95, 90, 80, and 70 % ethanol.
4. Xylene.
5. Paraffin: Paraplast X-TRA tissue embedding medium.
6. Plastic dish.
7. Slide glass: Poly-L-lysine coating slide glass.
8. Paraffin oven.
9. Rotary microtome.

2.4 Probe Labeling

1. 18S rDNA plasmid (0.25 μg/μL, GenBank accession no. AB239574).
2. Nick Translation Kit (Roche #976776, Mannheim, Germany): A mixture of dATP, dGTP, and dTTP, 10× buffer, and an enzyme mixture are included in this kit.
3. 1 mM Biotin-16-dUTP (Roche #1093070).
4. 4 M ammonium acetate.
5. tRNA and ssDNA mixture: 100 mg of *E. coli* tRNA (Roche, #10014) and 100 mg of denatured salmon sperm DNA (Sigma-Aldrich, #D-7656) are dissolved in 10 mL of Milli-Q water.
6. Deionized formamide.

2.5 FISH Procedure

1. Decolorization buffer: Prepare fresh 6 % hydrogen peroxide in PBS.
2. Permeabilization buffer: Prepare a solution containing 0.05 % Triton X-100 and 0.05 % saponin in 2× SSC.
3. RNase A solution: Prepare 10 μg/mL RNase A solution in 2× SSC.
4. Denaturing solution: Prepare 70 % formamide in 2× SCC.
5. 2× hybridization mix: 4× SSC, 4 mg/mL BSA (Roche #711454), 20 % dextran sulfate.
6. Wash buffer: 50 % formamide, 2× SSC.
7. TN buffer: 0.1 M Tris–HCl, pH 7.5, 0.15 M NaCl.
8. TNT buffer: 0.1 M Tris–HCl, pH 7.5, 0.15 M NaCl, 0.05 % Tween-20.
9. TNB buffer: 0.1 M Tris–HCl, pH 7.5, 0.15 M NaCl, 0.5 % blocking regent (Perkin Elmer, FP1020).
10. Sheep anti-DIG antibody (Roche Diagnostics).
11. Affinity-purified antibody against NS.

12. Hoechst solution: Prepare 8 μg/mL Hoechst 33258 in 2× SSC.
13. Fluorescence mounting medium.

3 Methods

3.1 Animal Maintenance

1. Up to ten newts are kept in a plastic container containing 3.5 L of rearing water at 22 °C.
2. Feed the newts with red bloodworms twice a week.
3. Change the rearing water everyday.

3.2 Lentectomy

1. Keep newts in the anesthetic solution until they are anesthetized (about 10 min).
2. Hold the newt neck between the thumb and forefinger.
3. Put the surgical blade into the cornea with the edge side up (Fig. 3c-1).
4. Carefully insert the blade between the cornea and iris until the tip of the blade reaches the opposite side (Fig. 3c-2).
5. Incision the cornea by drawing the blade upward. About two-thirds of the cornea diameter is cut in this operation (Fig. 3c-3).
6. Insert the forceps through the opening of the cornea (Fig. 3c-4, *see* **Note 3**).
7. Hold the surface of the lens with the forceps (*see* **Note 4**).
8. Carefully pull the lens out from the eyeball (Fig. 3c-5).
9. Return the newts to the rearing water. About 10–20 min after returning to the water, newts begin to awake from anesthesia.

3.3 Sampling and Tissue Sectioning

1. Anesthetize newts as described in, Subheading 3.2, **step 1**.
2. Cut off a head and remove a jaw from the head. The eyeballs are fixed with a head in methanol–acetic acid (3:1) solution at 4 °C for 12 h (*see* **Note 5**).
3. The fixed eyeballs are removed from the head.
4. Wash the eyeballs with 100 % methanol for 10 min, three times.
5. Cut ventral side of eyeball to create an opening to the posterior chamber in 100 % methanol (*see* **Note 6**).
6. Place the eyeballs in xylene for 10 min.
7. Place the eyeballs in xylene–paraffin (1:1) solution at 60 °C for 10 min.
8. Place the eyeballs in paraffin at 60 °C for 60 min, three times.

9. Embed the eyeballs in a plastic dish (*see* **Note 7**).
10. Paraffin-embedded eyeballs are sectioned at 20 μm.
11. Place the sections onto the surface of water dropped on a slide glass at 42 °C.
12. Dry the sections at 42 °C overnight (*see* **Note 8**).
13. Place the slides in xylene for 10 min, three times.
14. Rehydrate the sections through the following ethanol series for 5 min each, 100, 100, 95, 90, 80, and 70 % ethanol.
15. Wash the slides in PBS for 5 min, three times.

3.4 Probe Labeling

1. Mix following reagents: 2.0 μL of plasmid DNA (0.5 μg), 12 μL of a dATP, dGTP, dTTP mixture, 4 μL of 10× buffer, 2.2 μL of Biotin-16-dUTP, 12.8 μL of Milli-Q water, and 7 μL of enzyme mixture.
2. Incubate at 15 °C for 2 h.
3. Heat inactivation at 65 °C for 10 min.
4. Cool down to room temperature.
5. Add 4 μL of the tRNA and ssDNA mixture, 5.4 μL of 4 M ammonium acetate, and 66 μL of ethanol.
6. Mix by inverting tube.
7. Incubate at −80 °C for 20 min.
8. Centrifuge at 20,000 × *g* for 10 min.
9. Remove the supernatant and dry the precipitate.
10. Dissolve the precipitate in 40 μL of formaldehyde.
11. Incubate at 75 °C for 10 min.
12. On ice for 3 min (*see* **Note 9**).

3.5 Immuno-FISH

1. To decolorize PECs, put the deparaffinized slides (Subheading 3.3) in the decolorization buffer for 1 h under fluorescent light irradiation (*see* **Note 10**).
2. Wash the slides in PBS for 5 min, three times.
3. Permeabilization: Incubate in the permeabilization buffer for 60 min.
4. Rinse with 2× SSC for 15 min, three times.
5. RNA degradation: Incubate in the RNase solution at 37 °C for 60 min.
6. Rinse with 2× SSC for 15 min, three times.
7. Denaturing of dsDNA in nucleus: Incubate in the denaturing solution at 60 °C for 2 min (*see* **Note 11**).
8. Immediately place into the ice-cold denaturing solution.

9. Wipe the excess denaturing solution (avoid contacting tissue).
10. Add equal volume of 2× hybridization mix to the probe solution prepared in Subheading 3.4 and mix well (*see* **Note 12**).
11. Apply the hybridization mixture onto the specimens (10 μL for cm^2).
12. Cover solution with a piece of parafilm.
13. Incubate in a humid chamber with the wash buffer at 37 °C overnight.
14. Wash in the wash buffer at 37 °C for 30 min twice.
15. Wash in 1× SSC for 30 min three times at room temperature.
16. Wash in TN buffer for 5 min.
17. Block with TNB for 60 min.
18. Incubate with sheep anti-DIG antibody and affinity-purified anti-CpNS antibody diluted with TNB.
19. Wash in TNT for 15 min, three times.
20. Incubate with Alexa 488-conjugated anti-sheep antibody and Cy3-conjugated anti-rat antibody diluted with TNB.
21. Wash in TNT for 15 min, three times.
22. Wash in 2× SSC for 5 min.
23. Counterstain nuclei with the Hoechst solution for 5 min.
24. Wash in TN buffer for 5 min.
25. Mount with DACO fluorescence mounting medium.

4 Notes

1. During lentectomy, a lens epithelial cell layer is held with forceps. To hold it firmly and to avoid tissue damage, the tip of the forceps should not be sharp. With a whetstone, the tip of the forceps (e.g., INOX No. 5) is rounded and smoothen (Fig. 3a). When the forceps is closed, both ends of the tips should be attached together.
2. To insert a surgical blade into the eyeball through the cornea, the tip of the blade should be sharp. For this purpose Feather surgical blade #11 is suitable (the blade which has a rounded tip such as Kai #11 cannot be used).
3. Lens can be extruded by pushing the eyeball with forceps. However, this method may cause a damage on the iris. So the method presented in this chapter is highly recommended. Because of its transparency, the position of the lens in the eyeball is visually indistinguishable. Thus, the position of the lens must be realized based on the structure of the newt eye during the operation (*see* Fig. 3). Too much insertion of the forceps causes lens surface damage and makes it difficult to hold the lens.

4. At first, touch the eyeball with both tips of the forceps, and then close the forceps very carefully without any other movement.
5. The lens-regenerating eye is very fragile. To avoid tissue damage during its collection, eyeballs are fixed before dissection. Because a paraformaldehyde fixation causes strong signal reduction in newt FISH and immunohistochemical analysis, a methanol–acetic acid (3:1) solution is used.
6. This opening in the ventral side is necessary to avoid a crush of eyeballs during the xylene–paraffin exchange and serve as a mark indicating a dorsoventral orientation in sections.
7. The paraffin-embedded samples can be stored at 4 °C.
8. After drying the slides can be stored at 4 °C.
9. This probe solution can be stored at −20 °C for several months.
10. Because of the pigment granules in PECs, this treatment is performed.
11. To obtain FISH signals, the optimum temperature is 70 °C. However, to retain an antigenicity of NS, this denaturing treatment is performed at 60 °C. Denaturing condition must be determined for each antigen.
12. Because the 2× hybridization solution is viscous, widening the tip by cutting it is recommended.

Acknowledgments

This work is supported by JST PRESTO and a grant, KAKENHI (14514364), to N.M. I would like to thank Panagiotis A. Tsonis for critical reading and Rinako Maki for producing the illustrations.

References

1. Maki N, Kimura H (2013) Epigenetics and regeneration. Curr Top Microbiol Immunol 367:237–252
2. Maki N et al (2007) Rapid accumulation of nucleostemin in nucleolus during newt regeneration. Dev Dyn 236:941–950
3. Maki N, Tsonis PA, Agata K (2010) Changes in global histone modifications during dedifferentiation in newt lens regeneration. Mol Vis 16:1893–1897
4. Jullien J et al (2010) Characterization of somatic cell nuclear reprogramming by oocytes in which a linker histone is required for pluripotency gene reactivation. Proc Natl Acad Sci U S A 107: 5483–5488
5. Takahashi K, Yamanaka S (2006) Induction of pluripotent stem cells from mouse embryonic and adult fibroblast cultures by defined factors. Cell 126:663–676
6. Maki N et al (2010) Oocyte-type linker histone B4 is required for transdifferentiation of somatic cells in vivo. FASEB J 24:3462–3467
7. Maki N et al (2009) Expression of stem cell pluripotency factors during regeneration in newts. Dev Dyn 238:1613–1616
8. Murakami T et al (2007) Establishment of high-resolution FISH mapping system and its application for molecular cytogenetic characterization of chromosomes in newt, Cynops pyrrhogaster (Urodela, Amphibia). Chromosome Res 15:471–484
9. Eguchi G (1980) Lens regeneration: transdifferentiation of tissue cells. Iwanami shoten, Tokyo

Chapter 7

Studying Newt Brain Regeneration Following Subtype Specific Neuronal Ablation

Matthew Kirkham and Alberto Joven

Abstract

The realization that neuronal injury does not result in permanent functional or cellular loss in all vertebrates has fascinated regenerative biologists. Neuronal regeneration occurs in a subset of species, including lizards, teleost fish, axolotls, and newts. One tool for studying neuronal regeneration in the adult brain is intraventricular injection of selective neuronal toxins, which leads to loss of subpopulations of neurons. To trace cells involved in the regeneration process, plasmids encoding reporter proteins can be electroporated in vivo into the cells of interest. This protocol describes methods to label the ependymoglial cells of the brain of the red spotted newt *Notophthalmus viridescens* and follow their response after ablation of dopaminergic neurons.

Key words Neurogenesis, In vivo electroporation, 6-OHDA, Dopaminergic neurons, Neuronal regeneration

1 Introduction

A resurgence of interest in vertebrate brain regeneration has occurred over recent years, which has coincided with increased understanding of adult neurogenesis and stem cell biology within mammals [1]. The mammalian brain regenerates poorly after neuronal loss, but the persistent presence of neural stem cells and ongoing adult neurogenesis has offered hope that cell replacement treatments for neural injury and disease could be developed either via transplantation of cells or endogenous regeneration [2]. Hence, an increased understanding of how successful neuronal regeneration occurs could lead to improvements in cell regenerative therapies in humans.

Injury in adult neuronal regenerative models has been caused through two main methods. Mechanical injuries for example have included the removal of part of the optic tectum in *Notophthalmus viridescens*, severing the retinal axon in the lizard *Gallotia galloti*, cerebellar lesion paradigm in teleost fish and traumatic stab lesion

Anoop Kumar and András Simon (eds.), *Salamanders in Regeneration Research: Methods and Protocols*, Methods in Molecular Biology, vol. 1290, DOI 10.1007/978-1-4939-2495-0_7, © Springer Science+Business Media New York 2015

in the pallium of the zebrafish [3–6]. However, there are limited comparable injury models in mammalian systems, and due to the complexity of the removed tissue it is hard to determine the regeneration mechanisms related to any one neuronal type. The second method of neuronal injury utilizes neuronal toxins to ablate specific neuronal subtypes. 6-hydroxydopamine (6-OHDA) mediated ablation of dopaminergic neurons, which play a central role in Parkinson's disease, has been used to model Parkinson's disease in the mammalian brain for over 30 years [7]. Similarly, 6-OHDA has been demonstrated to ablate dopaminergic neurons in the newt (Fig. 1a, b) [8].

6-OHDA does not cross the blood–brain barrier, thus to ablate neurons in the brain, the drug should be injected directly in to the neuronal population of interest. Generally, a 6-OHDA injection is an inexpensive and reliable method for catecholaminergic neuron ablation. In this protocol the 6-OHDA is injected into the third ventricle of the newt and causes a bilateral ablation of tyrosine hydroxylase positive neurons in the ventral diencephalic/mesencephalic region [8, 9]. Tyrosine hydroxylase is a rate-limiting enzyme in the production of dopamine and is often used as a marker for catecholaminergic neurons.

In all brain regenerative models studied so far injury induced neurogenesis is driven by the proliferation and subsequent differentiation of ventricular localized glia cells, which in the newt are termed ependymoglial cells [1]. One method to effectively label these cells in vivo is electroporation of DNA plasmids (Fig. 1c). Plasmid preparation for electroporation is significantly simpler than creating transgenic lines or the production of virus used in viral based transfection methods. In vivo electroporation in amphibians has been used for a number of years including in the regenerating spinal cord and in juvenile axolotl and frogs [10, 11]. The protocol described here was adaptive from [12].

Electroporation generally results in multiple copies of the plasmids entering the cells resulting in a transient transfection or transient expression of the reporter protein, and like other chemical based methods of transient transfection expression of the reporter protein declines over time. Furthermore, as the plasmid DNA does not generally integrate into the genome there is no guarantee that progeny of the transfected cell will express the reporter protein. One method used to prolong expression in cells during regeneration has been to use the Gal4-VP16 enhancer system [9, 11, 13]. Here the GAL-VP16 gene is under the control of either a tissue specific promoter or a general promoter depending on the experimental design. The expression of the reporter protein is driven by the enhancer sequence Gal4UAS that Gal4-VP16 binds to. When the promoter is turned off, reporter gene expression is prolonged because of the presence of Gal4-VP16 protein that remains in the cell. An alternative strategy

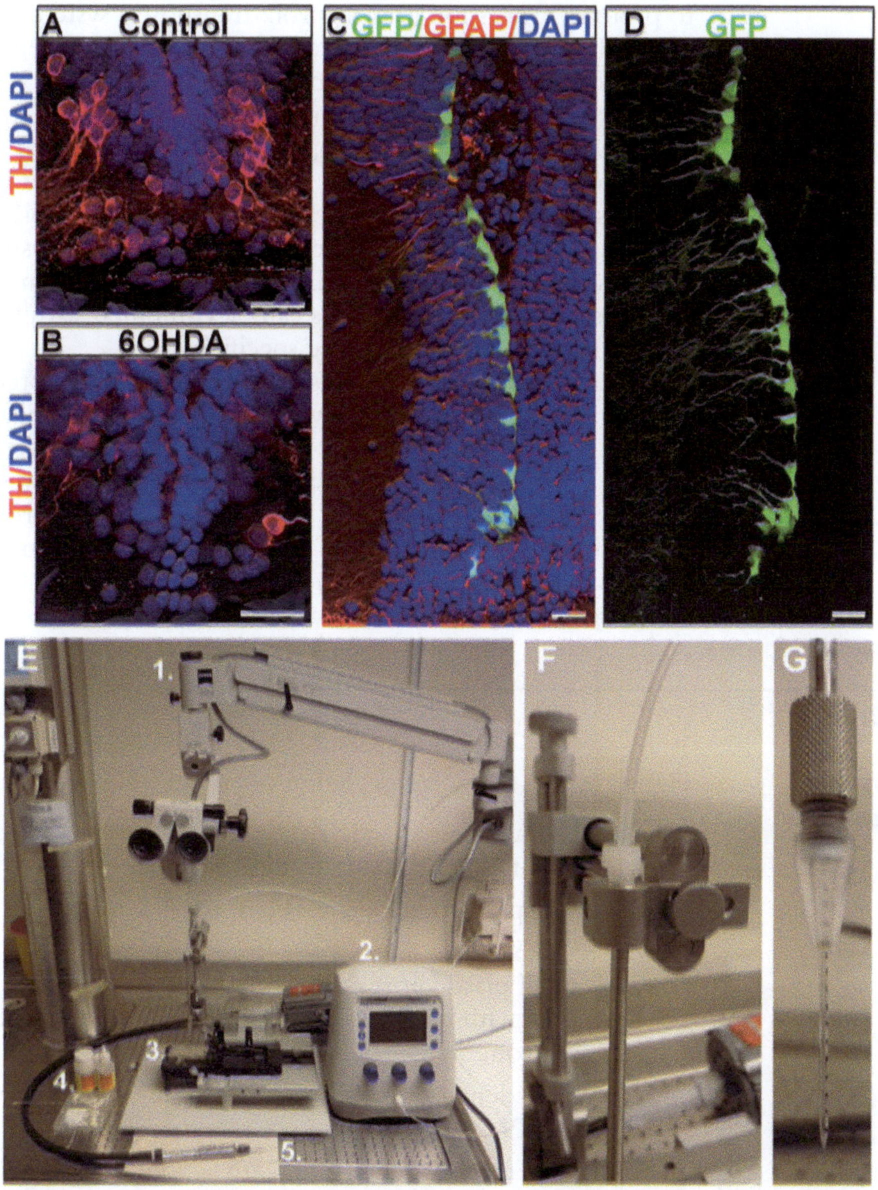

Fig. 1 Illustrations of the applications and equipment needed for intraventricular injection. (**a** and **b**) Midbrain dopaminergic neurons in a control newt, *Notophthalmus viridescens* (**a**) and 5 days after injection of 6-OHDA (**b**). The neurons are labeled with an antibody against Tyrosine hydroxylase (*red*) and the nuclei are labeled with the DNA dye DAPI (*blue*). (**c** and **d**) GFAP-positive ependymoglial cells (*red*) lining the third ventricle of a newt that has been electroporated with a plasmid encoding GFP (*green*) under a CMV promoter. (**e**) Overview of the equipment. (*1*) Stereomicroscope on bloom arm, (*2*) FemtoJet, (*3*) Stereotactic instrument, (*4*) Quick drying dental cement, and (*5*) Micro drill. (**f**) The universal capillary holder arm of the FemtoJet is attached to stereotactic instrument using a V clamp. (**g**) Glass needle in the universal capillary holder ready for injection. The barrel of the needle is marked with a marker pen. Scale bars 50 μM

to prolong expression uses the Tol2 transposon system [14]. This system has been employed to trace cells during limb regeneration for 2 months after electroporation [15], and has the advantage of integrating the plasmid DNA in to the genome of the transfected cell. Two plasmids are electroporated in a ratio of 1:1, the first encoding the gene for Tol2 transposase. The second plasmid encodes a promoter and reporter protein flank by Tol2 transposon sites. The sequence flanked by the Tol2 transposon sites is removed from the plasmid and inserted into the genome by Tol2 transposase. Both this methods can be combined with tissue specific promoters to give cell type specificity.

To specifically drive expression of a reporter protein in the ependymoglial cells in the newt, the human glial fibrillary acidic protein (GFAP) promoter has been used [9]. However, when a plasmid encoding GFP driven by the CMV promoter was injected into the ventricle of the newt brain followed by electroporation, the ependymoglial cells act as a barrier reducing the penetration of the plasmid to the brain parenchyma, resulting in 99 % of transfected cells being ependymoglial cells (Fig. 1c, d) [9].

2 Materials

2.1 Equipment for intraventricular Injection

Figure 1e–g is a graphic depiction of how the equipment is set up.

1. Small animal stereotactic device adapted for neonatal mouse and rodents with V clamp to hold a probe.
2. Small support rough 3×8×1 cm ideal a plastic lid or small plastic box.
3. Two small plastic boxes that are at least 4–5 cm high and that have a lid with air holes, small enough that the newts cannot escape through.
4. Electric dental-drill or micro-drill maximum speed 15,000 RPM.
5. Tungsten carbon dental bur with a round head 0.5 mm in diameter.
6. Eppendorf Femtotips (Eppendorf) or borosilicate glass capillaries (Harvard Apparatus: GC100F-10), and a Flaming/Brown type micropipette puller (Harvard Apparatus) (*see* **Note 1**).
7. Stereotactic dissection microscope on boom arm.
8. Microinjector (Eppendorf: FemtoJet or similar).
9. Microloader tips.
10. Microdissection tools: Dumont #4 Forceps and Spring Scissors with 6 mm curved blades.

2.2 Reagents for intraventricular Injection

1. Tank water. This can either be tap water or the salt solution used to house the animals.
2. 0.1 % MS-222 (Sigma). Dissolve 1 g MS-222 in 1,000 mL of tank water.
3. Quick drying dental cement (Harvard Dental).

2.3 Equipment for In Vivo Electroporation

1. Electroporator (Nepagene: NEPA21).
2. Electrodes: tweezers with round platinum plate electrodes 5 mm diameter (Nepagene CUY650P5).

2.4 Reagents Required for In Vivo Electroporation

1. Sterilized 10 mM Tris–HCl adjusted to a pH 8.5.
2. Plasmid purification Kit (OriGene: PowerPrep® HP Plasmid Maxiprep System).

2.5 Reagents for 6-OHDA Neuronal Ablation

1. Sterilized 0.9 % saline solution. 0.9 g NaCl dissolved in 100 mL water.
2. Ascorbic acid. Make fresh and keep on ice. 0.2 mg is dissolved in 1 mL of 0.9 % saline solution.
3. 6-OHDA (Sigma cat no: H4381). Dissolve 6 mg of 6-OHDA in ascorbic acid/saline solution and keep on ice until the time of injection. Alternatively aliquots can be stored at –20 °C for 6 months.

3 Methods

3.1 Intraventricular Injection of Plasmids

1. Prepare the plasmid solution for injection using PowerPrep® HP Plasmid Maxiprep System from OriGene (*see* **Note 2**) following the manufacturer's instructions with one exception. Resuspend the plasmid DNA in 50 μL of 10 mM Tris–HCl pH 8.5 to maximize DNA concentration. DNA concentration should be between 5 and 10 mg/mL.
2. Back fill a glass needle with microloader tips (*see* **Note 3**).
3. Setup the equipment (Fig. 1e–g). Attach the universal capillary holder arm of the FemtoJet to the stereotactic instrument using the same V clamp used to attach a probe. Load a needle containing DNA into the capillary arm and secure the correct dental bur in the micro drill. For needles created using micropipette puller the tip of the needle usually needs to be trimmed to allow fluid to come out (*see* **Note 4**).
4. Place a newt in 0.1 % MS-222 and wait for a minimum of 15 min before assessing the level of anesthesia. To determine if the newt is anesthetized pinch the leg of the newt or place it on its back. If the newt does not move within 5–10 s continue with the procedure. If the newt moves place it on its feet in

0.1 % MS-222 and wait for a further 5 min before repeating the assessment.

5. Once the newt is anesthetized, dry the head with a paper towel and place the newt on the plastic support and secure it using the ear bars and the snout bar to the stereotaxic instrument. The plastic support allows the ear bars to be firmly pushed against the temples of the newt.
6. Place a small piece of damp tissue over the newt to stop it drying out under the microscope.
7. Using the dental drill, make a 1–2 mm diameter hole on the skull of the newt at the junction between the parietal and frontal bones in the cranial midline. The drilling is performed under observation through a stereomicroscope. The hole does not need to go through the skull but split/crack the join between frontal bones to allow the glass needle to pass through (*see* **Note 5**).
8. Using the microscope, position the glass needle over the crack in the skull and push the glass needle through the meninges using the stereotaxic instrument to a depth of 1 mm (*see* **Note 6**).
9. Inject 500–600 nL of the plasmid solution into the brain by monitoring the drop in the meniscus of the fluid in the barrel of the needle between the pen marks (*see* **Note 1**). Wait for 30 s before removing the needle and then unfasten the newt from the stereotactic instrument. The newt should be electroporated within 5 min.

3.2 *In Vivo* Electroporation

1. Check that the electroporator is set up correctly and the electrodes are clean. The newt is electroporated in unidirectional manner at 150 mV/cm pore forming pulse and 110 mV/cm transfer pulse with a decay of 5 % between the pulses. The newt head is approximately 0.9 cm in width. To electroporate horizontally the machine is set to give one pore forming pulse at 135 mV 50 ms on 950 ms off, followed by five 50 ms pulses at 99 mV with 950 ms gap between each pulse.
2. Place the plastic support holding the newt on the bench. Dip the electrodes and the head of the newt into tank water. Place a lid of a 5 cm petri dish on top of the body of the newt to secure it during the electroporation.
3. Use the hole in the skull as a guide to position the electrodes on the side of the newt head, so that the current will pass through this region of the brain. Once the head is gripped firmly with the electrodes, start the electroporator.
4. Mix the fast drying cement and cover the drilled hole. The cement takes about 10 min to dry.
5. Allow the newt to recover in tap water or tank solution. Place the newt in a container with shallow water (1–5 cm in depth).

Use tissue paper to support the head of the newt so that it is out of the water, allowing the cement to dry and the newt to breathe freely while it regains consciousness.

6. Wait until the newt is moving freely before transferring it to a tank that has a water depth of at least 15 cm.
7. Protein reporter expression driven by a CMV promoter is visible after 15 h of electroporation.

3.3 Intraventricular Injection of 6-OHDA

Wait for at least 2–3 days after electroporation before proceeding to 6-OHDA injections.

1. Load the needle with 6-OHDA using the microloader tips.
2. Setup the intraventricular injections equipment and anesthetize the newts as described in Subheading 3.1, **steps 2–4**.
3. Anesthetize the newt and secure it to the stereotactic instrument as described Subheading 3.1, **step 5**.
4. Remove the dental cement covering the hole in the skull created previously. Use forceps to pull the edge of the cement away from the skull, and dry the head with tissue paper and cotton tips. Use fine forceps and cotton tips to remove the remaining cement with gentle scrapping action until the crack/hole used for the previous injection becomes visible (*see* **Note 7**).
5. Inject 200 nL of toxin through the same hole using the needle attached to the stereotactic device as described in Subheading 3.1, **step 9**.
6. Mix fresh cement and cover the hole. Allow the newts to recover in the shallow water as described in Subheading 3.2, **steps 5** and **6**.

4 Notes

1. The micropipette puller is used to manufacture glass needles. Mark the barrel of the needles every 2 mm with a marker pen. These marks can be used to gauge the volume of the fluid injected. The glass capillaries from Harvard Apparatus reference GC100F-10 have an inner diameter of 0.58 mm. Thus, the volume of fluid held in 1 mm of the barrel part of the needle is equal approximately to 200–300 nL.
2. The purity of the DNA is critical to the success of the electroporation. Plasmid preparation using other kits can have reduced efficiency and may have contaminates which can kill the ependymoglial cells.
3. Plasmid DNA with a concentration above 5 mg/mL can be very viscous and difficult to pipette. Patience is required, but the electroporation efficiency is substantially increased with high DNA concentration.

4. Once the needle is loaded test to see if any fluid comes out. If not, first check that the tubing is connected correctly. Next, try increasing the injection pressure and thirdly use scissors to cut the tip increasing the gauge of the needle. The injection pressure is dependent on how viscous the solution is and the gauge of the needle and can range from 100 to 2,000 hPa. The fluid should come out at a slow and steady rate, and to inject 500 nL can take up to 30 s.
5. To drill the hole in the skull maintain a continuous rpm and gently press the drill tip down on to the skull. Raise the drill tip at 1–2 s intervals to check the progress of the drilling.
6. The penetrance to the meninges is determined by the size and shape of the needle. Try several small sharp movements up and down with the needle. If this does not work, try a different needle. Care is needed not to break the needle at this point.
7. In some cases neuronal ablation is performed without electroporation. If this is the case, a hole is drilled in the skull as described in Subheading 3.1.

Acknowledgement

This work was supported by a grant from the Swedish Research Council to MK.

References

1. Tanaka EM, Ferretti P (2009) Considering the evolution of regeneration in the central nervous system. Nat Rev Neurosci 10:713–723
2. Kokaia Z, Martino G, Schwartz M, Lindvall O (2012) Cross-talk between neural stem cells and immune cells: the key to better brain repair? Nat Neurosci 15:1078–1087
3. Zupanc GKH, Sîrbulescu RF (2011) Adult neurogenesis and neuronal regeneration in the central nervous system of teleost fish. Eur J Neurosci 34:917–929
4. Lang DM, Monzón-Mayor M, Bandtlow CE, Stuermer CA (1998) Retinal axon regeneration in the lizard Gallotia galloti in the presence of CNS myelin and oligodendrocytes. Glia 23:61–74
5. Okamoto M, Ohsawa H, Hayashi T et al (2007) Regeneration of retinotectal projections after optic tectum removal in adult newts. Mol Vis 13:2112–2118
6. Kroehne V, Freudenreich D, Hans S et al (2011) Regeneration of the adult zebrafish brain from neurogenic radial glia-type progenitors. Development 138:4831–4841
7. Simola N, Morelli M, Carta AR (2007) The 6-hydroxydopamine model of Parkinson's disease. Neurotox Res 11:151–167
8. Parish CL, Beljajeva A, Arenas E, Simon A (2007) Midbrain dopaminergic neurogenesis and behavioural recovery in a salamander lesion-induced regeneration model. Development 134:2881–2887
9. Berg DA, Kirkham M, Beljajeva A et al (2010) Efficient regeneration by activation of neurogenesis in homeostatically quiescent regions of the adult vertebrate brain. Development 137: 4127–4134
10. Haas K, Jensen K, Sin WC et al (2002) Targeted electroporation in Xenopus tadpoles in vivo—from single cells to the entire brain. Differentiation 70:148–154
11. Echeverri K, Tanaka EM (2002) Ectoderm to mesoderm lineage switching during axolotl tail regeneration. Science 298:1993–1996

12. Barnabé-Heider F, Meletis K, Eriksson M et al (2008) Genetic manipulation of adult mouse neurogenic niches by in vivo electroporation. Nat Methods 5:189–196
13. Köster RW, Fraser SE (2001) Tracing transgene expression in living zebrafish embryos. Dev Biol 15:329–346
14. Kawakami K (2007) Tol2: a versatile gene transfer vector in vertebrates. Genome Biol 8(Suppl 1):S7
15. Sandoval-Guzmán T, Wang H, Khattak S et al (2014) Fundamental differences in dedifferentiation and stem cell recruitment during skeletal muscle regeneration in two salamander species. Cell Stem Cell 14:174–187

Chapter 8

The Accessory Limb Model: An Alternative Experimental System of Limb Regeneration

Tetsuya Endo, David M. Gardiner, Aki Makanae, and Akira Satoh

Abstract

Accessory limb model (ALM) was developed as an experimental model and functional assay for limb regeneration. The ALM provides several ways to identify pathways and test for signaling molecules that regulate limb regeneration. Here, we summarize the history of the ALM and describe the specific details involved in inducing ectopic blastemas and limbs from a skin wound on the side of the arm.

Key words Salamander, Axolotl, Nerve, Limb regeneration, Slow-release sustained bead

1 Introduction

Limb regeneration is one of the impressive biological phenomena that have been studied in two urodele amphibians, axolotls and newts. Limb amputation results in formation of a regeneration-specific structure called the blastema, which consists of undifferentiated cells. Blastema cells are derived from stump tissues and appear to be lineage restricted such that regenerated muscle is derived from myogenic progenitors, and connective tissue cells are derived from cells in the connective tissues of the stump. Once a blastema has formed, it starts restoring the original structure [1]. It is the early stages of blastema induction that are unique to limb regeneration. Therefore, understanding early phases of limb regeneration that lead to blastema formation is important in order to understand the regeneration potency in axolotls.

Historically, studies of limb regeneration were based on an amputation model, and the progression through the stages of blastema formation, growth, and differentiation was described. However, amputation causes wide-ranging injuries to the tissues at the amputation plane, and it is not possible to determine which of the multitude of responses that are triggered are required to induce limb regeneration. In addition, since regeneration is a default

Anoop Kumar and András Simon (eds.), *Salamanders in Regeneration Research: Methods and Protocols*, Methods in Molecular Biology, vol. 1290, DOI 10.1007/978-1-4939-2495-0_8, © Springer Science+Business Media New York 2015

response to total limb amputation, it is also difficult to assay candidate factors for the ability to induce a regenerative response. To address these problems, we revisited the phenomenon reported in earlier studies that a limb can be induced to form from non-regenerative skin wound without limb amputation [2], taking advantage from wounds which heal without forming a scar [3, 4]. We refined the surgical procedure and were able to induce ectopic blastemas and limbs reproducibly and at a high frequency [2]. As a consequence, studies of limb regeneration can be conducted using two experimental systems, the amputation model and the ALM.

As mentioned above, the ALM is based on earlier studies indicating that there are three essential sources of signaling in limb regeneration: the wound epidermis, nerves, and cells with positional information from the side of the limb that is opposite to the wound [5]. Since covering the amputation plane with uninjured, full-thickness skin (epidermis and dermis) inhibits limb regeneration [6], it appears that a direct interaction between the wound epidermis and underlying stump tissues is required. The wound epidermis in limb regeneration thickens to form the apical epithelial cap (AEC), which signals to the underlying mesenchymal cells, resulting in the formation and sustained growth of the blastema. The ability of the AEC to induce blastema growth is dependent on signaling from nerves, and this interaction does not occur if the limb is denervated, resulting in failure to regenerate the limb. When the limb is amputated either experimentally or in nature, the interactions between the wound epithelium, nerves, and blastema cells are very important [5].

In contrast to amputation-induced limb regeneration, the formation of an accessory limb de novo requires specific surgical manipulations in order to provide the necessary signals [5]. In the early experiments by Bodemer [7], the lateral side of the proximal limb was extensively traumatized, and brachial nerves were severed at the distal level and rerouted to the wounded site. It was subsequently discovered that traumatizing deep tissue could be replaced by grafting a limb/non-limb tissue to develop a limb structure, such as bone, muscle, or internal organs [8]. Later, it was shown that grafting a piece of skin from the opposite side of the contralateral limb to the nerve-deviated site is enough to induce a limb [9–12]. Egar [9] also reported that the removal of 1–2 mm square of skin at the site of a surgically deviated nerve increased the frequency of blastema induction which was interpreted to occur as a result of increasing the chance that the wound epithelium and the cut edge of the nerve interacted. This inference was consistent with the observation that the AEC is highly innervated, and the interaction between the nerve and wound epithelium is necessary to induce gene expression of an AEC marker [13, 14]. In addition to migration of epidermal cells and AEC formation, there is also a directed migration of dermal fibroblasts into the wounded site to form the blastema as well as to regenerate the dermis at the end of regeneration of the skin wound [15]. In amputation-induced limb regeneration, dermal fibroblasts are a major source of the blastema and redevelop into cartilages and connective tissues [16, 17]. As with amputations, dermal fibroblasts immediately migrate from the vicinity area of the skin window into the center of the wound in the ALM [2]. Given that the signals necessary to induce limb pattern formation in the ALM can be provided by grafting a piece of full-thickness skin from the side of the limb that is opposite to the nerve-deviated wound [9–12], it is likely that migration from the grafted skin and surrounding skin generates positional diversity in dermal fibroblasts, and such diversity is necessary for blastemas to undergo pattern formation [18]. Taken together, signals from both a deviated nerve and a grafted piece of skin are required to form a limb de novo. Nerves transition the initial phase of skin wound healing to the phase of blastema formation. If there is sufficient positional diversity derived from grafted fibroblasts from circumferentially different positions, there is sufficient information to transition to the phase of pattern formation. Therefore, the ALM enables us to examine functions of nerves and skin graft separately, leading to the description of the progressive steps of regeneration [2].

As the ALM allows us to assay for specific signals that regulate progression from one step to the next, it is now possible to investigate the function of the AEC, nerves, and positional interactions leading to limb regeneration. One way to do this is to test for the ability of proteins delivered on beads to influence the behavior of cells in wounds. The use of beads is widely accepted and has contributed in fields in developmental biology and regeneration. We describe the microsurgical techniques associated with the ALM and the preparation of gelatin beads for grafting in the context of regeneration.

2 Materials

2.1 Obtaining Axolotls and Husbandry

1. Animals: Axolotls are either purchased from colonies such as Ambystoma Genetic Stock Center at the University of Kentucky, Lexington, or spawned in the laboratory.
2. Animals are maintained at 25 °C on a 12-h light-dark cycle and fed every 2–3 days.
3. Axolotls of 5–12 cm (snout-to-tail tip; juveniles or young adults; *see* Fig. 1a, b) are preferred because they regenerate a

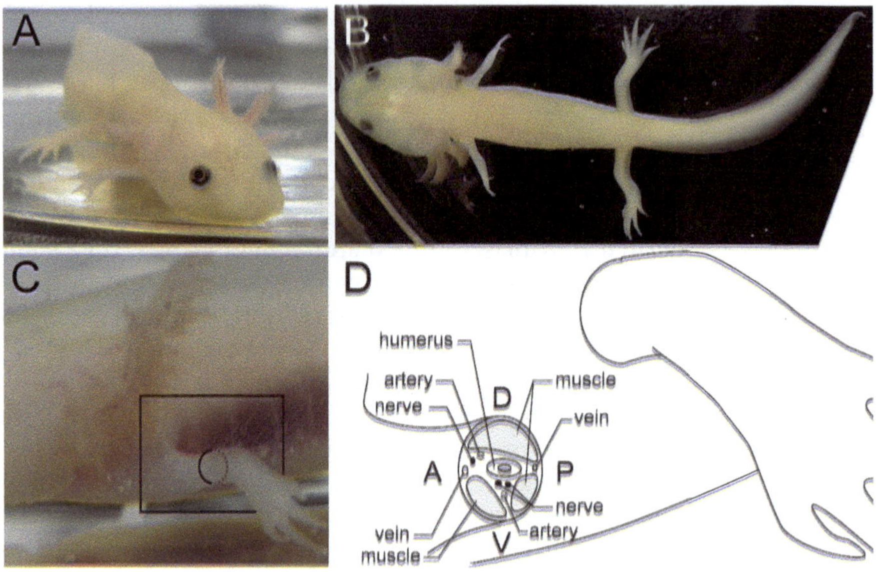

Fig. 1 The axolotl and the anatomy of the forelimb. (**a–c**) A juvenile leucistic axolotl. Front view (**a**) and dorsal view (**b**). (**c**) Left side view of the animal at the forelimb level. The left forelimb is seen from the anterior side. (**d**) The anatomy of the left forelimb at the middle of the forearm level as it would appear in transverse section at the level of the *black circle* indicated in (**c**) (modified from [20])

limb relatively fast and are easy to handle. Leucistic and albino mutants are useful since it is easier to visualize tissues through their unpigmented skin.

2.2 Axolotl Maintenance and Surgery

1. Holtfreter's solution buffered with Tris (HBT solution): Dissolve 0.46 g NaCl, 0.05 g KCl, 0.1 g $CaCl_2$, and 5.0 g Trizma preset crystals (pH 7.2) in 1 L of dechlorinated tap water (*see* **Note 1**).
2. 40 % HBT solution: Dilute the stock HBT in dechlorinated tap water.
3. Tricaine methanesulfonate (MS222): Prepare 0.1 % solution in 40 % HBT and adjust pH to 7.2–7.4 with Tris (pH 9.0).
4. Surgical supplies: Fine forceps (Dumont #5 or similar), angled ophthalmic scissors with 2.5 mm cutting edge, and tungsten needles. Sterilize the surgical instruments with 70 % ethanol or in a glass bead sterilizer.
5. Dissection and stereo fluorescence microscope.

2.3 Bead Preparation

1. Gelatin: Type B, ~225 g bloom, stored at room temperature and in dry air condition (*see* **Note 2**).
2. Olive oil.
3. Siliconized Petri dish: Add few drops of silicon solution (e.g., Sigmacote) and swirl the dish to coat evenly. Air-dry the dish

at room temperature. Wash the dish with deionized water (DDW) before use.

4. Glutaraldehyde fixative: 2–10 mM in deionized water containing 0.1 % Tween 20 (*see* **Note 3**).
5. Glycine: Prepare 10 mM in DDW.
6. Proteins (e.g., FGF2) (*see* **Note 4**).

3 Methods

The procedure described here induces an accessory limb on the anterior side of the left upper arm (for limb orientation, *see* Fig. 1c, d). An accessory limb can be induced at other proximal-distal levels or from any position around the circumference of the limb with appropriate modification of the procedure. Overview of the ALM surgeries and resulting phenotypes are shown in Fig. 2a–d.

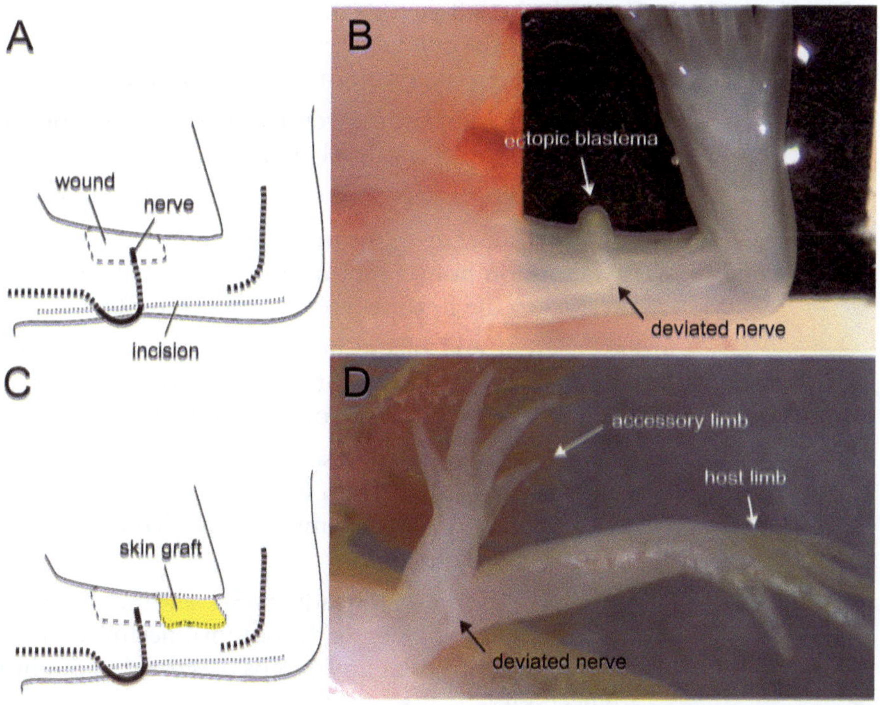

Fig. 2 Induction of an accessory limb in axolotls. (**a**) The surgical procedure that induces an ectopic blastema from an anterior skin wound by deviating a nerve. (**b**) An induced blastema by the (**a**) surgery. (**c**) The surgical procedure that induces an accessory limb from an anterior skin wound by deviating a nerve and grafting posterior skin. (**d**) An accessory limb induced by the (**c**) surgery. Host limbs in (**a**)–(**c**) and the accessory limb in (**d**) are seen from the ventral side

3.1 Skin Wounding

1. Put an animal into MS222 solution for anesthesia and leave it for 10–20 min.
2. Place a piece of tissue wipe on the surgical stage and moisten it with 40 % HBT solution.
3. Place the anesthetized animal on the surgical stage, turning the left side up (Fig. 3a).
4. Cover the animal with the moist wipe to protect from dehydration.
5. Orient the left forelimb for surgical manipulation.
6. Pull up the skin on the anterior side at the middle of the upper arm with forceps, and make a small incision carefully with ophthalmic scissors. If the animals are leucistic or albino, a thick blood vessel is visible through the skin (Fig. 3a, *see* also Fig. 1d).
7. Put the tip of the scissors into the incision and excise the piece of skin (1.0–1.5 mm × 1.0–1.5 mm for blastema induction without a skin graft, 1.0–1.5 mm × 2.0–3.0 mm for limb induction with skin graft) by expanding the incision (Fig. 3b) (*see* **Note 5**). Attention needs to be paid not to damage hypodermal muscles, nerves, and blood vessels.

3.2 Nerve Deviation

1. Raise up the left arm so that the ventral side of the limb is up (Fig. 3c). If the animals are leucistic or albino, two flexor nerve bundles are visible through the skin (*see* Fig. 1d).
2. Make an incision on the ventroposterior side from the shoulder level to the elbow level (Fig. 3d).
3. Bisect the flexor nerves distally (at the elbow level).
4. Pull out the free tip of the nerves carefully with forceps (Fig. 3e). Pay attention not to rupture the ventral brachial artery that is attached to the flexor nerves by connective tissues.
5. Pass the cut end of the nerve underneath the ventroanterior skin to the wound (Fig. 3f). Fix the skin back to the original position to close the ventroposterior incision.
6. Turn the left arm back to see the anterior wound. Position the cut edge of the nerve in the middle of the wound along the transverse plane of the limb (Fig. 3g). Trim the distal tip of the nerve, if it is too long.
7. These procedures (Subheadings 3.1 and 3.2) are sufficient to induce a blastema (Fig. 2a, b) but insufficient to induce a limb. To obtain a patterned limb, steps described in Subheading 3.3 are necessary.

3.3 Skin Grafting

1. To obtain a piece of skin for grafting, turn the animal left side down.
2. Hold the right arm up so that the posterior side of the limb is visible.

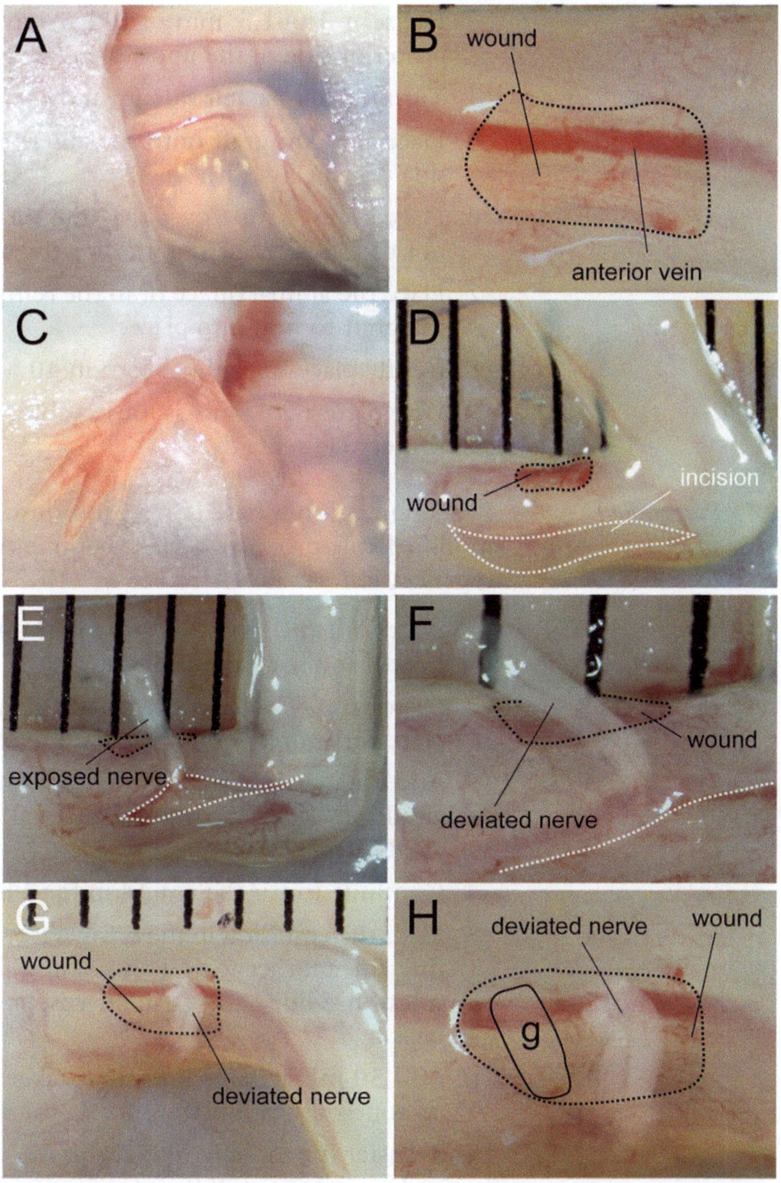

Fig. 3 A series of pictures of the surgery that induces an accessory limb. (**a**) For left forelimb surgery, the animal is laid down, turning the left side up. A *thick blood vessel* is visible on the anterior side of the forearm. (**b**) A *square skin* wound is made on the anterior side of the middle of the forearm. (**c**) The limb is turned up to expose the ventral side. (**d**) An *incision* is made in the ventroposterior skin to access the ventral nerve bundle. (**e**) The ventral nerve bundle is cut at the elbow level and pulled free of the connective tissues. (**f**) The cut end of the nerve bundle is passed underneath the ventroanterior skin to the wound. (**g**) The limb is turned down to see the anterior wound. The cut end of the nerve bundle is trimmed and placed in the middle of the wound. (**h**) A piece of skin from the opposite side of the limb (posterior skin is grafted to the anterior wound) is removed from the contralateral limb and grafted to the skin wound. *g* grafted skin

3. Excise a 1.0–1.5 mm × 1.0–1.5 mm piece of skin from the posterior side of the upper arm (*see* **Note 6**).
4. Turn the animal left side up and expose the anterior wound.
5. Place the skin graft onto the anterior wound with the dermal side down (Fig. 3h).
6. Place a moist wipe over the anterior wound, paying attention not to touch the skin graft directly with the wipe.
7. Place the animal on ice or in the refrigerator for 2–4 h to allow the graft to heal into place.
8. Gently place the animal back in 40 % HBT solution.
9. Well-patterned limb is observable about 60 days after the surgery (Fig. 2c, d).

3.4 Preparation of Beads

Bead generation is based on a procedure by Tabata et al. [19]. We modified this procedure for experimental use in amphibian limb regeneration.

1. Warm 50 mL olive oil up to 40–60 °C in a 200 mL glass beaker and stir with a stir bar.
2. Dissolve 0.5 g gelatin in distilled water by heating on a hot plate with stirrer or a water bath maintained at 100 °C.
3. Add the gelatin solution into the warm olive oil to generate an emulsion. Keep stirring the olive oil during this step (Fig. 4a) (*see* **Note 7**).
4. Continue to stir the emulsion for 2 h at room temperature.
5. Place the emulsion in the refrigerator at 4 °C and continue stirring for another 2 h. This cooling process makes the gelatin solution solidify and form microspheres that can be seen in the beaker.
6. Add 50 mL acetone into the beaker and continue stirring the acetone-oil-bead mixture for 1 h at 4 °C.
7. Transfer the mixture into 2× 50 mL centrifuge tubes (Fig. 4b) and centrifuge at 250 × *g* for 5 min; discard the supernatant.
8. Add 50 mL acetone and invert the tubes several times to mix the solution and centrifuge at 250 × *g* for 5 min.
9. Discard the supernatant and add 50 mL isopropanol to each tube, invert few times, and centrifuge at 250 × *g* for 5 min (Fig. 4c).
10. Wash the beads with DDW.
11. Fix the beads in 2–10 mM glutaraldehyde-Tween 20 solution overnight at 4 °C with constant stirring. Fixation results in hardening of the beads (*see* **Note 3**).
12. Centrifuge the solution in a 50 mL tube at 250 × *g* for 5 min and collect the beads.

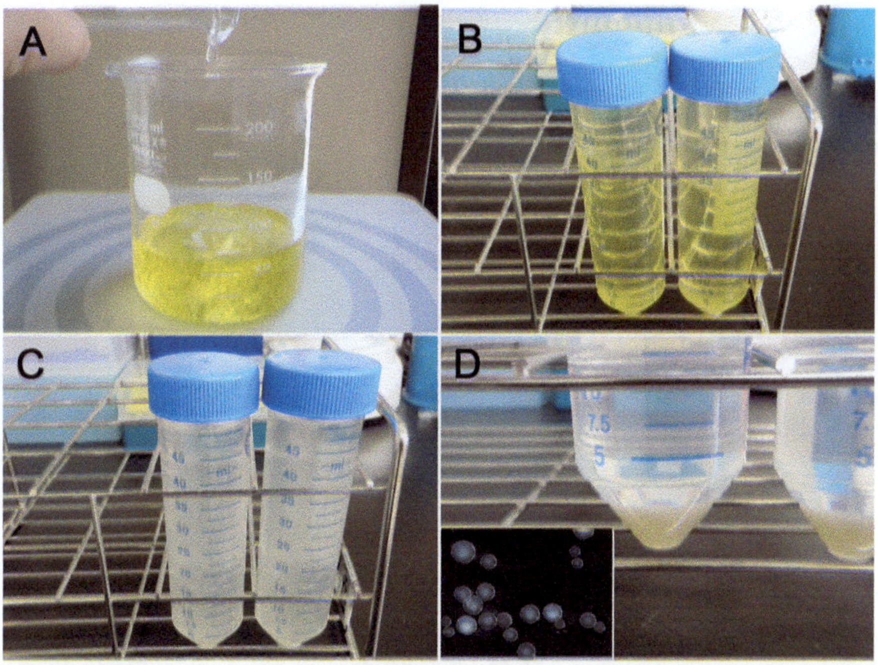

Fig. 4 Steps involved in making gelatin beads. (**a**) Gelatin solution is poured into warmed olive oil. (**b**) Acetone is added into the gelatin-oil emulsion. (**c**) Beads are washed by isopropanol. (**d**) Beads can be observed in the bottom of the tubes

13. Add deionized water and wash the beads by inverting few times.
14. Centrifuge again at 250 × *g* to remove the supernatant.
15. Transfer the beads in a glass beaker with 100 mL of 10 mM glycine/DDW solution to neutralize the remaining glutaraldehyde.
16. Stir the solution at room temperature for 1 h.
17. Centrifuge the solution in 50 mL tube at 250 × *g* for 5 min and discard the supernatant.
18. Wash the beads with DDW (shake well) and centrifuge as described in **step 17**. Repeat this procedure three times.
19. Store the beads in DDW at 4 °C (Fig. 4d). The beads are stable for 1 month.

3.5 Protein Absorption into the Beads

1. Pipette 3–5 mL of deionized water in a 35 mm Petri dish.
2. Transfer beads into the Petri dish by a plastic pipette. Do not transfer a large amount of beads. If an excess amount of beads is transferred, water in the Petri dish becomes cloudy, which makes bead selection difficult.

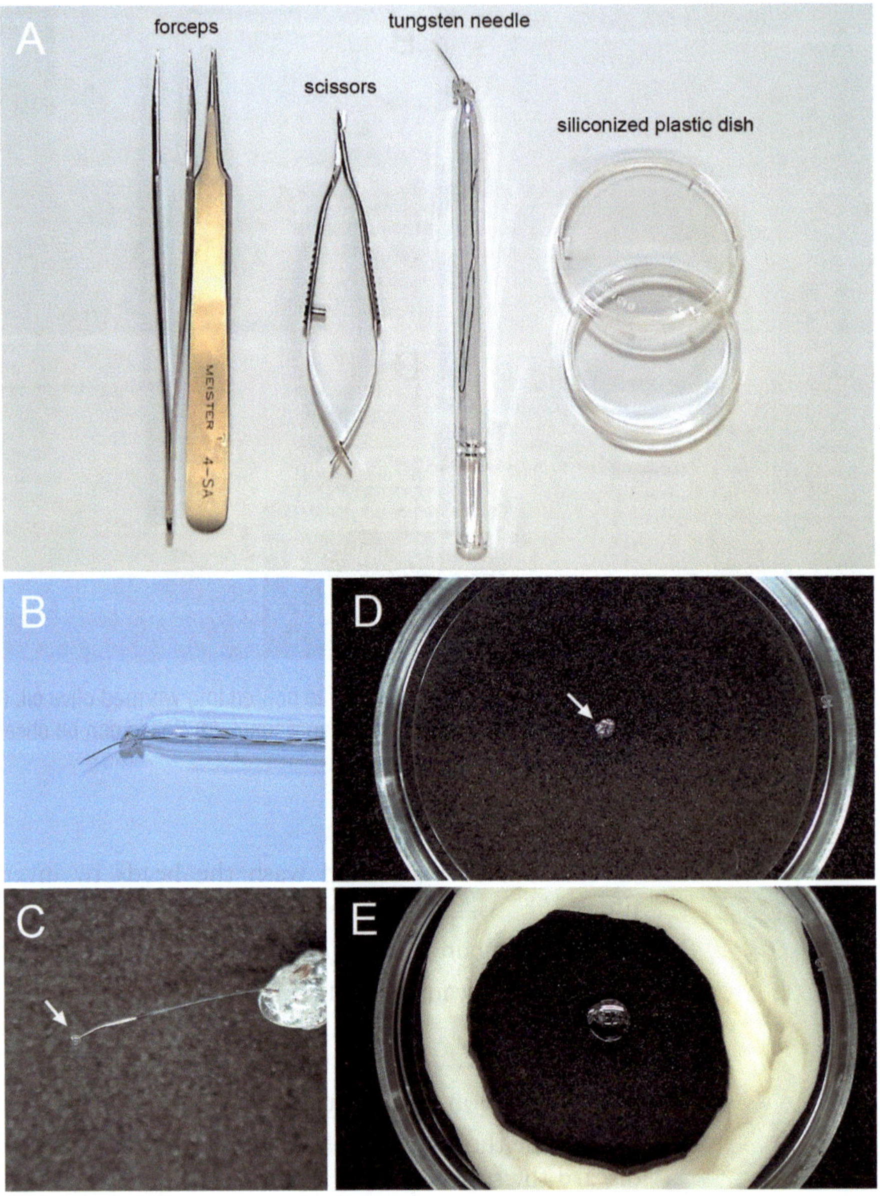

Fig. 5 Equipment and instruments required for grafting gelatin beads. (**a**) Surgical tools. Sizes of forceps and scissors depend on animal size. (**b**) A tungsten needle is melted into a glass Pasteur pipette. Tip of the needle is sharpened by electrolysis. (**c**) A bead is poked by the tungsten needle. (**d**) Beads are placed in the center of a dish and allowed to dry out, at which point they are shrunken and usually have a *yellow color*. (**e**) Proteins are dropped on the beads and absorbed by the beads. A damp wipe is placed along a margin of the dish to prevent dehydration of the beads

3. Select beads with the appropriate size for the experiment. Stick the bead with a tungsten needle (Fig. 5b, c) and transfer it into a new 35 mm siliconized Petri dish (Fig. 5d). Select beads of 200–400 μm diameter which are ideal for implantation in axolotls.

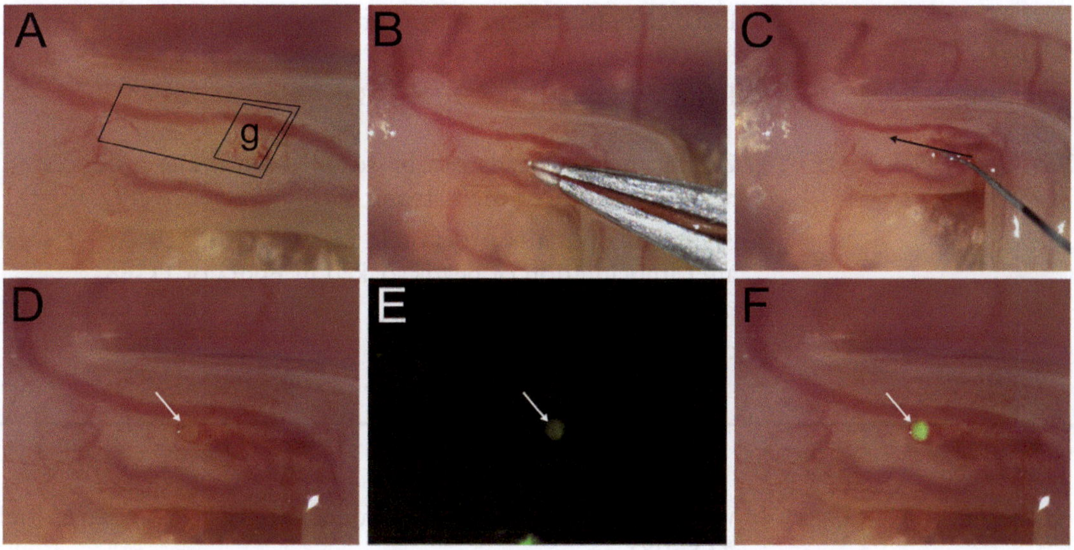

Fig. 6 A series of pictures of the surgery to graft a gelatin bead. (**a**) Skin is excised from a limb and a small piece of skin is grafted from the contralateral side of a limb. *g* grafted skin. *Boxes* indicate the border of the wound and the graft. (**b**) A small tunnel is created by forceps. (**c**) A bead is inserted underneath the wound epithelium through the tunnel using a tungsten needle. A *black arrow* indicates the direction of the tunnel. (**d–f**) The bead is left in the tunnel by removing the tungsten needle gently. *White arrows* indicate the grafted beads. (**d**) Bright field image. (**e**) Fluorescence image of (**d**) showing the autofluorescence of the grafted bead. (**f**) Merged image of (**d**) and (**e**)

4. Allow the beads to dry out completely by removing the lid on the plastic dish. Dried beads can be stored at room temperature for a couple of days.
5. Drop proteins of interest onto the dried beads. The beads will begin to swell as they rehydrate. Allow the beads to absorb the proteins for at least 2 h on ice or in a refrigerator. Place a wet tissue wipe along the brim of the dish (Fig. 5e) to prevent dehydration of beads.
6. The beads impregnated with proteins can be stored in 4 °C for 2–3 days (*see* **Note 8**).

3.6 Surgery and Bead Grafting

1. Anesthetize animals as described in Subheading 3.1.
2. Create a skin wound (usually 2 × 3 mm square).
3. Prepare and graft a skin graft as described in Subheading 3.3 (Fig. 6a).
4. Allow the graft to heal for 72 h.
5. Make a small incision by the wound (usually 2 mm long).
6. Create a small tunnel to the destination by insertion of fine-tip forceps (Fig. 6b).
7. Pick up a bead from the dish by a tungsten needle and insert it to the wound bed through the tunnel (Fig. 6c–f) (*see* **Note 9**).

8. Place the animals on ice for at least 2 h to allow the grafted beads to heal into place.
9. Return the animals to 40 % HBT.

3.7 Observation of Phenotypes

1. Grafted gelatin beads can be readily seen under a fluorescent microscope due to their strong autofluorescence (Fig. 6d–f).
2. A limb blastema is observable in an axolotl (10 cm) 1 week after implantation of beads soaked in FGF2 or FGF8.
3. Formation of a well-patterned limb induced by a skin graft is completed in about 2 months.

4 Notes

1. We prefer to use Trizma preset crystals, pH 7.2 (Sigma-Aldrich #T8508).
2. It might be appropriate to use type A gelatin depending on the electrical charge of the protein being used. For example, type A gelatin is for acidic FGF (FGF1).
3. This fixative needs to determine the stiffness of the beads. Usually, 2 mM of glutaraldehyde is sufficient for experiments in axolotls.
4. Proteins are purchased and reconstituted in accordance with the manufacturer's protocol. Usually the proteins are aliquoted into 1 μL each in siliconized microtubes. The aliquoted proteins are stored at −80 °C.
5. The size of the wound should be adjusted to approximately the same size as the transverse dimension of the limb (i.e., larger wounds on larger animals).
6. It is important not to disorient the skin graft when it is moved to the wound. To prevent that, prepare 2× 5–10 mm fine lens paper and place it next to the skin graft. Pick up the skin graft and flip it onto the lens paper, so that the epidermal side is facing the lens paper. When leucistic or albino mutants are used, put some dye (e.g., carbon ink) on the skin graft (the dermal side) to facilitate visualization.
7. The diameter of the expected beads cannot be predicted because of variability between researchers making the beads. The size of the bead depends upon the stirring speed and temperature of gelatin solution. Gelatin prepared at higher temperature appears to produce smaller beads. Therefore, the optimal speed of a stirrer to obtain the desired diameter should be determined empirically in each laboratory.
8. Proteins typically maintain activity for over 1 week, but usually maximum activities can be expected within the first couple of

days. It is strongly recommended that the beads are used on the day when the protein is initially added to the beads.

9. The presence of the grafted bead can be confirmed by its autofluorescence (Fig. 6d–f). However, depending on animal color (e.g., black or golden), autofluorescence of the bead can be hard to identify because of strong autofluorescence of the animals.

Acknowledgments

This work was supported by Grant-in-Aid for Scientific Research on Innovative Areas (KAKENHI #23124508 to AS and #22124006 to TE).

References

1. Bryant SV, Endo T, Gardiner DM (2002) Vertebrate limb regeneration and the origin of limb stem cells. Int J Dev Biol 46:887–896
2. Endo T, Bryant SV, Gardiner DM (2004) A stepwise model system for limb regeneration. Dev Biol 270:135–145
3. Seifert AW, Kiama SG, Seifert MG, Goheen JR, Palmer TM, Maden M (2012) Skin shedding and tissue regeneration in African spiny mice (Acomys). Nature 489:561–565
4. Levesque M, Villiard E, Roy S (2010) Skin wound healing in axolotls: a scarless process. J Exp Zool B Mol Dev Evol 314:684–697
5. Goss RJ (1969) Principles of regeneration. Academic, New York
6. Mescher AL (1976) Effects on adult newt limb regeneration of partial and complete skin flaps over the amputation surface. J Exp Zool 195: 117–128
7. Bodemer CW (1958) The development of nerve-induced supernumerary limbs in the adult newt, Triturus viridescens. J Morphol 102:555–581
8. Bodemer CW (1959) Observations on the mechanism of induction of supernumerary limbs in adult Triturus viridescens. J Exp Zool 140:79–99
9. Egar MW (1988) Accessory limb production by nerve-induced cell proliferation. Anat Rec 221:550–564
10. Lheureux E (1977) Importance of limb tissue associations in the development of nerve-induced supernumerary limbs in the newt Pleurodeles waltlii Michah (author's transl.). J Embryol Exp Morphol 38:151–173
11. Maden M, Holder N (1984) Axial characteristics of nerve induced supernumerary limbs in the axolotl. Roux's Arch Dev Biol 193: 394–401
12. Reynolds S, Holder N, Fernandes M (1983) The form and structure of supernumerary hindlimbs formed following skin grafting and nerve deviation in the newt Triturus cristatus. J Embryol Exp Morphol 77:221–241
13. Satoh A, Graham GM, Bryant SV, Gardiner DM (2008) Neurotrophic regulation of epidermal dedifferentiation during wound healing and limb regeneration in the axolotl (Ambystoma mexicanum). Dev Biol 319:321–335
14. Stocum DL (2011) The role of peripheral nerves in urodele limb regeneration. Eur J Neurosci 34:908–916
15. Gardiner DM, Muneoka K, Bryant SV (1986) The migration of dermal cells during blastema formation in axolotls. Dev Biol 118:488–493
16. Kragl M, Knapp D, Nacu E, Khattak S, Maden M, Epperlein HH, Tanaka EM (2009) Cells keep a memory of their tissue origin during axolotl limb regeneration. Nature 460:60–65
17. Muneoka K, Fox WF, Bryant SV (1986) Cellular contribution from dermis and cartilage to the regenerating limb blastema in axolotls. Dev Biol 116:256–260
18. Makanae A, Satoh A (2012) Early regulation of axolotl limb regeneration. Anat Rec 295: 1566–1574
19. Tabata Y, Hijikata S, Muniruzzaman M, Ikada Y (1999) Neovascularization effect of biodegradable gelatin microspheres incorporating basic fibroblast growth factor. J Biomater Sci Polym Ed 10:79–94
20. Bryant SV (1976) Regenerative failure of double half limbs in Notophthalmus viridescens. Nature 263:676–679

Acknowledgments

Chapter 9

High-Efficiency Electroporation of the Spinal Cord in Larval Axolotl

Aida Rodrigo Albors and Elly M. Tanaka

Abstract

Axolotls are well known for their remarkable ability to regenerate complex body parts and structures throughout life, including the entire limb and tail. Particularly fascinating is their ability to regenerate a fully functional spinal cord after losing the tail. Electroporation of DNA plasmids or morpholinos is a valuable tool to gain mechanistic insight into the cellular and molecular basis of regeneration. It provides among other advantages a simple and fast method to test gene function in a temporally and spatially controlled manner. Some classic drawbacks of the method, such as low transfection efficiency and damage to the tissue, had hindered our understanding of the contribution of different signaling pathways to regeneration. Here, we describe a comprehensive protocol for electroporation of the axolotl spinal cord that overcomes this limitations using a combination of high-voltage and short-length pulses followed by lower-voltage and longer-length pulses. Our approach yields highly efficient transfection of spinal cord cells with minimal tissue damage, which now allows the molecular dissection of spinal cord regeneration.

Key words Electroporation, Spinal cord, Regeneration, Axolotl

1 Introduction

In the late 1700s, Lazzaro Spallanzani found the salamander tail to be an eminent model to study regeneration. Due to the transparency of their regenerating tail tissue Spallanzani could describe for the first time the remarkable process of spinal cord regeneration [1]. The behavior of the cells lining the central canal of the spinal cord in response to injury is an integral aspect of spinal cord regeneration in salamanders. Following amputation of the tail, the spinal cord grows posteriorly as a simple neuroepithelial tube of self-renewing stem/progenitor cells. After a period of rapid expansion, the tube undergoes renewed neurogenesis to faithfully reconstruct tissue architecture and function [2–5]. Thus, in the view of current interest in regenerative medicine, salamanders are an unparalleled system to elucidate cellular and molecular mechanisms involved in spontaneous recruitment and expansion of neural stem/progenitor

Anoop Kumar and András Simon (eds.), *Salamanders in Regeneration Research: Methods and Protocols*, Methods in Molecular Biology, vol. 1290, DOI 10.1007/978-1-4939-2495-0_9, © Springer Science+Business Media New York 2015

cells in response to injury. Recently, significant efforts have been put into developing genetic, genomic, and transgenesis tools for a mechanistic understanding of regeneration [6–9] that have now made *Ambystoma mexicanum* (the axolotl) a premier organism for the molecular genetic analysis of spinal cord regeneration.

A key technical aspect for the molecular analysis of spinal cord regeneration is the ability to overexpress genes of interest. Electroporation has proven to be a highly valuable tool in this context due to the ability to rapidly target DNA plasmid constructs to the spinal cord. Electroporation is a technique which employs an electric field to transfer charged nucleic acids or other macromolecules into cells [10]. The exact mechanism is not fully known but, in essence, successful electroporation depends on two major processes. First, the application of a strong electric field induces the formation of transient and reversible pores in the otherwise hydrophobic cell plasma membrane. Then, electric pulses facilitate the entry of highly charged molecules such as DNA or RNA into the cell by electrophoresis [11]. Inside the cell, the electroporated DNA is transported into the nucleus, where it will be transcribed by the endogenous transcriptional machinery. Electroporation is a versatile tool and can be fine-tuned depending on the aim of the cellular and/or molecular analysis. For instance, in small and transparent axolotls, electroporation of GFP expression plasmids into single or small groups of spinal cord cells has been valuable to identify and track in vivo resident spinal cord cells during regeneration [6, 12, 13]. DNA plasmids and morpholinos [14] can also be used to test gene function, without going through the lengthy process of generating transgenic animals or using viral vectors, which are limited in insert size and can trigger immunological responses. The expression of regular expression plasmids and morpholinos is transient, but can be observed more than 21 days after electroporation depending on the cell type (our own observations). This is sufficient to cover the entire process of spinal cord regeneration in larval axolotls. For cell fate or lineage tracing studies, expression vectors such as the Tol2 transposon system can be used for stable gene expression by integrating the ectopic genes into the host genome [15].

Previous studies in the axolotl reported the electroporation of charged nucleic acids into small groups of spinal cord cells [6, 14]. However, electroporation as a method to study gene function in the axolotl spinal cord has remained elusive because often large numbers of electroporated cells are required within the tissue to detect an effect. Here we report a new strategy of in vivo spinal cord electroporation in the axolotl which yields highly efficient transfection of resident spinal cord cells with minimal tissue damage. Our strategy involves (1) microinjection of DNA encoding the gene of interest into the central canal of the spinal cord and (2) application of two sets of electric pulses, the so-called poring and transfer pulses, in which voltage and duration have been optimized to maximize efficiency and minimize tissue damage. Different from what has been previously described, our approach yields highly efficient transfection of resident spinal cord cells with minimal tissue damage opening a new door for simple and efficient gene function studies in the axolotl spinal cord.

2 Materials

1. Axolotls, 2.2–2.5 cm snout to tail tip (*see* **Note 1**).
2. 10× Phosphate buffered saline (PBS) stock: to prepare 1 L of 10× PBS, dissolve the following reagents in 800 mL of deionized water: 88 g of NaCl, 7 g of NaH_2PO_4, and 2.3 g of KH_2PO_4. Use HCl to adjust the pH to 7.4 and add deionized water up to 1 L. For 1× PBS solution, dilute 10× stock in deionized water. Autoclave to sterilize. Store PBS at room temperature.
3. DNA plasmid: we use the pCAGGs-Cherry-nls plasmid (ubiquitous chicken β-actin promoter with a CMV enhancer driving nuclear Cherry expression) diluted in PBS at a concentration of 1.5 μg/μL (*see* **Note 2**).
4. DNA Plasmid Maxi Kit (*see* **Note 3**).
5. Fast Green solution: make 5× Fast Green solution by dissolving 12.5 mg of Fast Green FCF powder in 10 mL of 1× PBS.
6. 10× TBS: to make 1 L, add 24.2 g of TRIS and 90 g of NaCl in 990 mL of distilled water. Mix well and adjust the pH to 8.0 by adding 10 mL of HCl 37 %. Store at room temperature for up to 12 months.
7. Holtfreter's solution 400 % (wt/vol): Mix 2.875 g of KCl, 5.36 g of $CaCl_2{\cdot}2H_2O$, 11.125 g of $MgSO_4{\cdot}7H_2O$, and 158.4 g of NaCl. Fill up to 10 L with distilled water. Store at room temperature for up to 6 months.
8. Benzocaine 10 % (wt/vol): To make benzocaine stock solution, dissolve 50 g of benzocaine in 500 mL of 100 % ethanol. Store at room temperature up to 12 months.
9. Benzocaine 0.03 % (vol/vol): To make 10 L, mix 500 mL of 10× TBS, 500 mL of 400 % Holtfreter's, and 30 mL of 10 % benzocaine. To anesthetize the animals, dilute the 0.03 % solution to 0.008 % in tap water.
10. Polymethyl methacrylate platform and SYLGARD silicone elastomer (Dow Corning), for stably holding the animal during microinjection (*see* **Note 4**).
11. Microloader tips, for loading the microinjection capillaries.

12. Sharp and ring forceps, for breaking the microinjection needles and handling the animals.
13. Borosilicate glass capillaries (1.2 mm O.D., 0.94 mm I.D.).
14. Pressure injector (e.g., PV830 Pneumatic PicoPump, World Precision Instruments).
15. A stereomicroscope with 3-axis micromanipulator attachment.
16. Electroporation plate: petri dish, UltraPure™ Agarose, PBS, scalpel.
17. Tweezer electrodes with round 10 mm platinum plate electrodes (CUY650P10, Nepa Gene).
18. NEPA21 electroporator (Nepa Gene).
19. Cups to house the animals.

3 Methods

3.1 Preparation for Microinjection and Electroporation

1. Make a 2 % (wt/vol) agarose solution in PBS. Heat in a microwave oven to dissolve, swirling to ensure proper mixing.
2. Pour the gel in a 60 mm petri dish and allow to cool down at room temperature. When the gel solidifies, cut out three wells with a surgical scalpel to make an electroporation plate (Fig. 1a). The two parallel wells will hold the electrodes for electroporating the tail; the third well, the body of the axolotl (Fig. 1b). Cut a thin line from the third well passing through the agarose between the two parallel wells. This cut will keep the tail straight and parallel to the electrodes during electroporation (Fig. 1c) (*see* **Note 5**).

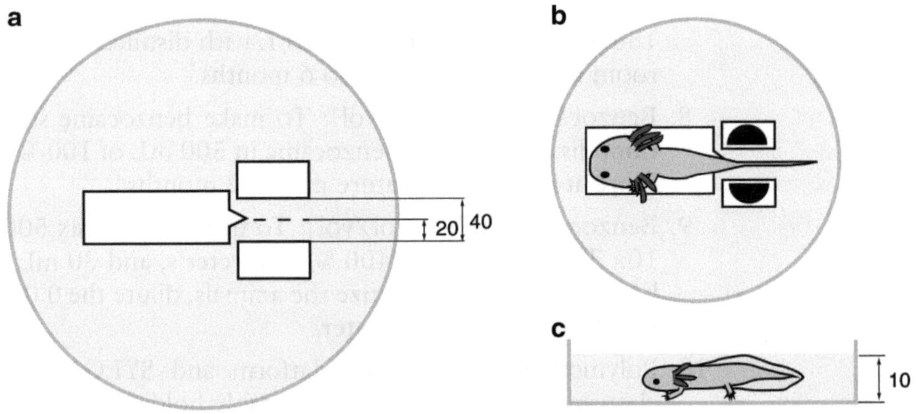

Fig. 1 Making of the electroporation plate for holding and protecting the tail during electroporation. (**a**) Cartoon illustrating the *top view* of the electroporation plate. Two wells, 40 mm spaced apart, will hold the electrodes. The thin cut, 20 mm away from each of these wells, will keep the tail of the axolotl in place and protected from the electric current, as shown in (**b**) and, in a *lateral view*, in (**c**)

3. Dilute DNA plasmid to 1.5 μg/μL in PBS (*see* **Notes 1** and **6**). Mix the diluted plasmid with Fast Green FCF solution at a 30:1 ratio (*see* **Note** 7).
4. Generate microinjection needles on a standard Flaming/ Brown micropipette puller or similar instrument (*see* **Note 8**).
5. Set the pressure injector at 2 psi eject pressure and the duration switch on gated mode (*see* **Note 9**).
6. Set up the electroporation parameters. The NEPA21 electroporator allows the configuration of two electrical pulses in one single electroporation: the poring pulse and transfer pulse. Use one bipolar poring pulse of 70 V and 5 ms length with a 50 ms interval and no voltage decay (total two pulses) followed by four bipolar transfer pulses of 40 V and 50 ms with 999 ms interval and 10 % voltage decay (total of eight pulses) (*see* **Note 10**).
7. Fix the electrode tweezers 7 mm apart (*see* **Note 11**). Connect the lead wires from the electrode tweezers to the electroporator power supply. Place a 100 mL glass bottle or dish filled with PBS near the work area to submerge the electrodes before and between each electroporation.
8. Fill the electroporation plate with ice-cold PBS before the start of the procedure (*see* **Note 12**).
9. Anesthetize axolotls in 0.008 % benzocaine. 2.2–2.5 cm axolotls will take 5–10 min to fall asleep.
10. Prepare individual cups with tap water.

3.2 Microinjection into the Spinal Cord

Make sure the microinjection and electroporation stations are set up before starting the experiment (Fig. 2a). For best outcome, operate swiftly.

1. Back-load the microinjection needle with a few microliters of DNA solution using a microloader tip. Insert the microinjection needle in the micropipette holder and mount on the micromanipulator (*see* **Note 13**).
2. Ensure that each axis of the coarse and fine drive unit of the micromanipulator is located at the midpoint position to maximize the working range. Set the micromanipulator arm holding the needle at an angle of 60° with respect to the stage plate of the stereomicroscope (Fig. 2b) (*see* **Note 14**).
3. Lay an axolotl on the silicone bed of the microinjection platform, parallel to the stage plate of the stereomicroscope. Overlay the tip of the needle with the spinal cord at the injection level and, using sharp forceps, carefully break the tip end (Fig. 2c) (*see* **Note 15**).
4. Insert the microinjection needle gently through skin and muscle to reach the central canal of the spinal cord (*see* **Note 16**).

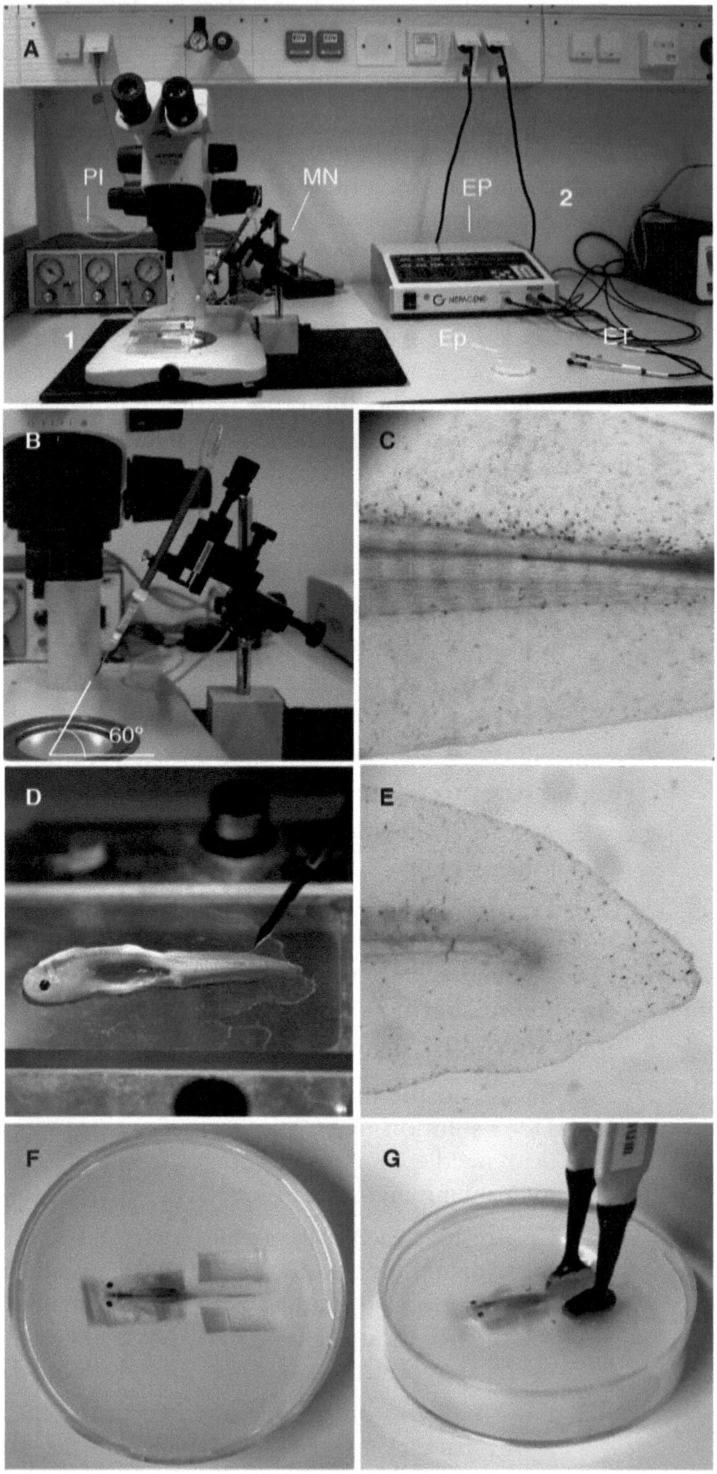

Fig. 2 Setup for microinjection and electroporation of the spinal cord of larval axolotls. (**a**) Microinjection (*1*) and electroporation stations (*2*) showing the pressure injector (PI), micromanipulator (MN), NEPA21 electroporator (EP), the custom-made electroporation plate (Ep) and commercial electrode tweezers (ET). (**b**) *Lateral view* of the manipulator arm showing the 60° angle respect to the stage plate of the microscope. (**c**) Image of the microinjection needle superposed to the spinal cord in the axolotl tail. The opening of the microinjection

5. Press the footswitch to pressure-inject the DNA solution into the central canal. Hold down the footswitch until the brain ventricles and the end of the spinal cord are filled with green dye (Fig. 2d) (*see* **Note 17**).

3.3 Electroporation of the Tail/Spinal Cord

1. Using round forceps, transfer the axolotl to the electroporation plate filled with ice-cold PBS (Fig. 2f) (*see* **Note 18**).
2. Electroporate the tail by applying one bipolar square-wave pulse of 70 V and 5 ms with 50 ms interval followed by four bipolar square-wave pulses of 40 V and 50 ms with 999 ms interval and 10 % voltage decay (Fig. 2g).
3. After electroporation, place the axolotl in its individual cup with tap water and allow it to recover (*see* **Note 19**).

3.4 Confirming Electroporation Efficiency

1. Four days post-electroporation, anesthetize the animals and examine electroporation efficiency and tissue integrity under a fluorescence microscope (Fig. 3a) (*see* **Note 20**).
2. Amputate the tail using a surgical scalpel, preferably at the same anterior–posterior level in each axolotl. The myotome number caudal to the cloaca can be used as a reference. Normally, we perform tail amputations within the 12th to 15th myotome (Fig. 3b) (*see* **Note 21**).
3. Return the axolotl to a cup for recovery.
4. Over the next days, follow up the process of spinal cord regeneration as intended (Fig. 3d).

4 Notes

1. This protocol has been carefully optimized to 2.2–2.5 cm axolotls. In these animals the electroporation efficiency is the highest, with the least tissue damage. The same procedure can be used in larger axolotls, although electroporation parameters will need to be modified to achieve the best results. Below we recommend some modifications to our original electroporation conditions that can be used as a starting point.

Fig. 2 (continued) needle should not be bigger than approximately half the thickness of the spinal cord. (**d**) View of an axolotl after a successful microinjection into the central canal of the spinal cord. The brain ventricles and spinal cord are filled with Fast Green. (**e**) View of the caudal end of the spinal cord after microinjection. The central canal must be filled with green dye. (**f**) *Top view* of an axolotl in the electroporation plate prior to the procedure. Importantly, the animal is immersed in ice-cold PBS and the tail protected by the agarose. (**g**) Image showing tweezer electrodes during the electroporation of the axolotl tail

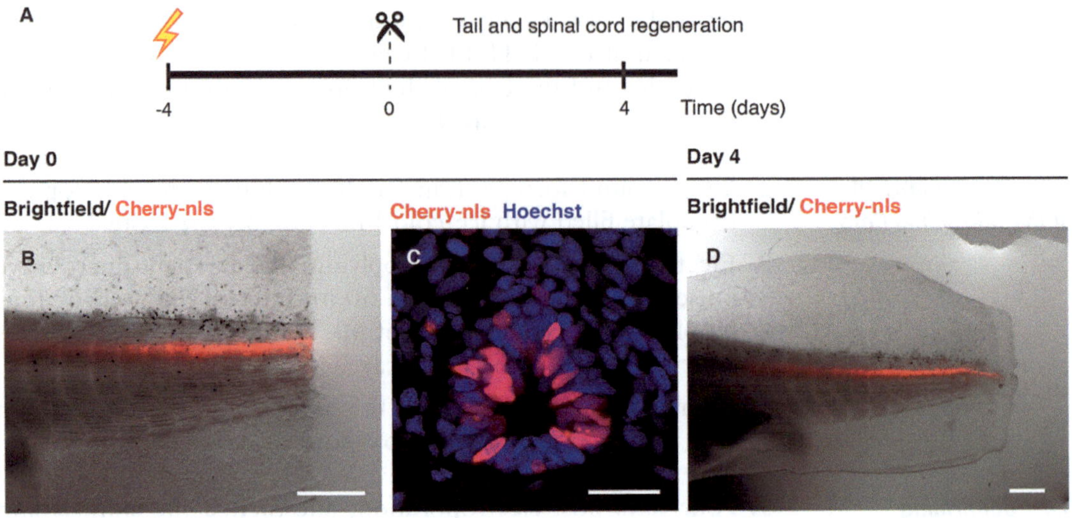

Fig. 3 Highly efficient spinal cord electroporation in a larval axolotl. (**a**) Schema placing the procedure of spinal cord electroporation in the context of a typical regeneration experiment in axolotl. (**b**) Combined bright-field and fluorescent image of the tail of an axolotl 4 days post-electroporation. The expression of the fluorescent reporter plasmid pCAGGS-Cherry-nls in the spinal cord suggests a high-efficient transfection of spinal cord cells. The tail has been amputated at the level of the myotome 12. (**c**) Cross-section of the electroporated spinal cord shown in (**a**), 4 days post-electroporation. (**d**) Combined bright-field and fluorescent image of the same animal depicted in (**a**) after 4 days of regenerative outgrowth. Scale bars in (**a**) and (**c**) are 0.5 mm; 50 μm in (**b**)

2. We suggest using a reporter gene with your plasmid of interest for visualizing electroporation efficiency. When preparing two or more plasmids, mix them with a molar ratio of 1:1. We recommend to keep the DNA solution close to 1.5 μg/μL. Higher DNA concentrations become too viscous and are difficult to inject, given the small opening of the microinjection needle required for injecting into the central canal of the spinal cord.
3. Normally, we purify plasmid DNA using QIAGEN Plasmid Maxi Kit and following manufacturer's instructions. It is important to dissolve the DNA at a relatively high concentration. Our DNA stocks have a concentration of 6–8 μg/μL in double-distilled water, and are stored at −20 °C.
4. The platform for microinjection we use is custom-made (MPI-CBG workshop, Dresden, Germany). However, a silicone bed in a petri dish will help keeping the animal in place during the microinjection procedure.
5. Besides maintaining the animal in place, the electroporation plate is important to allow the delivery of high-current electric pulses with minimal tissue damage. Tissue exposed directly to the electric field is more susceptible to overheat. Therefore, fill the dish with enough agarose to cover the entire axolotl tail

but do not fill it to the brim. For best results, change the electroporation plate after 60–80 electroporations.

6. Deionized water does not conduct the electrical pulses efficiently, therefore, dilute the DNA in PBS. For each 2.2–2.5 axolotl, calculate on 1.5 μL of DNA solution per spinal cord.
7. Fast Green FCF is an organic dye that is not taken up by living cells. Here it is used to visualize the injection of DNA into the central canal and gauge the amount of solution injected.
8. It is key to produce a microinjection needle tip thin enough to slip into the central canal of the spinal cord but sturdy enough to penetrate through the skin and muscle of the tail. Perform a "ramp test" as instructed by the manufacturer to determine the heat value inherent to the glass capillary tube. In our hands, the following pull cycle parameters (Heat: (ramp test value), Pull: 100, Velocity: 100, Time: 100) provide thinly tapered tips that will accommodate DNA concentrations up to 2 μg/μL.
9. Using gentle pressure and "manually" determining pulse length provides the best control over the injection.
10. The poring pulse is aimed to open pores in the cell membrane with minimum damage, and the transfer pulse to deliver charged DNA into the cells. Based on our experience, the above-mentioned parameters are the best compromise between electroporation efficiency and tissue damage in larval axolotls (2.2–2.5 cm snout-to-tail length). For larger animals, increase the voltage to 50 V and the number of pulses to 5.
11. For larger animals, start with a 6 mm gap between electrodes. If the tail tissue gets damaged, increase the distance to 7 mm.
12. The ice-cold PBS helps preventing excessive heating of the electroporated tail. Change PBS regularly, so it stays cold.
13. Bring the microloader tip as deep as possible into the microinjection needle, and then eject/push the DNA solution to avoid the formation of air bubbles. If bubbles form, gently tap the needle to get rid of them. The animals will not like an air bubble into the spinal cord or brain ventricles. If it happens, sacrifice the animal.
14. Bring the micromanipulator next to the stereomicroscope so that the tip of the microinjection needle appears in the middle of the field of view under low magnification. This will help to operate comfortably during microinjection.
15. In small and transparent axolotls, the spinal cord can be clearly identified as the brown tubular structure above the notochord. The opening of the tip should be in a 45° angle and be not bigger than half the thickness of the spinal cord at the level of injection (Fig. 2c). To avoid disrupting the spinal cord as much as possible, we recommend to inject the DNA solution 2–3 mm posterior to the planned amputation or injury site.

16. This is a critical step. The spinal cord sits only a few micrometers under the tail muscles, and a fine move of the fine drive unit of the micromanipulator will bring the tip to the central canal. Confirm this by gently depressing the footswitch and observing how the DNA solution spreads out. If the tip of the microinjection needle is inside the central canal the solution will flow inside the spinal cord. If it is not, make tiny movements bringing the needle tip back and forth while keeping the footswitch pressed. Stop when you see the green dye flowing inside the tube. If the DNA solution does not flow, dip the ring forces in PBS and bring the tip of the microinjection needle in the PBS "film" that forms on those. Then, activate the footswitch to unclog the needle without the need of breaking the tip.
17. Sometimes the DNA solution spreads out in a tube-like manner and gives the false impression that is inside the central canal of the spinal cord but is, instead, in the meningeal space. This will result in only the outer neuronal cells being electroporated. Inspect the caudal-most end of the spinal cord. It will be filled up with the green solution if the injection was successful (Fig. 2e).
18. It is important to carry out this step as quickly as possible to avoid the diffusion of injected DNA solution.
19. Examine the animals few hours after electroporation. If the tail appears visibly curled or burned, sacrifice the animal. We recommend not to change the water of the animals during the next 3–4 days to avoid damaging the tails.
20. The expression of the reporter plasmid can be readily observed 2 days post-electroporation. However, we recommend to wait 4 days before proceeding with the experiment for maximum expression, and to allow the animals to fully recover from the electroporation.
21. During the initial optimization/training electroporations, we recommend to fix the piece of amputated tail to quantify the efficiency of electroporation in tissue sections (Fig. 3c).

References

1. Spallanzani L (1766) Lettre to C. Bonnet dated A' Bomporto, près de Modene ce 21 Septembre 1766. Edizione Nazionale delle opere di Lazzaro Spallanzani edn.
2. Holtzer SW (1956) The inductive activity of the spinal cord in urodele tail regeneration. J Morphol 99:1–40
3. O'Hara CM, Egar MW, Chernoff EA (1992) Reorganization of the ependyma during axolotl spinal cord regeneration: changes in intermediate filament and fibronectin expression. Dev Dyn 193:103–115
4. Nordlander RH, Singer M (1978) The role of ependyma in regeneration of the spinal cord in

the urodele amphibian tail. J Comp Neurol 180:349–374
5. McHedlishvili L, Mazurov V, Grassme KS, Goehler K, Robl B, Tazaki A, Roensch K, Duemmler A, Tanaka EM (2012) Reconstitution of the central and peripheral nervous system during salamander tail regeneration. Proc Natl Acad Sci U S A 109:E2258–E2266
6. Echeverri K, Tanaka EM (2003) Electroporation as a tool to study in vivo spinal cord regeneration. Dev Dyn 226:418–425
7. Khattak S, Sandoval-Guzman T, Stanke N, Protze S, Tanaka EM, Lindemann D (2013) Foamy virus for efficient gene transfer in regeneration studies. BMC Dev Biol 13:17. doi:10.1186/1471-213X-13-17
8. Khattak S, Schuez M, Richter T, Knapp D, Haigo SL, Sandoval-Guzman T, Hradlikova K, Duemmler A, Kerney R, Tanaka EM (2013) Germline transgenic methods for tracking cells and testing gene function during regeneration in the axolotl. Stem Cell Rep 1:90–103
9. Khattak S, Murawala P, Andreas H, Kappert V, Schuez M, Sandoval-Guzman T, Crawford K, Tanaka EM (2014) Optimized axolotl (Ambystoma mexicanum) husbandry, breeding, metamorphosis, transgenesis and tamoxifen-mediated recombination. Nat Protoc 9: 529–540
10. Neumann E, Schaefer-Ridder M, Wang Y, Hofschneider PH (1982) Gene transfer into mouse lyoma cells by electroporation in high electric fields. EMBO J 1:841–845
11. Mir LM (2014) Electroporation-based gene therapy: recent evolution in the mechanism description and technology developments. Methods Mol Biol 1121:3–23
12. Echeverri K, Tanaka EM (2002) Ectoderm to mesoderm lineage switching during axolotl tail regeneration. Science 298:1993–1996
13. McHedlishvili L, Epperlein HH, Telzerow A, Tanaka EM (2007) A clonal analysis of neural progenitors during axolotl spinal cord regeneration reveals evidence for both spatially restricted and multipotent progenitors. Development 134:2083–2093
14. Schnapp E, Tanaka EM (2005) Quantitative evaluation of morpholino-mediated protein knockdown of GFP, MSX1, and PAX7 during tail regeneration in Ambystoma mexicanum. Dev Dyn 232:162–170
15. Kawakami K (2007) Tol2: a versatile gene transfer vector in vertebrates. Genome Biol 8(Suppl 1):S7

Chapter 10

Pseudotyped Retroviruses for Infecting Axolotl

Tzu-Hsing Kuo and Jessica L. Whited

Abstract

The ability to introduce DNA elements into host cells and analyze the effects has revolutionized modern biology. Here we describe a protocol to generate Moloney murine leukemia virus (MMLV)-based, replication-incompetent pseudotyped retrovirus capable of infecting axolotls and incorporating genetic information into their genome. When pseudotyped with vesicular stomatitis virus (VSV)-G glycoprotein, the retroviruses can infect a broad range of proliferative axolotl cell types. However, if the retrovirus is pseudotyped with an avian sarcoma leukosis virus (ASLV)-A envelope protein, only axolotl cells experimentally manipulated to express the cognate tumor virus A (TVA) receptor can be targeted by infections. These strategies enable robust transgene expression over many cell divisions, cell lineage tracing, and cell subtype targeting for gene expression.

Key words Axolotls, Salamander, MMLV, VSV-G, ASLV-A, Limb regeneration

1 Introduction

The astonishing regenerative abilities of axolotls and other salamanders have been the subject of intense experimentation over several centuries. However, a researcher's ability to unlock biological secrets is direct function of the power of the tools operational in the experimental system. In axolotls, a major hurdle has been the lack of a robust gene expression tool. Recent advances have permitted the creation of transgenic axolotls [1] and even transgenic axolotls with inducible gene expression [2, 3]. However, obtaining stable lines of transgenic animals is a lengthy endeavor perhaps best reserved for genes of special and intense interest. In other model systems, transgenesis has been complemented with faster methods for examining gene function. Axolotl researchers have long used electroporation to introduce exogenous DNA to cells in vivo. However, this method is transient as rapidly dividing cells, such as those in the regenerating limb blastema, partition the introduced DNA into daughter cells, resulting in an eventual diminishing of the expression across the tissue.

Anoop Kumar and András Simon (eds.), *Salamanders in Regeneration Research: Methods and Protocols*, Methods in Molecular Biology, vol. 1290, DOI 10.1007/978-1-4939-2495-0_10,

Additionally, electroporation in other salamanders has been documented to lead to deleterious and possibly confounding effects, such as muscle cell dedifferentiation [4]. Hence, alternative approaches for fast, robust, and potentially permanent exogenous gene expression are required.

An efficient means of permanent gene transfer is the use of a virus that incorporates its genetic material into the host genome, such as a retrovirus. Retroviruses have been used as very powerful tools to deliver genes into many mammalian cells and tissues that are difficult to transfect with other methods since the viral vector was established in early 1980s [5]. We have recently successfully applied this tool in axolotls [6]. Because retroviral genomes, including engineered elements, incorporate into the host genome, stable inheritance of these elements occurs as cells divide. Hence, all daughter cells continue to harbor the elements and, thus, retain the ability to express the transgene. This perdurant expression contrasts to other methods of expression such as plasmid electroporation [7]. Therefore, retroviral infections offer researchers a profound ability to track even highly proliferative cells and to robustly express transgenic elements of interest for prolonged periods of time without risk of diminishing expression.

This protocol provides details for generating a Moloney murine leukemia virus (MMLV)-based retrovirus that can be engineered to carry any gene of interest and can infect axolotl cells in vivo and in vitro. MMLV is gammaretrovirus that contains positive sensed, single-stranded RNA genome that replicates through DNA intermediate [8]. During the replication process, viruses randomly integrate their genetic information into the host genome. The viral genome contains three proteins, *gag*, *pol*, and *env*; one nucleic acid packaging signal, Ψ; and other *cis*-acting elements. Engineered retroviral vector maintains *cis*-elements but excludes sequences of proteins that can be supplied separately during the virus preparation process. This design results in generation of replication-incompetent MMLV. Pseudotyped retrovirus can also be generated by providing plasmids expressing a different envelope protein, which binds a particular cell surface receptor that mediates the specificity of cell that virus can infect. For example, VSV-G envelope protein enables pseudotyped retrovirus to infect almost any type of eukaryotic cell [9]. On the other hand, pseudotyped retroviruses with ASLV-A envelope protein infect only cells expressing avian TVA receptor [10]. The latter design enables tissue-specific expression of a gene of interest if TVA is expressed using a tissue-specific promoter. By creating batteries of transgenic animals expressing TVA in specific tissue types, researchers can easily produce viruses and test their effects in many different tissue types without altering the viral genome for each tissue of interest.

To generate pseudotyped retrovirus, mammalian 293T cells are transfected with three plasmids: (a) an engineered retroviral vector that contains a gene of interest, a packaging signal, and other *cis*-regulating elements; (b) a helper plasmid containing retroviral gag and pol open reading frames (ORFs); and (c) a plasmid containing the ORF of envelope protein such as VSV-G or ASLV-A. A concentration step is used to produce high-titer viruses. We further describe the methodology for infecting live axolotls with these retroviruses.

2 Materials

Before starting this protocol, please consult the safety department of your institution. Each institution/country may have specific regulations for handling pseudotyped retroviruses.

2.1 Plasmid Components

1. A pQCXIX (Clontech)-based retroviral vector (*see* **Notes 1** and **2**).
2. A helper plasmid that contains ORF of retroviral gag and pol [11].
3. A plasmid that contains ORF of VSV-G envelope protein [12] or ASLV-A envelope protein [13].

2.2 Virus Preparation Components

1. Cell culture medium for 293T cells: 10 % heat-inactivated fetal bovine serum (FBS) (*see* **Note 3**), 1× (100 U/mL) penicillin-streptomycin (P/S), in high glucose Dulbecco's Modified Eagle's Medium (DMEM). Open a new bottle of DMEM and take out 50 mL of media into a 50 mL conical tube. This serum-free DMEM can be used in the transfection described later. Add 50 mL heat-inactivated FBS and 5 mL 100× P/S into 450 mL DMEM. Store at 4 °C.
2. Poly-D-lysine (PDL)-coated tissue-culture plates: Prepare 0.1 mg/mL PDL by adding 50 mL of sterile tissue-culture grade water to a conical tube containing 5 mg PDL. Store reagent at −20 °C. Once thawed, PDL can be stored at 4 °C.
3. Polyethylenimine (PEI) transfection reagent: Make 1 mg/mL PEI solution in water, adjust the pH to 7.0 with HCl, and use a 0.22 μm filter to sterilize the solution. Make aliquots in sterile 1.5 mL microcentrifuge tubes and store the reagents at −80 °C. Once thawed, aliquots can be stored at 4 °C for 2–3 months.
4. Medium for virus collection: 10 % NuSerum and 1× (100 U/mL) penicillin-streptomycin in high glucose DMEM. Filter sterilize and store at 4 °C for 1 month.

2.3 Injection/Infection Components

1. Axolotl water, 4 L: Dissolve 7.2 g sodium chloride (NaCl), 960 mg magnesium sulfate heptahydrate ($MgSO_4 \cdot 7H_2O$), 328 mg calcium chloride dihydrate ($CaCl_2 \cdot 2H_2O$), and

192 mg potassium chloride (KCl) into 4 L deionized water. Sodium bicarbonate ($NaHCO_3$) is used to adjust the pH to the range of 6.75–7.25 (*see* **Note 4**).

2. 0.1 % Tricaine-S in axolotl water: Dissolve appropriate amount of Tricaine-S in axolotl water to equal 0.1 %, and then adjust pH up to 7.0 using sodium bicarbonate. Filter sterilize or use immediately.
3. 1 % fast green solution: Prepare by adding 0.1 g fast green in 10 mL H_2O. The solution is sterilized by passing through a 0.22 μm filter and then stored at 4 °C.
4. 10 % diluted bleach solution: Prepare in water and fill into squirt bottle or spray bottle for cleaning surfaces potentially contaminated with retrovirus.

2.4 Equipment and Supplies

1. 0.45 μm filtration system, Centricon® Plus-70 Centrifugal Filter Units (NMWL 10 kDa).
2. Ultracentrifuge tubes (Beckman Coulter Ultra-Clear Thinwall Tube Cat no. 344059 for SW41Ti rotor or Cat no. 344058 for SW28 rotor).
3. Ultracentrifuge (e.g., Beckman Coulter Optima L-90K and SW41Ti or SW28 bucket rotor and their accessories such as sleeves, caps, and sleeve holder).
4. Microcapillary needle puller.
5. Microinjection apparatus with accessories (microcapillary needles, microloader tips for loading virus into capillary needles).

3 Methods

3.1 Preparation of Plasmids

1. Clone gene of interest into engineered retroviral vector such as pQCXIX. A ubiquitous promoter such as CMV or CAGGs will drive robust gene expression in any infected cell. A fluorescent tag is also recommended so the infected cells can be visualized to (a) determine the location of infection and to (b) determine the titer of virus described in this protocol.
2. Purify all plasmids mentioned in Subheading 2.1 by commercial kit such as QIAGEN Plasmid Maxi Kit to reach sufficient amount.

3.2 Maintaining/Splitting 293T Cell Line (See Note 5)

1. Aspirate and discard old media.
2. Add 6–10 mL fresh complete 293T media to old plate.
3. Aspirate media against cell surface multiple times to detach cells.
4. Add 1 mL of cells to new plate containing 10 mL fresh media.
5. Place plate of 293T cells in a humidified incubator at 37 °C with 5 % CO_2.

3.3 Virus Preparation

DAY 1: Split 293T cells in the ratio of 1:20 from one confluent plate. For each gene of interest, six 10 cm culture plates are used for Day 1.

1. Aspirate and discard old media.
2. Add 10 mL fresh complete media to old plate.
3. Aspirate media against cell surface to detach cells.
4. Add 0.5 mL of cells to each new plate containing 10 mL fresh media.
5. Place all plates in a humidified incubator at 37 °C with 5 % CO_2.

DAY 4: Split one 10 cm plate from Day 1 to two PDL-coated 10 cm plates. For each gene of interest, twelve 10 cm plates are used for Day 4 (*see* **Note 6**).

6. Aspirate old media against cell surface to detach cells.
7. Split media to two PDL-coated (*see* **Note** 7) plates containing 5 mL fresh media.
8. Place all plates in a humidified incubator at 37 °C with 5 % CO_2.

DAY 5: Transfect 293T cells via PEI.

9. Mix plasmid DNA and PEI in serum-free DMEM in 15 mL conical tubes and vortex briefly. For each plate, 6 μg of retroviral vector, 3 μg of retro gag pol (helper) plasmid, 1 μg of coat protein plasmid (VSV-G or ASLV-A), 25 μL of PEI (1 mg/mL), and 250 μL of serum-free DMEM are used. We use 12 plates in this protocol; therefore, multiply the components by 12.5. Mix serum-free DMEM and DNA first, and then add PEI dropwise to the premixed DNA/DMEM solution.
10. Incubate the DNA/PEI mixture for 10–15 min.
11. Meanwhile, replace old media with 10 mL fresh complete medium for each plate.
12. Evenly apply DNA/PEI mixture dropwise onto the surface of cells.
13. Swirl plates gently and place plates back to incubator.

DAY 6: Replace media with 5 mL 10 % NuSerum medium for each plate.

14. Aspirate and discard old media (*see* **Notes 8** and **9**).
15. Add 5 mL fresh media containing 10 % NuSerum for each plate.
16. Place all plates in a humidified incubator at 37 °C with 5 % CO_2.

DAY 7: Collect and freeze first batch of media containing virus.

17. Collect media at 48 h post-transfection by tilting the plates to pull the media and using a tissue-culture pipette to gently aspirate the virus-containing media into sterile conical tubes. Collect a total of 60 mL media for each gene of interest and split the medium into two 50 mL conical tubes (*see* **Notes 10–12**).
18. Add 5 mL fresh media containing 10 % NuSerum to each plate.
19. Store collected media at −80 °C (*see* **Note 13**).

DAY 8: Collect media containing virus.

20. Collect medium at 72 h post-transfection as in **step 17**.
21. Plates containing 293T cells can be properly disposed after disinfection (*see* **Note 14**).
22. Store collected medium at −80 °C or start the concentration process with tubes collected from the previous day.

3.4 Concentrate Virus

1. Thaw frozen media containing virus in 37 °C water bath.
2. Place tubes on ice immediately after the last piece of frozen media are thawed.
3. Spin down at 4 °C and 840 × *g* for 5 min to pellet cell debris.
4. Apply all supernatant (~120 mL) to a sterile 0.45 μm filter system in tissue-culture hood (*see* **Note 15**).
5. Add 0.45 μm filtered media to Millipore Centricon Plus-70 centrifugal filter units.
6. Spin Plus-70 concentrator at 2,500 × *g* for 40 min at 4 °C.
7. Repeat **step 6** until all media were filtered (*see* **Note 16**).
8. Meanwhile, expose ultracentrifuge tubes, sleeves holding ultracentrifuge tubes, and caps of the sleeves to UV light for 20–30 min in a laminar flow tissue-culture hood (*see* **Note 17**).
9. Recover concentrated supernatant by inverting Plus-70 concentrator and spinning the unit for 5 min at 931 × *g* and 4 °C.
10. Collect the concentrated supernatant in the conical portion of the container and add the concentrated supernatant to ultracentrifuge tube.
11. Add the media that passed through the filter, collected in **step 7**, to ultracentrifuge tube for total volume of 13 mL (*see* **Note 18**).
12. Make a dummy tube for balancing if necessary.
13. Balance tubes by weighting each set of ultracentrifuge tube, sleeve holding the tube and its cap all together.
14. Ultracentrifuge at 4 °C, 12,300–13,200 × *g* for 3 h in SW41Ti or SW28 bucket rotor.

15. Discard supernatant by dumping out most of the medium with one movement. Keep the tube upside down at an angle (*see* **Note 19**).
16. Aspirate excess media by using vacuum aspirator. If the tube holds 13.2 mL, aspirate the droplet on the top of the tube. If the tube holds 38.5 mL, aspirate the media clinging to the wall by a zigzag motion up to half of the tube (*see* **Note 20**).
17. Place each tube back into the sleeve and cap each sleeve.
18. Put all capped sleeves containing tubes at 4 °C overnight to help with resuspension of virus (*see* **Note 21**).
19. The following day, shake the virus in the entire contraption (tubes/sleeves/caps) for 2 h on a platform/orbital shaker in a cold room until virus is well resuspended (*see* **Note 22**).
20. In the tissue-culture hood, gently mix virus by pipetting up and down using a P20 pipette. Avoid generating bubbles.
21. Aliquot (*see* **Note 23**) and freeze aliquots at −80 °C.

3.5 Test the Titer of Virus

To test the titer by using this protocol, an ORF of a fluorescent protein must be cloned into the retroviral vector (*see* **Note 24**).

DAY 1

1. Seed 293T cells in one 6-well plate per virus in the ratio of 1:20 per well from a 10 cm confluent plate (*see* **Note 25**).

DAY 2

2. Make serial dilution of virus from $1{:}10^3$ to $1{:}10^8$ in 1 mL complete medium in 1.5 mL microcentrifuge tubes. Add 1,200 μL complete medium to microcentrifuge tube containing 1.2 μL of concentrated virus (vial 1; $1{:}10^3$ dilution). Mix well. Take 120 μL medium from vial 1 to a microcentrifuge tube containing 1,080 μL medium (vial 2; $1{:}10^4$ dilution). Repeat the process till having virus dilution from $1{:}10^3$ to $1{:}10^8$.
3. Remove old media from each well of the 6-well plate. For each well, add 1 mL of media from each tube containing diluted virus.
4. Place plates in a humidified incubator at 37 °C with 5 % CO_2.

DAY 4

5. Identify and count the number of fluorescent colonies by observing infected cells under fluorescence microscope (*see* **Note 26**).
6. Calculate the titer from the number of observed colonies and the amount of virus applied to that well. Record this information.

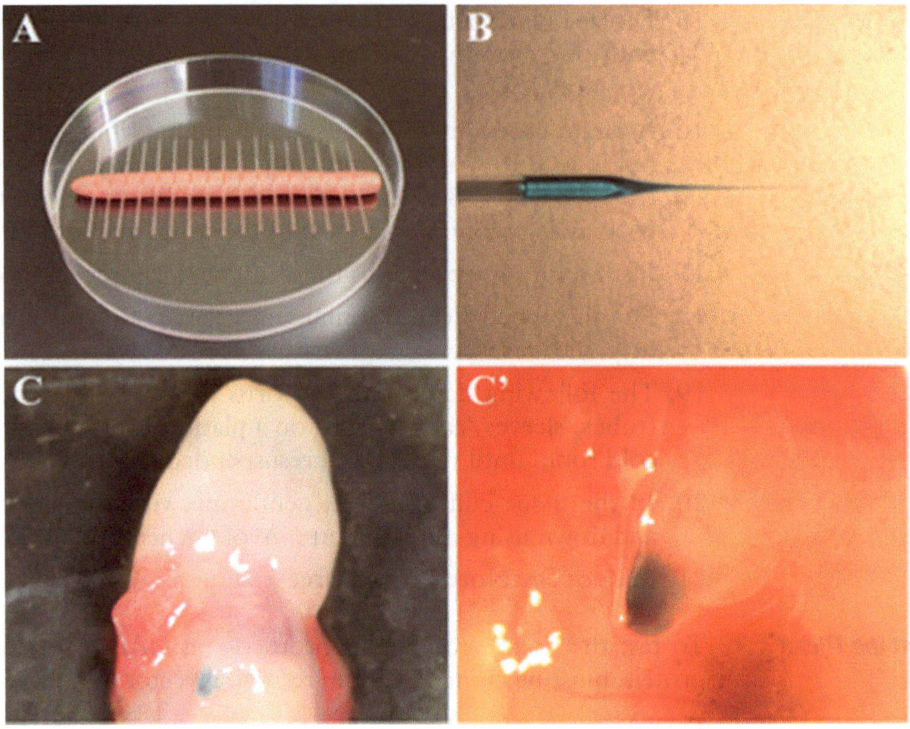

Fig. 1 Infecting axolotls with retrovirus. (***a***) Arrangement of unopened glass capillary needles in a 15 cm petri dish. Note that the needles are suspended from the surface by a strip of modeling clay. Lidding the dish keeps the needles from collecting dust. (**b**) Unopened needle filled with retrovirus and fast green, which allows for visualization during injection. Note the long taper of the needle. Filled needles can be broken with forceps at the tip to open. (**c**) Juvenile axolotl with a right upper forelimb blastema immediately following infection. Ventral side is up. (**c′**) Higher magnification of (**c**). Note that injection of the virus has been confined to the blastema and does not extend into the stump

3.6 Infect Limb Blastemas

1. Pull microcapillary needles (Fig. 1a, *see* **Note 27**).
2. Anesthetize axolotls by placing the animal in 0.1 % tricaine in axolotl water (*see* **Note 28**).
3. Thaw virus on ice (*see* **Note 29**).
4. Add 1/10 volume of 1 % fast green (final concentration 0.1 %) to thawed virus (*see* **Note 30**).
5. Backfill virus into microcapillary needle by microloader pipette tips (Fig. 1b, *see* **Note 31**).
6. Inject viral solution into blastema by microcapillary needle attached to pneumatic injector (Fig. 1c, c′, *see* **Notes 32** and **33**).
7. Place animal back in container with fresh axolotl water (*see* **Note 34**).
8. Monitor animal on subsequent days to confirm expression (Fig. 2).

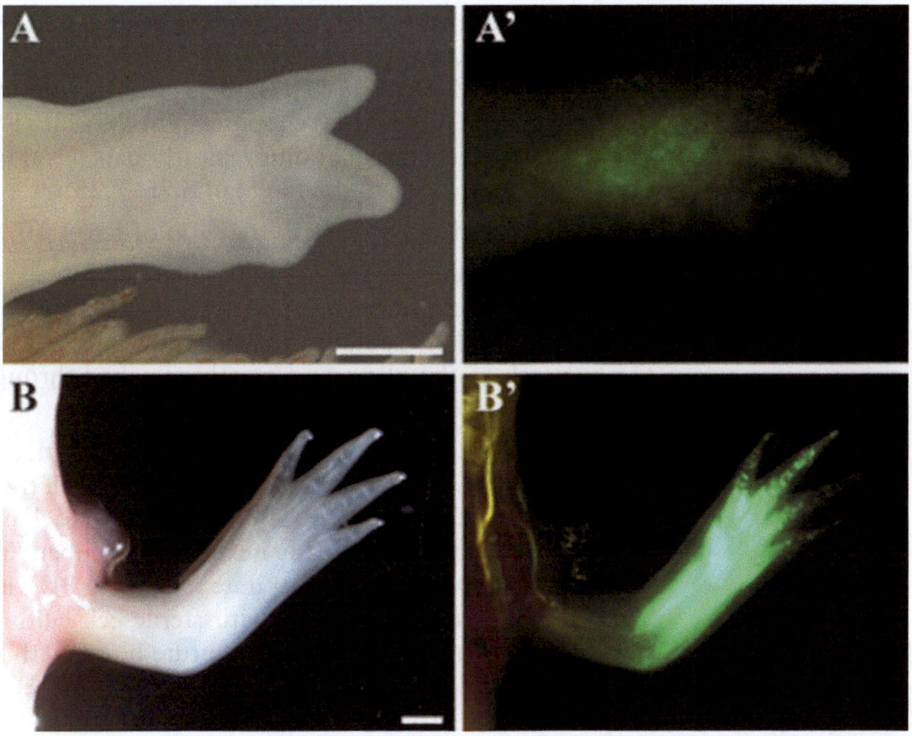

Fig. 2 In vivo expression of EGFP following infection with QC-EGFP retrovirus. (**a**, **a**′) *Bright-field* and *fluorescent images* of a right forelimb at 5 days postinfection, 19 days post-amputation. (**b**, **b**′) *Bright-field* and *fluorescent images* of same limb on live animal after full regeneration. Amputation plane was mid-humerus. Scale bar is 1 mm

4 Notes

1. The ORF of a gene of interest should be cloned after an appropriate promoter (for example, for ubiquitous expression, CMV can be used). For a single gene without a marker, a stop codon is included. However, we recommend using the 2A system to simultaneously allow for visualizing cells. In this strategy, the ORF can be cloned without a stop codon and instead followed directly by 2A and a fluorophore sequence. 2A encodes 18-amino acid self-cleaving peptide that is operational in axolotl [6]; hence, the engineered virus will express one transcript and a single polypeptide that is then cleaved into the protein of interest and the fluorophore.
2. A control virus should be prepared in parallel. For example, QCMV-EGFP, which expresses EGFP alone under the CMV promoter, can be used as a negative control.
3. To heat inactivate serum, incubate thawed FBS in 56 °C water bath for 30 min. Shake the bottle every 10 min.

4. Our laboratory makes a large volume of stock axolotl water by adding 720 g sodium chloride (NaCl), 96 g magnesium sulfate heptahydrate ($MgSO_4 \cdot 7H_2O$), 32.8 g calcium chloride dihydrate ($CaCl_2 \cdot 2H_2O$), and 19.2 g potassium chloride (KCl) into 400 L deionized water. Sodium bicarbonate ($NaHCO_3$) is used to adjust the pH to the range of 6.75–7.25.
5. Split 293T cells twice a week in the ratio of 1:6–1:10.
6. PDL (poly-D-lysine) coating is recommended beyond normal tissue-culture plate coating, especially if the 293T cells are not very adherent. This coating can be omitted if the 293T cells firmly attach to the plates and remain attached through virus collection/media change steps.
7. To coat plates with PDL, aspirate 5 mL or more PDL (0.1 mg/mL) into a plate. Gently rock the plate to ensure all the surface of plate is covered. Leave 1 mL PDL solution in each plate and aspirate the excess solution for next plate. Incubate all plates covered with PDL for 5 min. Aspirate remaining PDL in plates. Thoroughly rinse plate surface with sterile cell culture grade water. Allow plates to dry in tissue-culture hood, with lids ajar, for at least 30 min before use. PDL solution can be reused with proper storage.
8. The transfection rate can be inspected by visualizing the fluorescence if an ORF of fluorescent protein was cloned into the retroviral vector.
9. The medium will now contain live virus and should therefore be handled carefully and according to institutional guidelines.
10. Take caution as the media contains live virus. Use gentle aspiration and avoid bubbles as much as possible as they can have a deleterious effect on the virus.
11. Place tubes on ice temporarily.
12. Volume increases when medium freezes. Do not add more than 40 mL of media into one 50 mL tube to prevent leakage of medium and/or bursting of tube in freezer. Split the media evenly into two tubes so they are well balanced for centrifugation in Subheading 3.4, **step 3**.
13. Conical tubes can either be stored upright in −80 °C freezer or buried in a dry ice chest. To enable retrieval from a dry ice chest, place conical tubes inside plastic bags, and tie a label to each bag using a very long string; leave the label end of string outside of the dry ice chest.
14. Proper decontamination of materials and disposal of infected cells should be performed in pursuance of your institution's guidelines for biohazardous waste.

15. Be sure to use 0.45 μm instead of 0.22 μm filter. If a 0.22 μm filter is used, virus will be trapped in the filter and cannot be recovered.
16. Keep the pass through media for **step 11**.
17. This ensures the materials for concentrating virus will be sterile. Virus need not be sterile for use in axolotls (though sterility is preferred), but preparations with bacterial contamination will be impossible to titer in cultured cells.
18. The ultracentrifuge tube for SW41Ti rotor holds total volume of 13.2 mL and the tubes for SW28 rotor holds 38.5 mL.
19. Dump the supernatant in one swoop into a waste beaker in the hood. At the end of the protocol, decontaminate the contents pursuant to your institution's protocol and dispose.
20. Avoid aspirating off the viral pellet at the bottom of the tube. The viral pellet is usually visible as translucent or colored precipitate but may sometimes be difficult to discern from plastic tube.
21. At this point, it is not necessary or recommended to add media to the pellet. Residual media clinging to the wall of each tube following aspiration will drip down the insides of the tube and collect with the viral pellet overnight. Be careful to keep the tubes upright in the refrigerator or cold room; use a tube stand to ensure they do not fall over.
22. Shake tubes at a medium speed that is strong enough to resuspend virus but not to jostle the tubes out. Secure the tubes by placing them in racks inside a bucket. We advise using a rack inside a rectangular ice bucket and filling the bucket with ice to minimize jostling and ensure the virus is kept cold.
23. Around 50 μL of concentrated virus is recovered, though the final volume depends on how much residual media was aspirated from the sides of the tube. Set aside 1.2 μL of concentrated virus in one microcentrifuge tube for testing of the viral titer later. Multiple freeze-thaw cycles decrease viral titers. Estimate the required volume of virus for each experiment and make aliquots. The best infections may be achieved with fresh virus prepared without freezing.
24. Other methods may be designed to detect infected cells. These methods may include detection via histological procedures such as enzymatic activity or antibody stains or qRT-PCR.
25. When first learning to prepare virus, it is advisable to reserve and freeze some unconcentrated virus so that it can be tested for activity when the concentrated preparations are titered. To do this, simply save a small aliquot (100–150 μL) of fresh virus-containing media after the initial filtering step removing cell debris. Prepare separate plates or wells for infecting

293T cells with unconcentrated virus. Apply 10–100 μL of unconcentrated virus, and count infected colonies to compare to concentrated titer. If virus is produced but lost in concentration, this will become apparent in this comparison.

26. Infected cells form colonies from single cells; therefore, identify number of colonies instead of cells for each well. The duration of viral infection required to detect fluorescence may vary depending on co-expressed proteins. Be consistent between preparations so viral titers prepared from different batches are comparable. Do not wait so long that the number of cells/colony becomes very large. The ideal colony size to count is 4–8 cells. To determine the titer of virus, find a well with a countable number of colonies (i.e., there are many colonies but not too many to count, and the colonies are far enough apart that they can be individually discerned). For example, if there are eight colonies in the well of $1{:}10^8$ dilution, the titer for the virus is approximately 8×10^8.

27. Practice and instrument calibration are necessary to produce perfect glass capillary needles. We use a Sutter Instruments P-87 machine. This machine pulls two closed-end needles from each capillary tube. Following pulling, we line the needles up inside a large petri dish with a bar of modeling clay to keep them from breaking.

28. The time required for axolotls to be fully anesthetized depends on the size of the animal. Handle animals pursuant to your institution's animal care guidelines.

29. Avoid having the virus thawed for lengthy amounts of time before injection as this could diminish infectivity.

30. Virus is sensitive to bubbles. Therefore, mix virus and fast green by gently stirring pipette tip inside microcentrifuge tube instead of pipetting solution up and down, which generates bubbles easily.

31. Take the upmost care to squirt the virus as close to the tip of the glass capillary needle as possible. Eject the virus in one pipette movement without releasing. Avoid bubbles within the virus column in the glass capillary as they can make injecting with low pressure impossible. Insert the filled needle into the holder of the pneumatic device. A forceps is used to break a piece of the needle tip off under a stereodissection microscope. Be careful not to break too much of the tip off; needles can always be broken again, but if too much has been broken off, the virus will be ejected too quickly and can be wasted or can lead to air being injected into the animal.

32. Place the animal on a silicone pad. For blastemas, the animal can be placed on its ventral side and the limb pulled away from the animal atop the pad. Alternatively, animals can be placed on their sides with limbs resting atop the bodies. Be sure to keep axolotls moist throughout the procedure. This can be accomplished by performing the procedure quickly or by dabbing the body with axolotl water/tricaine while on the silicone pad.
33. Practice with your specific pneumatic device will be necessary to become adept at delivering virus. The idea is to fill the area of interest with the virus-containing fast green solution and to do it with minimal damage to the animal's tissue, minimal waste/leakage, and minimal carryover to other tissues. Low pressure is advisable, but too low can result in clogged needles from slimy tissue becoming impacted at the ejection site or from virus clogging the needle. Too high will result in premature ejection of the virus, too much virus, and possibly in the injection of air (which is not good for the animals).
34. Normally, axolotls can be immediately returned to housing containers and fresh axolotl water. There is no worry that virus will "leak out" provided the needles and injection entries were quite small. However, when first learning the procedure, axolotls can be kept asleep postinfection for 1 h to ensure the injection site is not disrupted. To do this, gently transfer the injected animals to a prepared hydration chamber (for example, a large petri dish with tricaine-soaked filter paper). Following the resting period, transfer the axolotls to fresh water in normal housing. Throughout the procedures, proceed in compliance with your institution's biosafety guidelines. This may involve temporarily housing the infected animals in a "BL2-area" until the virus is no longer infectious or until a predetermined time dependent upon your institution. Decontaminate all areas exposed to the virus. Wear appropriate personal protective gear (PPE) as VSV-G-coated retroviruses are capable of infecting human cells.

Acknowledgements

This work was supported by start-up funds from Brigham and Women's Hospital (J.L.W.). The authors would like to thank Christina DeMaso in Dr. Connie Cepko's lab (Harvard Medical School) for sharing similar protocols used in infecting mammalian cells.

References

1. Sobkow L et al (2006) A germline GFP transgenic axolotl and its use to track cell fate: dual origin of the fin mesenchyme during development and the fate of blood cells during regeneration. Dev Biol 290:386–397
2. Whited JL, Lehoczky JA, Tabin CJ (2012) Inducible genetic system for the axolotl. Proc Natl Acad Sci U S A 109:13662–13667
3. Khattak S et al (2013) Germline transgenic methods for tracking cells and testing gene function during regeneration in the Axolotl. Stem Cell Reports 1:90–103
4. Atkinson DL et al (2006) Cellular electroporation induces dedifferentiation in intact newt limbs. Dev Biol 299:257–271
5. Cone RD, Mulligan RC (1984) High-efficiency gene transfer into mammalian cells: generation of helper-free recombinant retrovirus with broad mammalian host range. Proc Natl Acad Sci U S A 81:6349–6353
6. Whited JL et al (2013) Pseudotyped retroviruses for infecting axolotl in vivo and in vitro. Development 140:1137–1146
7. Echeverri K, Tanaka EM (2003) Electroporation as a tool to study in vivo spinal cord regeneration. Dev Dyn 226:418–425
8. Maetzig T et al (2011) Gammaretroviral vectors: biology, technology and application. Viruses 3:677–713
9. Yee JK, Friedmann T, Burns JC (1994) Generation of high-titer pseudotyped retroviral vectors with very broad host range. Methods Cell Biol 43:99–112
10. Young JA, Bates P, Varmus HE (1993) Isolation of a chicken gene that confers susceptibility to infection by subgroup A avian leukosis and sarcoma viruses. J Virol 67:1811–1816
11. Ory DS, Neugeboren BA, Mulligan RC (1996) A stable human-derived packaging cell line for production of high titer retrovirus/vesicular stomatitis virus G pseudotypes. Proc Natl Acad Sci U S A 93:11400–11406
12. Beier KT et al (2011) Conditional expression of the TVA receptor allows clonal analysis of descendents from Cre-expressing progenitor cells. Dev Biol 353:309–320
13. Landau NR, Littman DR (1992) Packaging system for rapid production of murine leukemia virus vectors with variable tropism. J Virol 66:5110–5113

Chapter 11

Thyroxine-Induced Metamorphosis in the Axolotl (*Ambystoma mexicanum*)

Peggy S. Coots and Ashley W. Seifert

Abstract

The axolotl (*Ambystoma mexicanum*) has remained an important model for regeneration and developmental biology for over a century. Although axolotls in captive-bred colonies usually exist in an aquatic form, they retain the ability to undergo metamorphosis following exposure to thyroid hormone. Here we present a robust method for inducing metamorphosis in adult axolotls that results in high survivability and produces terrestrial animals that can be maintained in long-term captivity.

Key words Metamorphosis, Thyroxine, T_4, Thyroid hormone, Axolotl, *Ambystoma*, Salamander, Regeneration

1 Introduction

Amphibian metamorphosis produces a suite of behavioral, morphological, physiological, biochemical, and genetic changes associated with the transition from a larval to adult form [1–3]. These changes are mediated by alterations in circulating thyroid hormones (T_3—triiodothyronine and T_4—thyroxine) and their interaction with thyroid hormone receptors in target cells [3, 4]. Among urodeles, however, paedomorphosis has evolved in several lineages to produce species that fail to undergo an obvious metamorphosis in natural populations [5]. Instead, these species become sexually mature while retaining many larval traits. Although the biochemical changes that mediate the failure to undergo metamorphosis in these paedomorphic lineages have not been identified, some urodeles, including *Ambystoma mexicanum* (axolotl) and *Ambystoma tigrinum* (tiger salamander), retain the ability to undergo metamorphosis following exposure to thyroid-stimulating hormone (TSH), T_3 and T_4 [2, 3, 6–9].

Anoop Kumar and András Simon (eds.), *Salamanders in Regeneration Research: Methods and Protocols*, Methods in Molecular Biology, vol. 1290, DOI 10.1007/978-1-4939-2495-0_11, © Springer Science+Business Media New York 2015

In order to induce metamorphosis in axolotls, two basic methods using hormones are usually employed: peritoneal injection of the hormone [7] or addition of the hormone to the rearing water [2]. While the majority of studies using induced metamorphosis of the axolotl have focused on aspects of metamorphosis itself, some recent studies have sought to use captive terrestrial axolotls to explore mechanisms regulating regeneration [10, 11]. For these studies, animals must be healthy, actively eating, and able to survive in captivity for more than a year. Although more time intensive than single peritoneal injections, using the method presented here, >50 % of completely metamorphosed animals survive in long-term captivity.

2 Materials

1. Axolotls (*see* **Note 1**).
2. Round plastic containers (diameter = 21.5 cm, height = 15 cm).
3. Plastic shoeboxes with lids (width = 33.5 cm, depth = 19.5 cm, height = 9 cm).
4. L-Thyroxine (100 μM) stock T_4 solution: Add 100 mg L-thyroxine (T_4) (Sigma; Cat. no: T2376) in 10 mL 0.4 M NaOH in a 15-mL conical tube. Vortex the tube for 30 s and incubate the tubes in a water bath maintained at 37 °C for 5 min. Repeat vortex and heat cycle until L-thyroxine is completely dissolved. This solution is now 13 mM T_4. Take 5 mL of T_4 solution and add to 645 mL of sterile water to make 100 μM stock solution. The remaining 5 mL of 13 mM T_4 can be wrapped in foil and stored at 4 °C.
5. L-Thyroxine (T_4) working solution (50 nM): Make 50 nM T_4 as needed by mixing 100 μM T_4 stock solution with 40 % modified Holtfreter's solution at 1:2,000 dilution (*see* **Note 2**).
6. 40 % modified Holtfreter's solution: For a 1,136-L (300-gallon) container, add 1,589.7 g NaCl, 90.8 g $NaHCO_3$, 22.7 g KCl, 181.7 g $MgSO_4$–$7H_2O$, and 90.8 g $CaCl_4 \cdot 2H_2O$. Add all the dry components into the container and fill to volume with sterile water. Mix with circulation until all components have gone into solution. Add 75.7 mL Amquel and 75.7 mL Novaqua to this solution. Solution is now ready to use (*see* **Note 3**).
7. California blackworms (*Lumbriculus species*).
8. Night crawlers (earthworms).

3 Methods

1. To facilitate successful metamorphic transition, feed animals California blackworms every day for 1 week prior to the first addition of T_4 working solution (normal feeding schedule is every 3 days).
2. After 1 week, transfer axolotls into plastic containers with working T_4 solution (one animal per container). The amount of T_4 solution should fill the container (*see* **Note 4**).
3. Remove and replace working T_4 solution every 3 days.
4. Feed animals after each changing with a generous portion of California blackworms (approximately 20–25 worms). Since they will stop eating during metamorphosis, the goal is to get them to eat as much as possible.
5. Obvious morphological changes are rarely visible until approximately 2 weeks after T_4 exposure. The first visible changes occur in the third week and include noticeable weight loss, dorsal ridge recession, tail fin recession, and eye protrusion (Fig. 1) (*see* **Note 5**).
6. Concomitant with the above morphological changes, the gills begin to disappear. It is critical to begin lowering the water level as this happens. The water level should be reduced an inch per water changing. When the gills have nearly disappeared,

Fig. 1 Pictorial representation of morphological changes occurring during thyroxine-induced metamorphosis in adult axolotls. The first obvious changes are visible 2–3 weeks following thyroxine exposure. The dorsal fin begins to regress in a cranial to caudal direction, as does the ventral portion of the tail fin (*red arrows*). Four weeks after exposure, the gills have begun regressing and the dorsal and anal fins continue to disappear. The eyes begin to protrude from the head. The gills, dorsal, and anal fins have almost entirely disappeared in the fifth week (*red arrows*). By the sixth week, animals have metamorphosed to a completely terrestrial form

place two pieces of bunched-up paper towel at the bottom of the containers (*see* **Note 6**).

7. After 4–5 weeks, the gills have completely disappeared and T_4 administration is no longer required. Transfer each axolotl to a plastic shoebox container with lid, paper towel, and 1 in. of 40 % modified Holtfreter's solution. Feed each animal ½ or ¼ of an earthworm (*see* **Note 7**).
8. From this point on, animals can be maintained on paper towels in 1 in. of 40 % Holtfreter's solution. Animals should be offered food every 3 days (*see* **Note 7**).

4 Notes

1. In our experience, wild type and albino color morphs exhibit higher survivability than white mutants. We have used this method successfully on animals ranging in size from 7 cm TL to >20 cm TL adults.
2. We typically add 50 mL of T_4 stock solution to a 44-gallon container (e.g., large Rubbermaid trash bin) to make 100 L working solution (with 40 % modified Holtfreter's solution). The working T_4 solution can be used for several days when morphing a large number of animals.
3. Although we maintain our colony of axolotls at a relatively constant water temperature of 18–20 °C, for metamorphosis we use Holtfreter's solution that is at room temp (~23–25 °C).
4. It is imperative that plastic containers used for induced metamorphosis are kept separately from normal housing containers because T_4 can induce metamorphosis at extremely low concentrations. Once containers are exposed to L-thyroxine, they should be clearly labeled.
5. As the skin matures to an adult phenotype, it is shed frequently and this will cause the water to appear dirty. In our experience, this does not require an increase in water changing frequency.
6. Paper towels provide a surface to help the axolotls reach the surface and acquire oxygen along with helping to keep the containers moist. Axolotls will eat significantly less at this point during metamorphosis and may refuse to eat at all. To help encourage consumption, place 5–10 blackworms directly in front of their mouth.
7. Resumption of feeding is critical for metamorphs to survive in captivity. A behavioral change is associated with transition to a terrestrial mode of feeding. Every animal must be offered food, although every axolotl may not eat at every feeding. This is normal. Patience is required while feeding metamorphs.

Earthworms should be offered with a tweezer and placed to the left or right of the axolotl's nostril to encourage eating. A light tap in front of the eyes can sometimes encourage the animal to bite for food. When animals are reluctant to eat earthworms, we offer blackworms.

Acknowledgments

This work was supported by the Office of the Vice President of Research at the University of Kentucky.

References

1. Dodd M, Dodd J (1976) The biology of metamorphosis. In: Lofts B (ed) Physiology of the Amphibia, vol 3. Academic, New York, pp 467–599
2. Page RB, Monaghan JR, Samuels AK, Smith JJ, Beachy CK, Voss SR (2007) Microarray analysis identifies keratin loci as sensitive biomarkers for thyroid hormone disruption in the salamander Ambystoma mexicanum. Comp Biochem Physiol C Toxicol Pharmacol 145:15–27
3. Rosenkilde P, Ussing AP (1996) What mechanisms control neoteny and regulate induced metamorphosis in urodeles? Int J Dev Biol 40:665–673
4. Brown DD (1997) The role of thyroid hormone in zebrafish and axolotl development. Proc Natl Acad Sci U S A 94:13011–13016
5. Shaffer HB, Voss SR (1996) Phylogenetic and mechanistic analysis of a developmentally integrated character complex: alternate life history modes in ambystomatid salamanders. Am Zool 36:24–35
6. Norris DO, Platt JE (1974) T3-induced and T4-induced rates of metamorphosis in immature and sexually mature larvae of Ambystoma tigrinum (Amphibia caudata). J Exp Zool 89:303–310
7. Prahlad KV, DeLanney LE (1965) A study of induced metamorphosis in the axolotl. J Exp Zool 160:137–145
8. Taurog A (1974) Effect of Tsh and long-acting thyroid-stimulator on thyroid I-131-metabolism and metamorphosis of Mexican Axolotl (Ambystoma mexicanum). Gen Comp Endocrinol 24:257–266
9. Taurog A, Oliver C, Eskay RL, Porter JC, McKenzie JM (1974) The role of TRH in the neoteny of the Mexican axolotl (Ambystoma mexicanum). Gen Comp Endocrinol 24:267–279
10. Monaghan JR, Stier AC, Michonneau F, Smith MD, Pasch B, Maden M, Seifert AW (2014) Experimentally induced metamorphosis in axolotls reduces regenerative rate and fidelity. Regeneration 1(1):2–14. doi:10.1002/reg2.8
11. Seifert AW, Monaghan JR, Voss SR, Maden M (2012) Skin regeneration in adult axolotls: a blueprint for scar-free healing in vertebrates. PLoS One 7:e32875. doi:10.1371/journal.pone.0032875

Chapter 12

Generation of Aneurogenic Larvae by Parabiosis of Salamander Embryos

Anoop Kumar and Jean Paul Delgado

Abstract

Limb regeneration of salamanders is nerve dependent, and the removal of the nerves in early stages of limb regeneration severely curtails the proliferation of the blastemal cells and growth of the regenerate. The removal of the neural tube from a developing salamander embryo results in an aneurogenic larva and the aneurogenic limb (ANL) develops independently without innervation. Paradoxically, the limb in an ANL is capable of regeneration in a nerve-independent manner. Here, we describe a detailed method for the generation of ANL in the spotted salamander, *Ambystoma maculatum*, for regeneration studies.

Key words Newt, Axolotl, Limb regeneration, Nerve dependence

1 Introduction

Innervation of tissues and organs is a fundamental requirement in vertebrates as well as invertebrates. In a classical perspective, the nerves are associated with neurotransmission, whereas in a context like organ regeneration, the nerves exert dependence on the tissues. The nerve dependence is established during embryonic development and it persists throughout adult life. In tissue regeneration of mammals, the recent findings indicate that regeneration of hair follicle [1], digit tip [2], bone marrow [3], and earlobe [4, 5] requires the presence of nerves. The nerve dependence of tissues was discovered in relation to salamander limb regeneration [6] and has been subjected to extensive experimental studies spanning several decades. These studies revealed several salient features of nerve dependence during salamander limb regeneration [7–9].

Transection of the salamander limb in an experimental context invokes rapid wound healing, followed by dedifferentiation and proliferation of progenitor cells from the local stump tissues, leading to formation of a limb blastema. The limb blastema is a heterogeneous collection of cells and the origin and nature of these cells varies among salamander families [10–12]. Despite these

Anoop Kumar and András Simon (eds.), *Salamanders in Regeneration Research: Methods and Protocols*, Methods in Molecular Biology, vol. 1290, DOI 10.1007/978-1-4939-2495-0_12,

differences in the origin of the blastemal cells, in all cases, blastemal cell proliferation requires the presence of nerves in its immediate proximity during the early stages of regeneration [8]. The growth of nerve axons into the regenerating tissues of the limb occurs very early, and the nerve branches reach both epithelial and mesenchymal populations [13–15]. The cellular components of the nerve which invade the tissues originate from ensheathing Schwann cells and associated fibroblasts [16]. Transection of the nerve prior to limb amputation does not prevent the formation of an initial cohort of blastemal cells [17, 18], but the nature of the wound epithelium is different, and the proliferation of blastemal cells is severely curtailed, leading to failure of the limb regenerate [16, 19]. A minimum threshold of nerve fibers is also required within the limb tissues for regeneration [20]. Recently we have shown that the Schwann cell-derived newt anterior gradient protein (AGP) is able to largely substitute for the presence of the nerve in adult newts [21]. The AGP is a ligand for the taxon-specific protein Prod1 [22–24], which is expressed as a cell surface GPI-anchored protein predominantly by dermal fibroblasts and blastemal cells [25]. In a denervated early limb regenerate, ectopic expression of *AG* gene by electroporation leads to robust expression of the protein in the regenerate and this acts as a growth factor for proliferating blastemal cells [21]. Proliferation of the blastemal cells leads to restoration of the secretory functions of the wound epithelium and induction of glandular cells. Remarkably, this cascade of events leads to completion of the regenerative process with complete autopodial elements, though lacking in skeletal muscle.

During development of salamander embryos, the limb bud develops at a later stage compared to other amniotes, and it evolves semiautonomously [26]. The salamander embryos are easily amenable to experimental manipulation, a feature which was established by experimental embryology in the early twentieth century [27]. Therefore, the entire neural tube from the base of the brain to the tail bud can be removed from a salamander embryo by microsurgical methods, and the limb bud develops and differentiates into a limb in an autonomous manner [28–30]. This limb, which has never had peripheral innervation, is called an "aneurogenic limb" (ANL). The ANL has exceptional properties in that, it can regenerate without innervation. However, if the ANL is transplanted to the location of the normal limb in a host larva, the brachial nerves grow into the transplanted ANL and establish connections [31]. This leads to imprinting of the nerve dependence in the ANL, which fails to regenerate in a denervated context. Our recent experiments on the nature of ANL have identified the cellular and molecular events associated with ANL development and regeneration [32]. The AGP is expressed at high levels in the larval epidermis during development and regeneration, but decreases when it is transplanted to a normal larval host due to innervation. In an ANL limb, the high level of secreted AGP acts in lieu of the nerve as a source of growth factor during regeneration [32].

The autonomous nature of the limb bud development in salamander embryos is advantageous for experimental manipulation. The limb bud is amenable to extensive surgical intervention, electroporation, orthotopic, and ectopic transplantation. Therefore, it is possible to develop functional assays to delineate gene functions and tissue interactions which occur during limb development and regeneration. We have modified the previous protocol [28, 33, 34] and describe here a detailed procedure for the generation of aneurogenic larvae during embryonic development of the spotted salamander, *Ambystoma maculatum*.

2 Materials

2.1 Obtaining and Maintenance of Embryos

1. The embryos of spotted salamander (*Ambystoma maculatum*) are available seasonally from January to March. The supplier is Charles D. Sullivan & Co., Nashville, TN, USA. Parabiosis is possible using embryos of other salamanders like axolotl and *Pleurodeles waltl* (*see* **Note 1**).

2.2 Reagents for Embryo Storage and Manipulation

1. Solution A: Dissolve 34 g NaCl in 200 mL sterile water.
2. Solution B: Dissolve 1.0 g KCl in 200 mL sterile water.
3. Solution C: Dissolve 1.6 g $Ca(No_3)_2$. H_2O in 200 mL sterile water.
4. Solution D: Dissolve 4.1 g $MgSo_4{\cdot}7H_2O$ in 200 mL sterile water.
5. Steinberg solution: Mix 20 mL solution A, 10 mL each of solution B, C, and D 750 mL of deionized water. Add 0.56 g Tris buffer and 1 mL of 50 mg/mL gentamicin sulfate. Adjust the pH to 7.8–8.0 using 5 N NaOH. Make up the solution to 1 L, sterilize using a 0.45 μm filter, and store at 4 °C.
6. Autoclaved tap water: Autoclave filtered tap water available in an amphibian facility and store at room temperature.
7. 1 % KOH solution: Dissolve 10 g KOH in 1 L of deionized water and at room temperature.
8. Saturated potassium nitrate (KNO_3) solution: Prepare 250 mL of saturated solution of KNO_3 in deionized water and store at room temperature.
9. MS 222 (tricaine methanesulfonate): Prepare 0.02 % solution of MS 222 in Steinberg solution and store at 4 °C.

2.3 Microsurgery Tools

1. Dumont #5 forceps, tungsten needle holder, and microspatula with flat tip (e.g., Fine Science Tools, cat no. 10091-12) (Fig. 1d).

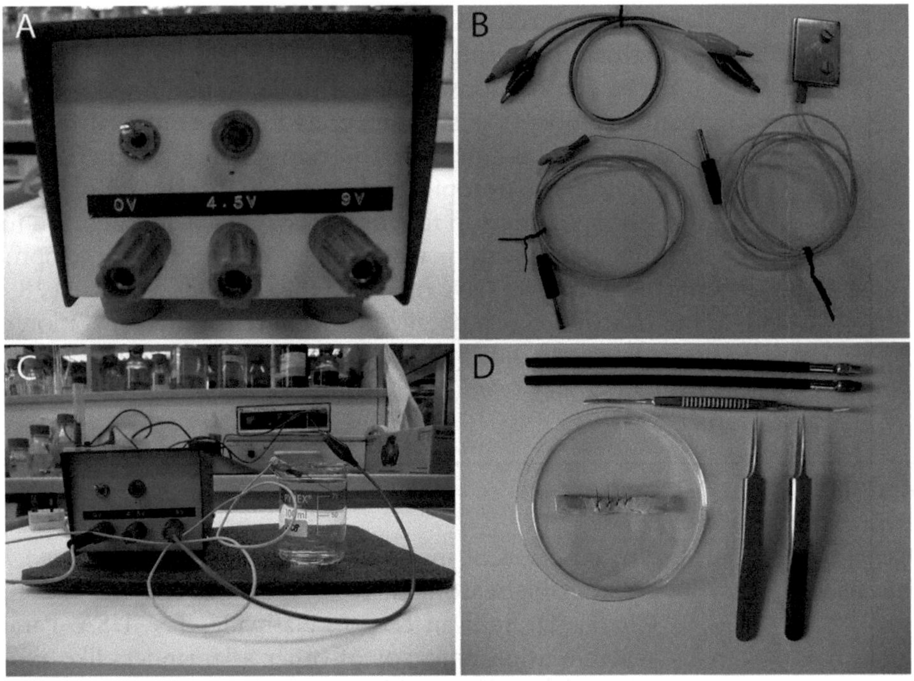

Fig. 1 Equipment and tools required for microsurgical methods. (**a**) Custom-built power supply unit with a 4.5 and 9 V D.C. supply. (**b**) Electrodes with connecting leads. A custom-made electrode holder for multiple tungsten needles (*right*) and single electrode holder (*left upper panel*) and negative electrode with platinum wire for immersion in the electrolyte solution (*left, bottom panel*). (**c**) Electrolysis setup. The platinum electrode is immersed in the beaker containing saturated solution of potassium nitrate, and the terminal is connected to 0 V connector. The tungsten wire is held by the crocodile plug, and the end is connected to desired voltage terminal. Supply of higher voltage results in faster electrolysis. The low voltage is desirable for fine sharpening of the needles. (**d**) Microsurgery tools: tungsten electrode holder, spatula for reshaping the FIMO clay, tweezers, and sharpened tungsten needles in its storage container

2. FIMO soft modeling clay (Eberhard Faber, Germany) (*see* **Note 2**).
3. Tungsten wires: thickness 0.2–0.5 mm, available from fine metal suppliers in 10 cm length. Cut the tungsten wire to 2.5 cm length using a diamond cutter.
4. Benchtop power supply unit: a custom-made or commercially available variable power supply unit with 4.5 and 9 V power output terminals (Fig. 1a). Custom-fabricated terminal leads with small crocodile plugs. One end of the negative terminal is fitted with fine platinum electrode (*see* **Note 3**).
5. Small gas torch.
6. Plastic Pasteur pipettes with varying tip diameter. This can be prepared by cutting the tip of a 3.5 mL Pasteur pipette to make it wider and quickly fire polishing it under the flame using a gas torch. This procedure will produce a smooth pipette which is essential to prevent any damage to the embryo.

7. Agar Petri dishes: Weigh 2.5 g agar in a conical flask, add 100 mL deionized water, and heat in a microwave oven to dissolve. Allow the solution to cool, and add 0.1 mL 50 mg/mL solution of gentamicin sulfate. Pour the solution into 60 mm Petri dishes to fill about 2/3 of the dish. Allow the agar to solidify at room temperature. Add 1 mL Steinberg solution to each dish and store at 4 °C.
8. Tungsten needles: Tungsten needles are prepared by electrolysis [35]. Connect the terminals to the benchtop power supply unit. Place a 2.5 cm tungsten wire to the crocodile clip at the 9 V positive terminal. Fill a 100 mL glass beaker with 80 mL saturated KNO_3 solution, and immerse the platinum electrode from the negative terminal in this solution. The terminal can be affixed to the side of the beaker using blue tack or a tape. Immerse the tip of the tungsten wire in the solution to etch the tip of the wire. Use a constant vertical motion at the tip of the wire to create a tapered end to the tungsten needle. This process takes about 15–20 min, and the sharpness of the needle can be examined under a microscope (*see* **Note 4**). Rinse the needles in deionized water and store in a box. Multiple needles can be created simultaneously using a multi-pin holder (Fig. 1b).

3 Methods

3.1 Maintenance of Embryos

1. *A. maculatum* lays the eggs in clutches and each clutch usually holds about 100 embryos. Transfer each clutch of eggs into a plastic container containing 1 L of autoclaved tap water. The embryos can be stored at 10–15 °C in a cold cabinet for 1–2 weeks. Change the water every 4–5 days (*see* **Note 5**).
2. To release the embryos from a clutch, immerse a clutch of eggs for 30 s in 1 % KOH solution (*see* **Note 6**).
3. Quickly transfer the clutch into another container with autoclaved tap water to remove the KOH solution.
4. Repeat **step 3** two times to remove any remaining KOH.
5. Transfer the clutch of eggs into a 15 cm Petri dish.
6. Using a blunt-tipped forceps, carefully separate the egg mass into 2–3 workable portions. Move the forceps between the embryos so that the embryos are not punched. Each embryo is encapsulated in a separate capsule.
7. Work through each small portion of the clutch to release individual embryo from the jelly. A pair of student quality #5 forceps is sufficient for this purpose.
8. Transfer embryos to another large Petri dish.
9. Release each embryo from its individual jelly capsule using a pair of sharp #5 forceps (*see* **Note 7**).

10. Transfer the embryo to a well in a 24-well microplate containing 1× Steinberg solution. Keep only one embryo per well.
11. Repeat **steps 9** and **10** and transfer the microplates to an incubator maintained at 14–16 °C. Change the Steinberg solution every 3–4 days. Periodically check the embryos, and remove any dead or abnormal embryos (*see* **Note 8**).

3.2 Parabiosis of Embryos

The embryos between stages 24 and 26 are suitable for parabiosis (Fig. 2a).

1. Place a pair of embryos of identical developmental stage in the parabiosis chamber containing 1× Steinberg solution.
2. Using a sharp tungsten needle, remove a rectangular piece of flank epithelium from each larva (Fig. 2a). Care should be

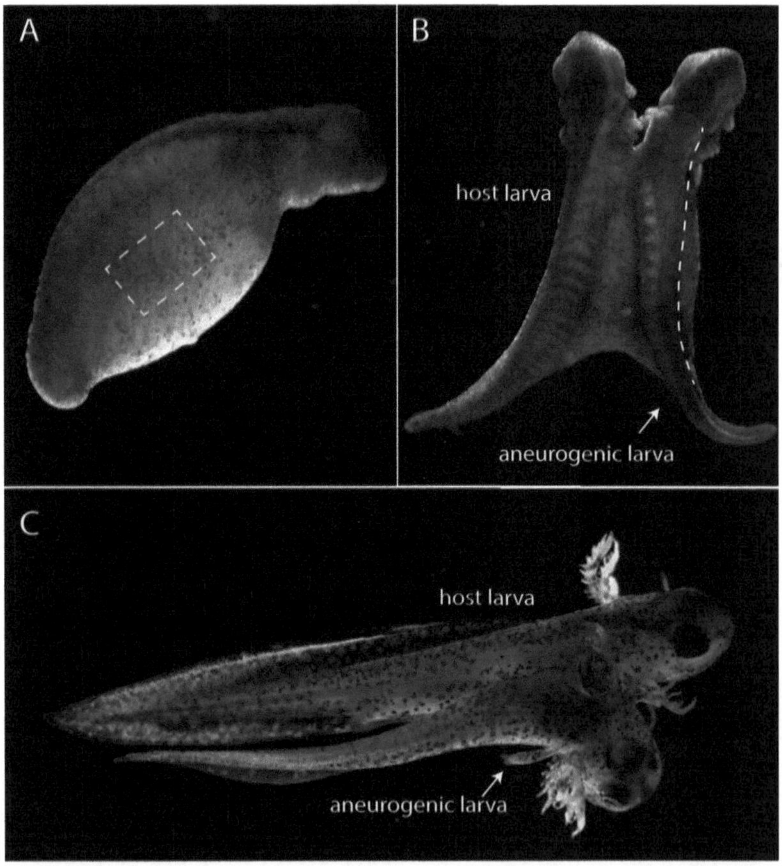

Fig. 2 *A. maculatum* embryos at various stages of development. (**a**) A stage 26 embryo suitable for parabiosis. The epidermis from the marked area on the flank can be removed for parabiosis. (**b**) A pair at stage 36–37 of development. Note the complete fusion, along the flank area. The *dotted line* on the aneurogenic larva (*right*) indicates the area lacking the neural tube (**c**) Parabiotic pair at late stage of growth. The host larva is at stage 46, whereas the aneurogenic larva is at stage 43–44. The forelimb of the aneurogenic larva is at 2-digit stage (*arrowed*)

taken not to remove large areas of the epithelium which will result in the collapse of the embryo and leakage of the yolk (*see* **Note 9**).

3. Bring both the embryos together using a flat spatula, and adjust the embryos so that the wound surfaces face each other.
4. Bring edges of the modeling clay using a spatula from both sides of the clay chamber. This procedure will make the chamber narrow, resulting in stabilizing the embryos together at their wound surface. The edges of the molding clay should touch the embryos so as to prevent any ciliary movement of the embryos.
5. Repeat the **steps 1–4**.
6. Leave the parabiosis chamber at room temperature (18 °C) overnight to allow the fusion of the pairs. The chamber can be covered with a lid and placed in a plastic box with moistened tissue to prevent evaporation of the medium.
7. Gently remove the edges of the clay from all the chambers and carefully add more Steinberg solution to the wells in excess to float the parabiotic pairs.
8. Transfer the parabiotic pairs to a 12-well multiplate chamber containing 1× Steinberg solution.
9. Raise the parabiotic pairs at 16–18 °C. Change the medium every 3–4 days.
10. Transfer the larvae to a 6-well multiplate for growth.

3.3 Aneurogenic Larvae

Aneurogenic larvae are generated by removal of the entire neural tube between stages 29 and 32. The neural tube is removed from one of the parabiotic pairs (Fig. 2b). The microsurgery is also possible in individual embryos. The surgical procedures are performed under a stereo microscope.

1. Remove a rectangular piece of agar, slightly larger than the size of a parabiotic pair from the middle of the Petri dish.
2. Pick a pair of embryos with a wide-mouth pipette and place in agar plate. The embryos are immersed in the Steinberg solution during the procedure.
3. Using a tungsten needle, gently make a vertical incision at the base of the brain to remove the epithelium (*see* **Note 10**).
4. Run the needle all the way toward the base of the tail to remove the epithelium.
5. Removal of the epithelium exposes the neural tube which lies in the grove, and the epithelium is visible under the microscope as a faint opaque line. Run the needle through the line—from the base of the brain to the last somite to remove the neural tube (*see* **Note 11**).

6. Remove the parabiotic embryos and place them in fresh Steinberg solution in a 6-well multi-dish. Return the embryos to an incubator maintained at 7 °C to complete wound healing for 5–7 days (*see* **Note 12**).
7. Transfer the embryos to 12–14 °C for growth.

3.4 Maintenance of Aneurogenic Larvae

The host larvae are able to feed on brine shrimp during larval growth (Fig. 2c). The brine shrimps can cause contamination and fungal infections in the larvae if left for longer periods. Therefore, it is essential to change the Stenberg solution few hours after feeding. Change the multi-well plates every 4–5 days.

4 Notes

1. The rate of limb development varies among different amphibian taxa. The limb development from bud stage to digit stage is very slow in axolotl larvae compared to *A. maculatum*. Therefore, in aneurogenic axolotl larvae, reinnervation of the limb is possible due to slow growth.
2. In our experience, the modeling clay from FIMO works well. Other brands which we tested were non-malleable or leaches color and appear to be toxic to the developing larvae. It is also possible to make parabiosis chambers from agar, gelatine, or Dow Corning 184 elastomer. However, these chambers are not as flexible as FIMO clay to hold the embryos together. This drawback can be ameliorated by making small bridging plates made of the same components or thin glass cover slips.
3. The negative platinum electrode is immersed in the electrolyte. Platinum helps to prevent corrosion and prolong the longevity of the electrode.
4. The voltage can be reduced to 4.5 V to create a sharp needle with 2–3 μm tip diameter. It takes longer to produce a fine tip with reduced voltage, but the low voltage gives better control in shaping the needle.
5. Long-term storage of the embryos at lower temperature is not recommended. The embryos develop deformity, and many embryos in the inner mass of the clutch eventually perish. In our experience, the length of storage also depends on the stage of development. Early stage embryos (stage 10–15) withstand lower temperature (7–10 °C), whereas preferred temperature for late stage embryos is above 10 °C. At this storage, developmental progression occurs in the embryos.
6. This step is critical; do not prolong incubation in the KOH solution. The embryos will perish or deformity may occur during development. If the clutch of eggs contains late stage

embryos (stage 30 or above), immersion may not be necessary. This also depends on the size of the clutch.

7. Removal of jelly capsule from the embryo requires practice and training is necessary to develop these skills before embarking on large-scale experiments. Early stage embryos are more difficult to handle. Late stage embryos get released from the capsule with relative ease. Embryos can also maintain individually without the removal of the capsule. The embryos hatch from the capsule at late stage of the development.
8. Usage of multi-well plates eliminates cross contamination and easy stacking of plates in the incubator. Aeration is not necessary for the development of salamander embryos. Periodical changes of the Steinberg solution is necessary to avoid fungal (most common) and bacterial contamination. In our experience, gentamicin is better than penicillin-streptomycin. The larval growth is dependent on the temperature, and for synchronized growth, an incubator is essential (e.g., cooled incubator manufactured by LMS Ltd., United Kingdom). The incubator can be fitted with light and timer for photoperiodic control (12L:12D).
9. This is the most critical step in fusion of the embryos. Smaller rectangular window results in separation of the embryos at later stages of development, whereas large epithelial wound leads to leak of the yolk and collapse of the embryo. The sharpness of the tungsten tip is critical for creating a rectangular wound, and the epithelium can be removed by a fine no. 5 forceps or using the tip of another tungsten needle. The embryos should be kept under the culture medium throughout the procedure.
10. The epithelium is very thin, and usually the tissue sticks to the tungsten needle. A pair of needles is ideal for this procedure, so that one can be used to remove the tissue from the tip of the other needle.
11. Use a pair of tungsten needles during this procedure. Use one needle to cut open the tissue and the other to remove the tissue debris from the tip of the needle. Very sharp tungsten needles are required for this procedure. Periodical resharpening of the needles is recommended. The needle for removing tissue debris can be can be blunt ended to avoid causing any damage to the embryo.
12. In our experience, storage of the aneurogenic embryos at a lower temperature after the surgery considerably decreases the mortality. About 65 % of embryos survive by adopting this procedure. The temperature is later increased to 12 °C for growth of the larvae.

Acknowledgements

We thank Professor Jeremy Brockes for advice and support. Our work is funded by a Program Grant and MRC Research Professorship to Jeremy Brockes. JP Delgado is funded by Programa-Sostenibilidad 2013–2014, University of Antioquia, Colombia.

References

1. Brownell I, Guevara E, Bai CB, Loomis CA, Joyner AL (2011) Nerve-derived sonic hedgehog defines a niche for hair follicle stem cells capable of becoming epidermal stem cells. Cell Stem Cell 8:552–565
2. Takeo M, Chou WC, Sun Q, Lee W, Rabbani P, Loomis C, Taketo MM, Ito M (2013) Wnt activation in nail epithelium couples nail growth to digit regeneration. Nature 499: 228–232
3. Yamazaki S, Ema H, Karlsson G, Yamaguchi T, Miyoshi H, Shioda S, Taketo MM, Karlsson S, Iwama A, Nakauchi H (2011) Nonmyelinating Schwann cells maintain hematopoietic stem cell hibernation in the bone marrow niche. Cell 147:1146–1158
4. Buckley G, Metcalfe AD, Ferguson MW (2011) Peripheral nerve regeneration in the MRL/MpJ ear wound model. J Anat 218: 163–172
5. Buckley G, Wong J, Metcalfe AD, Ferguson MW (2012) Denervation affects regenerative responses in MRL/MpJ and repair in C57BL/6 ear wounds. J Anat 220:3–12
6. Todd TJ (1823) On the process of reproduction of the members of the aquatic salamander. Q J Sci Lit Arts 16:84–96
7. Kumar A, Brockes JP (2012) Nerve dependence in tissue, organ, and appendage regeneration. Trends Neurosci 35:691–699
8. Singer M (1952) The influence of the nerve in regeneration of the amphibian extremity. Q Rev Biol 27:169–200
9. Stocum DL (2011) The role of peripheral nerves in urodele limb regeneration. Eur J Neurosci 34:908–916
10. Brockes JP, Kumar A (2002) Plasticity and reprogramming of differentiated cells in amphibian regeneration. Nat Rev Mol Cell Biol 3:566–574
11. Simon A, Tanaka EM (2013) Limb regeneration. Wiley Interdiscip Rev Dev Biol 2: 291–300
12. Sandoval-Guzman T, Wang H, Khattak S, Schuez M, Roensch K, Nacu E, Tazaki A, Joven A, Tanaka EM, Simon A (2014) Fundamental differences in dedifferentiation and stem cell recruitment during skeletal muscle regeneration in two salamander species. Cell Stem Cell 14:174–187
13. Lentz TL (1967) Fine structure of nerves in the regenerating limb of the newt Triturus. Am J Anat 121:647–669
14. Salpeter MM (1965) Disposition of nerve fibers in the regenerating limb of the adult newt, Triturus. J Morphol 117:201–211
15. Singer M (1949) The invasion of the epidermis of the regenerating forelimb of the urodele, Triturus, by nerve fibers. J Exp Zool 111:189–209
16. Brockes JP (1984) Mitogenic growth factors and nerve dependence of limb regeneration. Science 225:1280–1287
17. Singer M, Craven L (1948) The growth and morphogenesis of the regenerating forelimb of adult *Triturus* following denervation at various stages of development. J Exp Zool 108: 279–308
18. Gordon H, Brockes JP (1988) Appearance and regulation of an antigen associated with limb regeneration in Notophthalmus viridescens. J Exp Zool 247:232–243
19. Mescher AL (1976) Effects on adult newt limb regeneration of partial and complete skin flaps over the amputation surface. J Exp Zool 195: 117–128
20. Singer M, Rzehak K, Maier CS (1967) The relation between the caliber of the axon and the trophic activity of nerves in limb regeneration. J Exp Zool 166:89–97
21. Kumar A, Godwin JW, Gates PB, Garza-Garcia AA, Brockes JP (2007) Molecular basis for the nerve dependence of limb regeneration in an adult vertebrate. Science 318:772–777
22. Garza-Garcia A, Harris R, Esposito D, Gates PB, Driscoll PC (2009) Solution structure and phylogenetics of Prod1, a member of the three-finger protein superfamily implicated in salamander limb regeneration. PLoS One 4: e7123

23. Garza-Garcia A, Driscoll PC, Brockes JP (2010) Evidence for the local evolution of mechanisms underlying limb regeneration in salamanders. Integr Comp Biol 50:528–535
24. Morais da Silva S, Gates PB, Brockes JP (2002) The newt ortholog of CD59 is implicated in proximodistal identity during amphibian limb regeneration. Dev Cell 3:547–555
25. Kumar A, Gates PB, Brockes JP (2007) Positional identity of adult stem cells in salamander limb regeneration. C R Biol 330: 485–490
26. Galis F, Wagner GP, Jockusch EL (2003) Why is limb regeneration possible in amphibians but not in reptiles, birds, and mammals? Evol Dev 5:208–220
27. Harrison RG (1907) Experiments in transplanting limbs and their bearing upon the problems of the development of nerves. J Exp Zool 4:239–281
28. Fekete DM, Brockes JP (1988) Evidence that the nerve controls molecular identity of progenitor cells for limb regeneration. Development 103:567–573
29. Yntema CL (1959) Blastema formation in sparsely innervated and aneurogenic forelimbs of ambystoma larvae. J Exp Zool 142:423–439
30. Tassava RA, Olsen-Winner CL (2003) Responses to amputation of denervated ambystoma limbs containing aneurogenic limb grafts. J Exp Zool 297:64–79
31. Thornton CS, Thornton MT (1970) Recuperation of regeneration in denervated limbs of ambystoma larvae. J Exp Zool 173: 293–301
32. Kumar A, Delgado JP, Gates PB, Neville G, Forge A, Brockes JP (2011) The aneurogenic limb identifies developmental cell interactions underlying vertebrate limb regeneration. Proc Natl Acad Sci U S A 108:13588–13593
33. Hamburger VA (1966) A manual of experimental embryology. University of Chicago Press, Chicago, IL
34. Johnson LG (2001) Patterns & experiments in developmental biology. McGraw Hill, London
35. Hubel DH (1957) Tungsten Microelectrode for recording from single units. Science 125: 549–550

Chapter 13

In Vivo Modulation and Quantification of microRNAs During Axolotl Tail Regeneration

Jami R. Erickson and Karen Echeverri

Abstract

The ability to regenerate diseased, injured, or missing complex tissue is widespread throughout lower vertebrates and invertebrates; however, our knowledge of the molecular mechanisms that regulate this amazing ability is still in its infancy. Many recent papers have shown important roles for microRNAs in regulating regeneration in a number of species. The ability to detect and quantify miRNA expression fluctuations at a single cell level in vivo in different cell types during processes like regeneration is very informative. In this chapter, we describe how to use a dual-fluorescent green fluorescent protein (GFP)-reporter/monomeric red fluorescent protein (mRFP)-sensor (DFRS) plasmid to quantitate the dynamics of specific miRNAs over time following miRNA mimic injection as well as during regeneration. In this bicistronic vector, the mRFP allows for verification of miRNA expression, while the GFP functions as an internal control to normalize miRNA expression and thus obtain quantitative results. In addition, we demonstrate how this technique revealed dynamic miR-23a expression and function during tail regeneration.

Key words microRNAs, Axolotl, Sensor plasmids

1 Introduction

The ability to regenerate multiple body parts is a fascinating phenomenon that remains poorly understood at a molecular level [1–12]. While model systems like axolotls are desirable since they can regenerate these tissues, lack of a full axolotl genome sequence hinders classical genetic studies. Recent studies have shown that small noncoding microRNAs (miRNAs) are highly conserved and may play an important role in regulating gene expression after injury in many species including axolotls [13–19].

miRNAs are small noncoding RNAs that are approximately 22 nucleotides long [20–22]. The major function of miRNAs is post-transcriptional regulation of gene expression. The canonical mechanism of action for miRNAs is to bind to a region of complementary sequence in the 3′ untranslated region (UTR) of the target messenger RNA (mRNA). The binding of the miRNA to its target

Anoop Kumar and András Simon (eds.), *Salamanders in Regeneration Research: Methods and Protocols*, Methods in Molecular Biology, vol. 1290, DOI 10.1007/978-1-4939-2495-0_13, © Springer Science+Business Media New York 2015

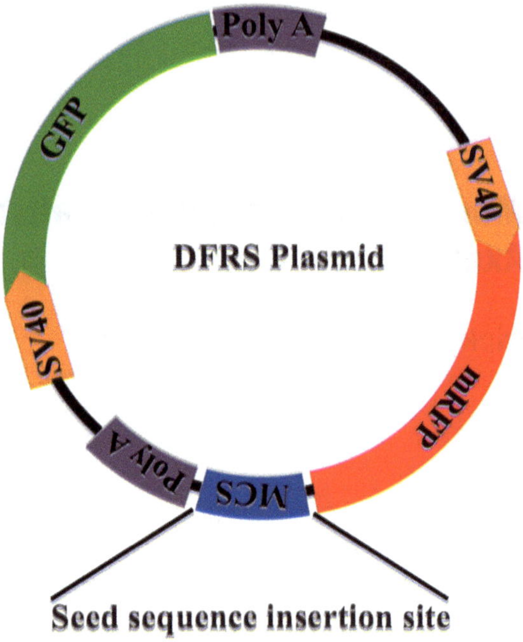

Fig. 1 Schematic of DFRS plasmid. This dual miRNA reporter construct can be used to analyze and quantify expression of endogenous miRNAs and to quantitatively evaluate the effect of modulators of microRNAs, like inhibitors or mimics. The DFRS plasmid contains both GFP (control) and mRFP (reporter), each under the control of their own SV40 promoter. Following the open read frame of mRFP is a multiple cloning site (MCS) to allow addition of seed sequences for an miRNA of interest

mRNA often leads to message degradation [23]. miRNAs play an essential role in regeneration [19, 24–26], yet dynamics of miRNA expression during regeneration remains to be delineated. One tool that allows us to explore temporal fluctuation of specific miRNAs in a tissue of interest is the dual-fluorescent green fluorescent protein (GFP)-reporter/monomeric red fluorescent protein (mRFP)-sensor (DFRS) plasmid (Fig. 1) [27, 28]. In addition to normal bacterial expression components, the DFRS plasmid is a bicistronic GFP (reporter) and mRFP (sensor) transiently expressed under the control of two independent SV40 promoters. Sequences that are complementary to an miRNA of interest can be cloned into the 3′UTR of mRFP in the DFRS plasmid using a variety of restriction enzymes for digestion followed by ligation [21].

Using microinjection and electroporation, the DFRS plasmid can be injected into a tissue of interest. The expression of the DFRS plasmid will not only confirm miRNA mimic uptake into a specific cell but will also allow observation of changes in expression levels of miRNA in a single cell in vivo. By live animal imaging, the amount of mRFP can be normalized to GFP, with the changes in mRFP negatively correlate with activity of the given miRNA, such that the relative decrease in mRFP expression correlates with increased miRNA expression in that cell. This can be used to model temporal changes in expression of an miRNA of interest as regeneration progresses as well as spatial changes as miRNA expression fluctuates relative to the regenerating tissue.

2 Materials

1. Fluorescent dissecting scope.
2. Pressure injector.
3. Micromanipulator.
4. Electroporator.
5. Electrodes.
6. Needle puller.
7. Axolotls.
8. miRNA DFRS plasmids.
9. miRNA mimic.
10. Anesthesia (0.01 % *p*-aminobenzene).
11. Fast Green.
12. Glass capillaries.
13. Sylgard-dissecting pad.
14. Insect pins.
15. Microloader pipette tips.
16. 1× PBS.
17. Tweezers.
18. Image J.

3 Methods

3.1 Preparation of Plasmid Solution for Injection

1. To prepare the plasmid solution for injection, dilute the dual-fluorescent green fluorescent protein (GFP)-reporter/monomeric red fluorescent protein (mRFP)-sensor (DFRS) plasmid to a final concentration of 0.5 μg/μL with 0.2 mg/mL Fast Green and 1× phosphate buffered saline (PBS) (*see* **Note 1**).

3.2 Preparation of Needles

1. Pull glass capillaries to form injection needles (*see* **Note 2**). Following a ramp test, place a glass capillary into the needle puller and input the following settings: Heat (Ramp value + 10), Pull 60, Velocity 60, Time 250, and Pressure 500.
2. Load 2–4 μL of plasmid solution from Subheading 3.1 into the injection needle using a pipette with a microloader tip.

3. Insert and secure the loaded injection needle into the pressure injector nozzle attached to the micromanipulator.
4. To open the needle, very gently break the very tip using a forceps. To ensure the needle is open, press the injection pedal to observe release of the plasmid solution.

3.3 Injection and Electroporation of Animals

1. Set the pressure injector to 20 psi ejection pressure and 100 ms timed ejection.
2. Place axolotls into anesthesia (*see* **Note 3**).
3. Once the axolotl is motionless and unresponsive to touch, place the axolotl on a Sylgard-dissecting pad and stabilize the axolotl by pinning the fin down with four insect pins. Assure the axolotl remains moist by adding a drop of the anesthetic on the axolotl's head.
4. Visualize the target tissue using a dissecting microscope.
5. Using the micromanipulator, insert the injection needle into the tissue of interest, such as muscle or spinal cord, and press the injection pedal to release the plasmid solution (*see* **Note 4**).
6. While the axolotl is still pinned on the Sylgard-dissecting pad, cover the axolotl with 1× PBS. Place the electrodes on either side of the tail ensuring that either electrode is not in direct contact with tissue.
7. Electroporate the axolotl with 5 pulses of 50 V, 50 ms each (*see* **Note 5**).
8. Remove the insect pins and return the axolotl to water (*see* **Note 6**).

3.4 Tail Amputation

1. Place axolotls into anesthesia (*see* **Note** 7).
2. Once the axolotl is motionless, place the axolotl in a petri dish under a fluorescent dissecting microscope.
3. Visualize the DFRS plasmid-expressing cells by GFP expression. Align a scalpel immediately posterior to GFP-expressing cells. Amputate the tail using a scalpel.
4. Image the axolotl about the amputation, and return to water.
5. Image the axolotl at desired time points during regeneration. Be sure to use the same exposure time as well as other microscope settings as the first imaged time point (*see* **Note 8**).

3.5 Mimic Injection

1. Prepare the mimic solution so the final concentration of the miRNA mimic is 10 μM with 0.2 mg/mL Fast Green and 1× PBS (*see* **Notes 9–11**).
2. Load a new injection needle with 2–4 μL of mimic solution using a pipette with a microloader tip. Insert and secure the needle into the pressure injector nozzle attached to the

micromanipulator. To open the needle, very gently break the very tip using a forceps. To ensure the needle is open, press the injection pedal to observe release of the plasmid solution.

3. Place axolotls into anesthesia.
4. Once the axolotl is motionless and unresponsive to touch, place the axolotl on a Sylgard-dissecting pad and stabilize the axolotl by pinning the fin down with four insect pins. Assure the axolotl remains moist by adding a drop of the anesthetic on the axolotl's head.
5. Place the Sylgard-dissecting pad under a fluorescent dissecting scope. Visualize the DFRS plasmid-expressing cells by GFP expression, and adjust the position of the needle to inject immediately adjacent to the DFRS plasmid-expressing cells as to not damage the cell by puncture.
6. Release the mimic solution by pressing the pedal of the injector using the settings in Subheading 3.1 (*see* **Note 12**).
7. Cover the axolotl in 1× PBS and electroporate as in Subheading 3.3.
8. Remove the pins and image the injected area of the axolotl (Fig. 3a).
9. Return the axolotl to water.
10. Image the axolotl at desired time points (*see* **Note 13**).

3.6 Quantification

1. Open RFP and GFP channel images in separate windows in image J (or similar program).
2. Synchronize windows by going to Analyze → Tools → Synchronize Windows → Synchronize All.
3. Select the freehand selection tool and trace the outline of the DFRS plasmid-expressing cell.
4. Measure the fluorescence intensity of the GFP window by going to Analyze → Measure. Be sure the measurement output contains "Area," "Integrated Density," and "Mean." Repeat the measurement on the RFP window (*see* **Note 14**).
5. Once each DFRS plasmid-expressing cell is measured, select a region that has no fluorescence and measure the region from each window. This will serve as a background.
6. Copy and paste the Image J results window into an excel worksheet (or similar program) for further analysis.
7. Calculate the corrected total cell fluorescence (CTCF) for both RFP and GFP channels:

$$\text{CTCF} = \text{Integrated Density} - (\text{Area of selected cell} \times \text{Mean of background}).$$

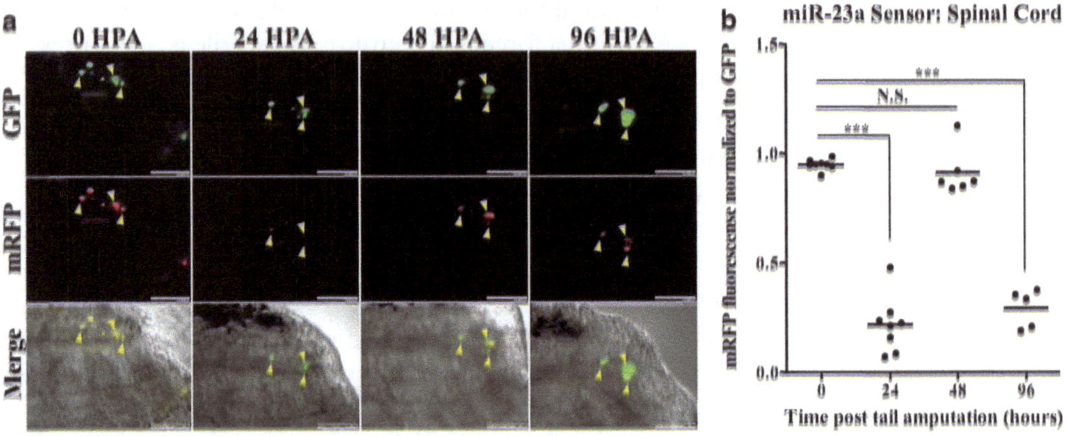

Fig. 2 Quantifying miR-23a dynamics in single cells during tail regeneration. (**a**) Control GFP expression (*green*) and reporter mRFP expression (*red*) shown at 0, 24, 48, and 96 hour post amputation (hpa). *Arrowheads* indicate examples of cells that are expressing the DFRS plasmid and are used for quantification. Scale bar is 200 μm. (**b**) Graphical representation of mRFP expression modulated by miR-23a in the spinal cord during regeneration. mRFP expression is dynamic during the regenerative process, revealing that miR-23a expression is increased at 24 and 96 hpa, while miR-23a levels at 48 hpa are similar to steady state levels of miR-23a (***$P<0.001$, *NS* not significant)

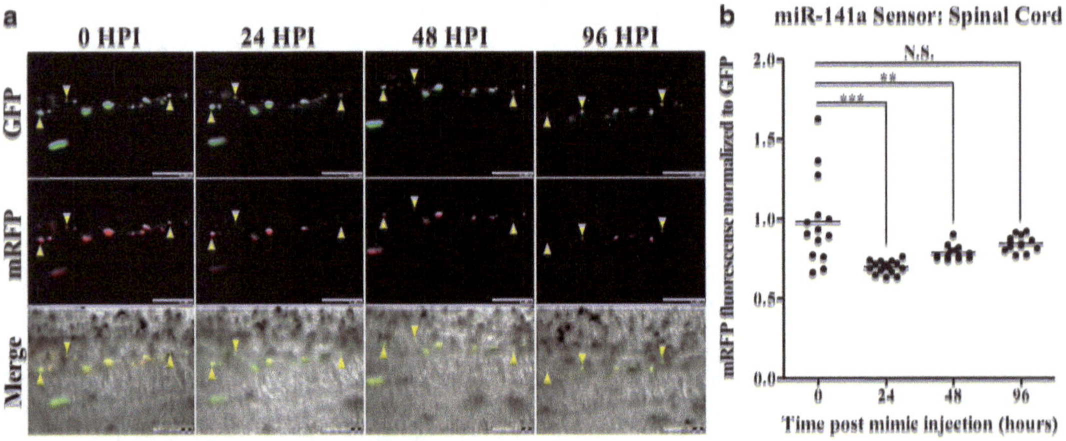

Fig. 3 Determining the effect of microRNA mimic injection using miR-141a sensor. (**a**) Control GFP expression (*green*) and reporter mRFP expression (*red*) shown at 0, 24, 48, and 96 hour post mimic injection (hpi). Arrowheads indicate examples of cells that are expressing the DFRS plasmid and are used for quantification. Scale bar is 200 μm. (**b**) Graphical representation of mRFP expression modulated by miR-141a in the spinal cord following mimic injection. Overall there is a general decrease in the amount of mRFP expression until 96 hpi, indicating that most DFRS plasmid-expressing cells also took up and processed injected miR-141 mimics (***$P<0.001$, **$P<0.01$, *NS* not significant)

8. Normalize expression levels by dividing the RFP CTCF by the GFP CTCF.
9. Use the normalized RFP values to produce a graph (Figs. 2b and 3b).

4 Notes

1. Adding Fast Green will allow visualization of the injection under visible light and is an ideal way to localize injections to the correct tissue site.
2. Needles can also be pulled manually or pre-pulled needles can be ordered.
3. Animals used here are 2–3 cm in length, injection volumes and electroporation conditions may need to be varied depending on the size of animals used. For further details on how microinjection and electroporation are performed, refer to refs. 29 and 30.
4. The injection is successful if the Fast Green color can be visualized in the tissue of interest. If not, adjust the injection needle using the micromanipulator and reinject.
5. The orientation of the electric field can be used to direct plasmid uptake to specific regions, or the electrodes can be inverted and the axolotl can be electroporated a second time for bidirectional plasmid uptake [29, 30].
6. Fluorescence of mRFP and GFP can be visualized within 24 h and can confirm a successful injection.
7. After visualization of mRFP and GFP, the tail can be amputated adjacent to the DFRS-expressing cells to examine the expression patterns of the miRNA of interest (Fig. 2a).
8. If the expression of the miRNA of interest increases during the regenerative process, a decrease in mRFP expression relative to GFP will be observed (Fig. 2a, b).
9. Here we have demonstrated the effect of a mimic on microRNA levels, and this same technique can also be used to assay the quantitative effect of microRNA inhibitors in vivo.
10. For each individual microRNA mimic or inhibitor, the optimum concentration must be determined by testing different concentrations.
11. After visualization of mRFP and GFP, an miRNA mimic can be injected and targeted to the DFRS-expressing cells.
12. If Fast Green color appears in the desired location, remove the needle. If the injection was not administered in the correct location, adjust the position of the injection using the micromanipulator and reinject.
13. Be sure to use the same exposure time as well as other microscope settings as the first imaged time point. We have found that a significant decrease in relative mRFP fluorescence can be observed at 24 h after mimic injection. If the injected mimic is targeting the binding site at the 3′ UTR of mRFP, a decrease in the mRFP expression relative to GFP can be observed (Fig. 3a, b).
14. Synchronizing the windows will allow the same area to be measured in each channel.

Acknowledgements

We thank Davide De Pietri Tonelli and Antonio Giraldez for the kind gift of the DFRS plasmids.

The authors are grateful for support from the Stem Cell Training Grant (T32 HD060536 04) for this project as well as funding from the Department of Genetics, Cell Biology and Development at the University of Minnesota.

References

1. Kumar A, Brockes JP (2012) Nerve dependence in tissue, organ, and appendage regeneration. Trends Neurosci 35:691–699
2. Brockes JP, Kumar A (2008) Comparative aspects of animal regeneration. Annu Rev Cell Dev Biol 24:525–549
3. Brockes JP, Kumar A (2005) Appendage regeneration in adult vertebrates and implications for regenerative medicine. Science 310:1919–1923
4. Tanaka EM, Reddien PW (2011) The cellular basis for animal regeneration. Dev Cell 21(1): 172–185
5. Tanaka EM (2003) Regeneration: if they can do it, why can't we? Cell 113:559–562
6. Elliott SA, Sanchez Alvarado A (2013) The history and enduring contributions of planarians to the study of animal regeneration. Wiley Interdiscip Rev Dev Biol 2:301–326
7. Sanchez Alvarado A, Tsonis PA (2006) Bridging the regeneration gap: genetic insights from diverse animal models. Nat Rev Genet 7:873–884
8. Gemberling M, Bailey TJ, Hyde DR, Poss KD (2013) The zebrafish as a model for complex tissue regeneration. Trends Genet 29: 611–620
9. Poss KD (2010) Advances in understanding tissue regenerative capacity and mechanisms in animals. Nat Rev Genet 11:710–722
10. Reddien PW, Sanchez Alvarado A (2004) Fundamentals of planarian regeneration. Annu Rev Cell Dev Biol 20:725–757
11. Reddien PW (2013) Specialized progenitors and regeneration. Development 140:951–957
12. Bryant SV, Endo T, Gardiner DM (2002) Vertebrate limb regeneration and the origin of limb stem cells. Int J Dev Biol 46:887–896
13. Wang J, Martin JF (2014) Macro advances in microRNAs and myocardial regeneration. Curr Opin Cardiol 29:207–213
14. Yin VP, Poss KD (2008) New regulators of vertebrate appendage regeneration. Curr Opin Genet Dev 18:381–386
15. Yi R, Fuchs E (2012) A miR image of stem cells and their lineages. Curr Top Dev Biol 99:175–199
16. Nakamura K, Maki N, Trinh A, Trask HW, Gui J, Tomlinson CR, Tsonis PA (2010) miRNAs in newt lens regeneration: specific control of proliferation and evidence for miRNA networking. PLoS One 5(8):e12058. doi:10.1371/journal.pone.0012058
17. Diaz Quiroz JF, Tsai E, Coyle M, Sehm T, Echeverri K (2014) Precise control of miR-125b levels is required to create a regeneration-permissive environment after spinal cord injury: a cross-species comparison between salamander and rat. Dis Model Mech 7:601–611
18. Thatcher EJ, Patton JG (2010) Small RNAs have a big impact on regeneration. RNA Biol 7:333–338
19. Sehm T, Sachse C, Frenzel C, Echeverri K (2009) miR-196 is an essential early-stage regulator of tail regeneration, upstream of key spinal cord patterning events. Dev Biol 334: 468–480
20. Bartel DP (2004) MicroRNAs: genomics, biogenesis, mechanism, and function. Cell 116: 281–297
21. Aravin A, Tuschl T (2005) Identification and characterization of small RNAs involved in RNA silencing. FEBS Lett 579:5830–5840
22. Bartel DP (2009) MicroRNAs: target recognition and regulatory functions. Cell 136:215–233
23. Cai Y, Yu X, Hu S, Yu J (2009) A brief review on the mechanisms of miRNA regulation. Genomics Proteomics Bioinformatics 7:147–154
24. Yin VP, Thomson JM, Thummel R, Hyde DR, Hammond SM, Poss KD (2008) Fgf-dependent depletion of microRNA-133 promotes appendage regeneration in zebrafish. Genes Dev 22:728–733

25. Eulalio A, Mano M, Dal Ferro M, Zentilin L, Sinagra G, Zacchigna S, Giacca M (2012) Functional screening identifies miRNAs inducing cardiac regeneration. Nature 492: 376–381
26. Gonzalez-Estevez C, Arseni V, Thambyrajah RS, Felix DA, Aboobaker AA (2009) Diverse miRNA spatial expression patterns suggest important roles in homeostasis and regeneration in planarians. Int J Dev Biol 53:493–505
27. De Pietri Tonelli D, Calegari F, Fei JF, Nomura T, Osumi N, Heisenberg CP, Huttner WB (2006) Single-cell detection of microRNAs in developing vertebrate embryos after acute administration of a dual-fluorescence reporter/sensor plasmid. Biotechniques 41:727–732
28. Mishima Y, Abreu-Goodger C, Staton AA, Stahlhut C, Shou C, Cheng C, Gerstein M, Enright AJ, Giraldez AJ (2009) Zebrafish miR-1 and miR-133 shape muscle gene expression and regulate sarcomeric actin organization. Genes Dev 23:619–632
29. Diaz Quiroz JF, Echeverri K (2012) In vivo modulation of microRNA levels during spinal cord regeneration. In: Michael Conn P (ed) Laboratory methods in cell biology: biochemistry and cell culture, vol 112. Elsevier, Amsterdam, pp 235–246
30. Echeverri K, Tanaka EM (2003) Electroporation as a tool to study in vivo spinal cord regeneration. Dev Dyn 226: 418–425

Part III

Salamander Cells in Culture

Chapter 14

Derivation and Long-Term Culture of Cells from Newt Adult Limbs and Limb Blastemas

Patrizia Ferretti and Anoop Kumar

Abstract

Notwithstanding the key importance of in vivo models for understanding patterning and cellular interactions in the regenerating tailed amphibian (salamander) limb, dissection of molecular mechanisms, as in other species, can be greatly aided by robust in vitro models. This chapter focuses on derivation and maintenance of cell lines from adult post-metamorphic salamanders and in particular cells derived from normal and regenerating limbs. We also describe a protocol for nucleofecting newt cells that can be used both to investigate the gene function in short-term studies and to establish stable cell lines.

Key words Axolotl, Blastema, Cell culture, Limb, Nucleofection, Regeneration, Transfection, Salamander

1 Introduction

Cultures of mesenchymal cells from the late-bud stage blastemas of 9–15-month-old axolotls were used by Boilly and Albert [1], but they were not established as lines available to the scientific community.

Long-term cultures of cells derived either from limb blastemas or normal limbs from adult urodeles (tailed amphibian) with the ability to undergo myogenic differentiation were first established in 1988 [2]. The cell lines derived from the newt, *Notophthalmus viridescens*, that have been more extensively used are a limb blastema cell line, originally named as BlH1 (blastema hind limb 1) and more recently called B1H1, and two lines from normal hind limbs derived either from ankle (A1) or thigh (TH4B) tissue [3, 4]. Both cell lines derived from normal limbs and limb blastemas contain thin, well-spread, and large cells as well as smaller, thicker, and often bipolar cells (Fig. 1a, b). All cells present in these cultures are much larger than mammalian cells, with nuclei up to 50 μm in diameter. They have an average duplication time of 2 days, which can vary depending on the line and culture

Anoop Kumar and András Simon (eds.), *Salamanders in Regeneration Research: Methods and Protocols*, Methods in Molecular Biology, vol. 1290, DOI 10.1007/978-1-4939-2495-0_14, © Springer Science+Business Media New York 2015

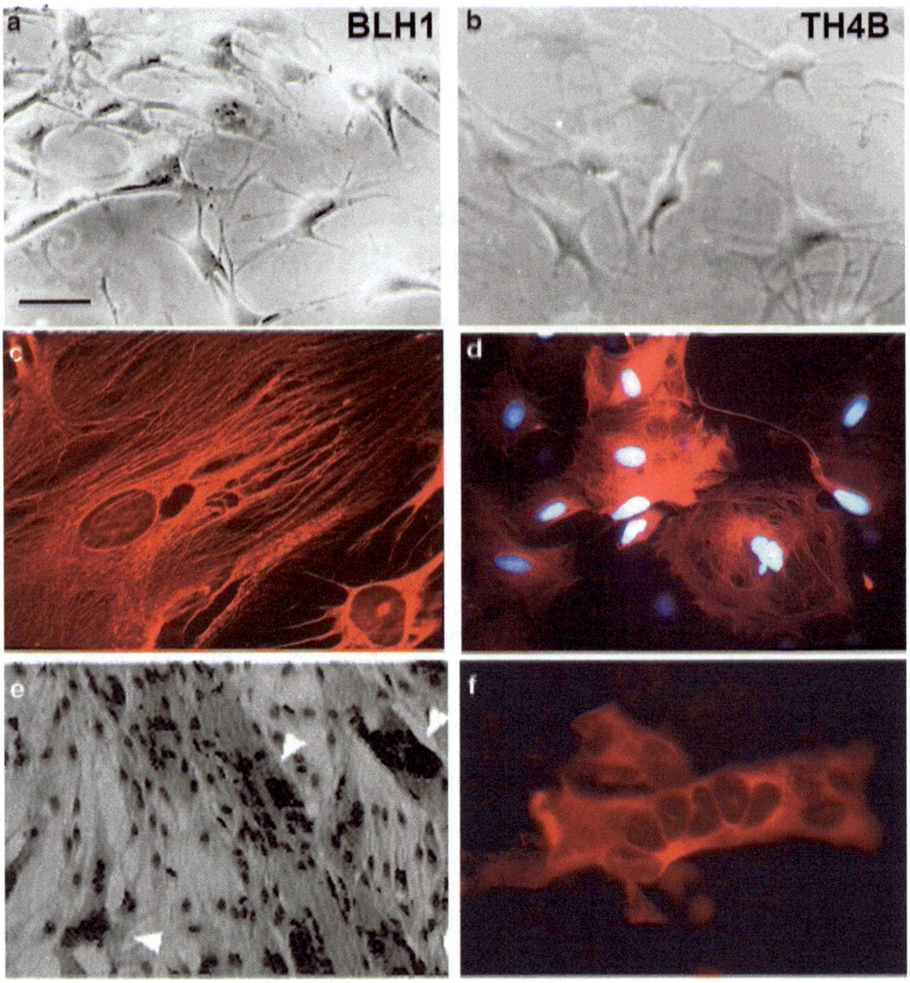

Fig. 1 Newt cells cultured from normal limb and limb blastema maintained either at low/medium density (**a–c**) or at high density to induce myogenic differentiation (**e**, **f**) express blastemal (**c**, **d**) and muscle markers (**f**), respectively. (**a**, **b**) Live images of BlH1 and TH4 cells. (**c**, **d**) Newt cells stained for blastemal markers, 22/18 (**c**) and K8 (**d**). (**e**) Histological staining of cells grown at high density; some myotubes are indicated by arrowheads. (**f**) Myotube stained with the muscle marker, 12/101

conditions (e.g., batch of fetal calf serum). A correlation between the larger size of newt cells and their relatively slow cell cycling time, as compared to mammalian cells, and the size of the newt genome (estimated to be up to tenfold larger than that of mammals) has been suggested [5].

Attempts at cloning the newt cell lines were not successful as, at a clonal density, these cells would survive for a long time but not divide. In contrast, satellite cells derived from cultured isolated myotubes could be expanded at clonal density [6]. It is not known at this stage whether the culture conditions used to grow satellite cell clones would allow one to obtain also clonal BlH1,

A1, or TH4B cell lines. Nonetheless, although the established newt cell lines have not been cloned, they appear to be relatively homogeneous, as indicated by markers of hypomethylation, immunoreactivity profile, and differentiation potential [2, 7, 8]. All the newt lines generated express regeneration-associated molecules such as the protein detected by the *22/18* monoclonal antibody and keratins *8* and *18* [2, 7] (Fig. 1c, d). These proteins are down-regulated in limb and blastemal cultures following differentiation into muscle, mimicking their regulation in vivo. These cell lines, though untransformed, have shown no sign of senescence in long-term cultures. In this respect, their behavior is similar to those of newt cells derived from the iris pigment epithelium that can also be expanded long-term [9, 10]. In contrast, cells derived from adult newt heart and liver were found to have a relatively short life in culture [2, 11]. Newt cells undergo myogenic differentiation (Fig. 1e, f) when grown at high density [12]. However, cells with self-renewing capability are always recovered upon passaging even after being left for prolonged periods at high density, consistent with the ability of newt muscle to dedifferentiate. This is not the case when mammalian satellite cells or myoblasts are allowed to grow at high density and undergo terminal differentiation, which results in the loss of the progenitor cell population. Furthermore, newt cells that have undergone myogenic differentiation can reenter the cell cycle upon serum stimulation [13, 14]. The newt cell lines have been extensively used to study responses to retinoids and growth factors [3, 4, 15, 16], cell differentiation and dedifferentiation [12, 13, 17, 18], and molecular mechanisms underlying patterning [3, 4, 19, 20]. In addition, A1 cells have been used to generate hybrid myotubes by fusing them with a mouse myoblast cell line [21].

When starting to culture amphibian, and in general nonmammalian vertebrate cells, it is important to keep in mind that these are cold-blooded animals. Therefore, their cells need to be cultured at a lower temperature (24–26 °C) than their mammalian counterparts. It is also important to be aware of the fact that the osmolarity of the amphibian blood is lower than that of mammalian blood (approximately 225–250 mOsm depending on the species) [2, 22]; hence, culture media require dilution to match the osmolarity of the blood. When media utilizing a sodium bicarbonate buffer are used, the percentage of CO_2 in the tissue culture incubator needs to be 2 % to maintain the pH at 7.4–7.6. The newt cells do not attach to glass or plastic cell cultureware. Their adhesive properties also differ from those of axolotl mesenchymal cells derived from skin that can attach to the plastic in the absence of a substrate [23]. Newt cells can grow both on collagen and gelatine. Gelatine, however, has become the preferred method as it provides a cheaper and more convenient alternative to collagen [2, 4].

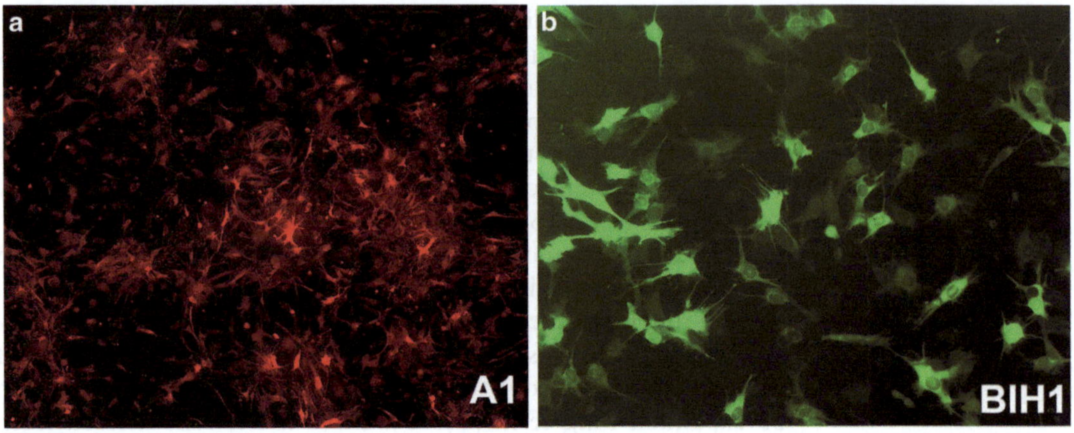

Fig. 2 A1 and BIH1 newt cells nucleofected with either RFP (**a**) or GFP (**b**) constructs. Note that the high levels of expression of the nucleofected gene are observed in both cell lines

This chapter describes how to establish cell lines from newt blastemas and normal limbs and to maintain them as long-term cultures. It is challenging to introduce plasmid DNA into newt cells efficiently using lipid-based transfection reagents. Here we describe a protocol for nucleofecting newt cells (Fig. 2) that can be used both to investigate the function of a transfected gene in short-term studies and to establish stable cell lines expressing genes of interest or markers genes for cell tracking after cell grafting in vivo [20, 24].

2 Materials

Here we indicate the materials required and the sources we use, but unless otherwise specified, equivalent reagents obtained from other suppliers will work as well. Tissue culture flasks from Corning, Nunc, or BD Falcon have all proven suitable for culturing newt cells (plastic from other sources has given variable results).

2.1 Sources of Adult Newts

Suppliers of adult newts can be purchased from Charles D. Sullivan & Co., USA, NASCO Biology, USA, or Connecticut Valley Biological Supply Company, USA. Newts may also be obtained from local suppliers. Follow the national and institutional guidelines for the import and use of animals for laboratory research.

2.2 Surgical Tools, Anesthetics, and Reagents for Limbs and Limb Blastema Harvest

All the surgical instruments are sterilized by autoclaving, by using a glass bead sterilizer, or by flaming after dipping the tip of the instrument in 80 % alcohol.

1. Dumont #5 forceps.
2. Surgical scissors.

3. 12 cm spring scissors.
4. Bone cutter.
5. Dissection board.
6. Sterile water: Aliquot commercially available HPLC grade water in 250 mL bottles and autoclave. Store at room temperature (*see* **Note 1**).
7. Sterile glass Pasteur pipettes.
8. Sterile Petri dishes and 50 mL tubes.
9. Sodium hypochlorite solution: Prepare 1 % sodium hypochlorite solution in sterile water and adjust the pH to 8.8 with 1 N HCL.
10. Tricaine (ethyl 3-aminobenzoate methanesulfonate) (MS-222): Prepare a 0.1 % solution using filtered water available in the amphibian facility. Adjust the pH between 7.2 and 7.5 using 1 N NaOH. Prepare fresh before use.
11. Recovery solution for newts: Dissolve 10 g sulfamerazine (Sigma, Cat. no. S0800) in 2 L of filtered water available in an amphibian facility. Add 0.1 mL analgesic solution (Alvegesic® Vet, stock 10 mg/mL) and adjust the pH between 7 and 7.5 with 5 N NaOH.

2.3 Reagents and Equipment for Newt Cell Isolation and Maintenance

Preparation of tissue culture solution and media and handling of cells are carried out under sterile conditions in a class II biological safety cabinet swiped with 70 % ethanol and turned on at least 15 min before starting the procedure. Use tissue culture grade plastic and sterile glassware for all reagent preparation.

1. Sterile water (*see* Subheading 2.2, **item 6**).
2. Amphibian phosphate buffered saline (APBS). Add 25 mL of sterile water to 100 mL of sterile tissue culture grade PBS. This is to adjust the osmolarity of the medium to a level comparable to that of amphibian blood (225 mosmol ± 5). Store at room temperature.
3. Ca^{2+}/Mg^{2+}-free amphibian phosphate buffered saline (Ca^{2+}/Mg^{2+}-free APBS). Use the procedure described in **item 2**, substituting Ca^{2+}/Mg^{2+}-free PBS to standard PBS.
4. 100× penicillin/streptomycin (10,000 units/mL of penicillin and 10,000 μg/mL of streptomycin): Defrost the stock solution and aliquot as convenient (e.g., 2 mL aliquots if routinely preparing 200 mL of culture medium). Store at −20 °C.
5. 1,000× gentamicin stock solution (35 mg/mL): Dissolve 350 mg of gentamicin sulfate salt in 10 mL of sterile water. Filter-sterilize the solution using a 0.2 μm syringe filter. Store as 100 or 500 μL aliquots at −20 °C.

6. 100× glutamine stock solution (200 mM): Aliquot the stock solution (e.g., 2 mL aliquots if routinely preparing 200 mL of culture medium) and store at −20 °C.
7. 100× insulin stock solution (0.5 mg/mL stock): Weigh 50 mg insulin in a beaker and dissolve it by adding 50 mL of 0.1 M HCl (1 mg/mL). Make up the solution to 100 mL with Ca^{2+}/Mg^{2+}-free APBS to avoid precipitation. Sterilize the solution by filtering it through a 0.2 μm filter, aliquot (e.g., 2 mL aliquots if routinely preparing 200 mL of culture medium), and store at −20 °C.
8. Fetal calf serum (FCS): Defrost the stock bottle and place it in a water bath at 56 °C for 30 min to inactivate the complement. Cool the bottle to room temperature, aliquot (e.g., 20 mL aliquots if routinely preparing 200 mL medium), and store at −20 °C. There is great variability in the ability of FSC batches to support newt cell growth. Hence, batch testing to select the FCS is needed (*see* **Note 2**).
9. Amphibian minimum essential medium (AMEM; for growth in a CO_2 incubator): Dispense 130 mL minimum essential medium (1× MEM), 20 mL of FCS, 2 mL of penicillin/streptomycin, 2 mL of glutamine, and 2 mL of insulin in a sterile bottle (e.g., 75 cm^2 tissue culture flask; sterile glass bottle with a phenolic cap). Bring the volume up to 200 mL with sterile water. This dilution gives amphibian cell culture medium of the appropriate osmolarity (*see* **Note 3**).
10. Dissection medium: Dispense 65 mL of L-15, 1 mL of penicillin/streptomycin, and 100 μL of gentamicin in a sterile bottle. Bring the volume up to 100 mL with sterile water (*see* **Notes 4** and **5**).
11. Gelatine solution (0.75 %): Weigh 3.75 g of gelatine and add it to 500 mL of sterile water pre-warmed in a glass bottle at 65 °C in a water bath. Allow the gelatine to dissolve slowly at 65 °C in a shaking water bath or by shaking the bottle manually from time to time. As soon as the gelatine has solubilized, filter it while warm using a 0.45 μm filter flask. Dispense 50 mL aliquots (e.g., Falcon tubes) and store at 4 °C.
12. Collagen (5 mg/mL). Collagen can be purchased already in solution for convenience (First Link, Cat. no. 60-35-807, replacing Vitrogen originally used). Store at 4 °C in 5–10 mL aliquots to reduce the possibility of contamination.
13. Gelatine-coated culture flasks/dishes: Warm the gelatine solution at 37 °C in a water bath to liquefy. Pipette the gelatine solution into the tissue culture flask or dish to be coated, and swirl it to spread the gelatine evenly. Remove excess gelatine solution and use it to coat the next flask/dish if required. As an

indication, 5–10 mL of gelatine solution can be used to coat several 75 cm^2 flasks. The gelatine-coated flasks/Petri dishes can be used immediately without drying.

14. Collagen-coated culture flasks/dishes: Pipette the collagen solution and coat as for gelatine coating (**item 13**). Dry the collagen-coated dishes in a tissue culture cabinet (overnight is usually sufficient). The collagen-coated dishes can be stored wrapped in cling film for at least 2 weeks at 4 °C.
15. Trypsin stock solution (2.5 %) for cell passaging (Invitrogen, Cat. no. 15090-046). Aliquot the stock solution in 2 mL and store at −20 °C.
16. Trypsin working solution for cell culture (0.25 %): To prepare the working solution, thaw a 2 mL aliquot and bring the volume up to 20 mL with Ca^{2+}/Mg^{2+}-free APBS. The solution can be stored at 4 °C for approximately 1 week.
17. Trypsin working solution for cell dissociation (0.05 %): To prepare the working solution, dilute the stock solution (**item 15**) to 1:50 with Ca^{2+}/Mg^{2+}-free Puck's saline (original protocol) or Ca^{2+}/Mg^{2+}-free APBS.
18. EDTA solution: Prepare 0.1 % EDTA in Ca^{2+}/Mg^{2+}-free APBS.
19. Sterilizing filters (syringe filters and filter flask; 0.2 μm/0.45 μm).
20. Cell strainer or nylon mesh (100 μm size).
21. Freezing medium: Dispense 5 mL of dimethyl sulfoxide (DMSO; sterile-filtered DMSO supplied in 5 mL vials is convenient; Sigma, Cat. no. D2650) into a 50 mL sterile tube (e.g., Falcon tube). Bring the volume up to 50 mL with AMEM and mix well. Store as 5 mL aliquots in sterile tubes at −20 °C. The freezing medium can be stored for several months.
22. 1 °C freezing container (Nalgene, Cat. no. 5100-0001); alternatively a cardboard or plastic box for microtube storage with bottom and lid lined with cotton wool can be used.
23. Cryogenic vials.
24. Amphibian cell culture incubator: Set the incubator for amphibian cell cultures at 25 °C and 2 % CO_2. The incubator is humidified with sterile water containing copper sulfate (e.g., one teaspoonful of copper sulfate). Ideally, a CO_2 incubator with a cooling unit should be used to ensure a stable temperature and avoid any risk of overheating. If a standard incubator is used, the tissue culture room needs to be maintained at 20 °C. Amphibian cells can also be cultured in L-15 medium that does not require maintaining the cells in a CO_2 incubator. However, for long-term cultures, a CO_2 incubator is preferable.

2.4 Equipment and Reagents for Nucleofection of Newt Cells

1. Lonza Nucleofector 2b Device (basic model).
2. Ingenio 0.2 cm cuvettes (Cat. no. 50121).
3. Cell Line electroporation kit V (Lonza, Cat. no. VCA-1003) or Ingenio electroporation solution (Mirus Bio LLC).
4. Fine-tip disposable pipettes.
5. Plasmid DNA.
6. 50 mL centrifuge tubes.

3 Methods

3.1 Newt Limb Regeneration Blastemas

1. Anesthetize the newts in tricaine solution for 7–10 min (20–30 blastemas are usually sufficient for establishing blastemal cultures).
2. Amputate the forelimb (amputation can be unilateral or bilateral depending on your animal license) at mid-humerus level using 12 cm surgical spring scissors.
3. Gently pull the stump tissue slightly back with forceps to expose the bone stump and trim the protruding bone with a bone cutter or strong scissors. Bone protruding from the stump will hamper wound healing and regeneration.
4. Trim the stump soft tissues to obtain a neatly cut surface using a 12 cm spring scissors.
5. Move the amputated newts to the recovery solution.
6. After 24 h, transfer the animals to an aquarium set up at 24 °C. Limbs will regenerate also at the lower maintenance temperature (18–20 °C), but the process will be slower. In animals kept at 24 °C, the mid-bud blastema stage (conical mound of cells at the tip of the stump) is reached by 12–15 days. As there is variability among batches of newts, animals must be closely monitored and care taken to provide abundant live food and maintain the temperature constant.

3.2 Tissue Preparation for Culturing

1. Kill the newt with an overdose of anesthetics (according to your animal procedure approved directives).
2. Dip the whole body in 1 % sodium hypochlorite, pH 8.8, for a couple of minutes (50 mL tubes are quite convenient for small-/medium-sized newts). This time can be increased for preparation of explants from normal limbs, as normal skin provides a strong barrier to the underlying tissue.
3. Rinse the body 4–5 times in sterile Ca^{2+}/Mg^{2+}-free APBS.
4. Remove the normal limb or the limb blastema with sterile dissection tools.
5. Place the limb or limb blastema in cold dissection medium (*see* **Notes 4** and **5**).

3.3 Dissociated Cultures from Newt Limb Blastemas

1. Follow **steps 1–5** in Subheading 3.2.
2. Transfer the mid-bud blastemas into a dish containing EDTA solution.
3. Incubate in the EDTA solution for 10–15 min at room temperature.
4. Mechanically remove the wound epidermis with fine sterile forceps. It should peel off quite easily if sufficient time in EDTA is allowed.
5. Cut each blastema in 4–5 pieces.
6. Transfer the blastema fragments into a screw cap vial (e.g., a 7 mL Bijou vial) containing 2–5 mL of 0.05 % trypsin solution. Always use a freshly prepared solution.
7. Digest the blastema fragments overnight at 8 °C.
8. Inactivate the trypsin by adding an equal volume of L-15.
9. Gently triturate the tissue suspension by passing it through progressively smaller plastic pipettes (from 5 to 0.5 mL). An autoclaved nylon mesh or cell strainer (100 μm size) can be used to remove undigested tissue fragments; however, it should be noted that some of these fragments can attach as explants and increase the cell yield.
10. Centrifuge the cell suspension for 5 min at $176 \times g$ in conical centrifuge tubes.
11. Resuspend the pellet in AMEM.
12. Remove fragments of undigested tissue by passing the cell suspension through the 100 μm pore cell strainer or nylon mesh. This step can reduce the yield and can be skipped, particularly when starting with a small number of blastemas.
13. Plate the cells onto either collagen- or gelatine-coated multiwell plates (e.g., 24- or 6-well plates) keeping the medium volume as low as possible to allow more rapid conditioning (*see* **Note 6**).
14. Place the dishes in a humidified incubator with 2 % CO_2 at 25 °C.
15. Wait for the cells to attach. Do not remove the medium even if there is abundant cell debris until adhering cells are observed. This can take 2–3 days or much longer, as notwithstanding high viability after dissociation, initial attachment is poor and can take several days for cell growth to become apparent (10–20 days).

3.4 Explant Cultures from Normal Newt Limbs and Limb Blastemas

1. Remove the normal limb skin keeping the limb immersed in L-15 (*see* **Note 7**).
2. Remove wound epidermis from blastemas following **steps 1–4** in Subheading 3.3.

3. Cut the limb tissue into small cubes of approximately 3–5 mm^3 in L-15.
4. Gently place each fragment onto either collagen (as in the original protocol) or gelatine-coated Petri dishes (30–35 explants/10 cm dish) or multiwell plates (e.g., 24-well plates, 1–3 explants per well) and cover it with a drop of AMEM or L-15 if using an incubator without CO_2 (*see* **Note 6**).
5. Add just enough AMEM to cover the explants and put the dish back in the incubator.
6. Leave the explants for a few hours or overnight in a humidified incubator with 2 % CO_2 at 25 °C to promote attachment ensuring they do not dry out.
7. Wait for cells to migrate out. This can take several days. After 3–4 days, remove floating explants (usually about 40 %), but do not replace the whole medium.
8. Once cells have started to migrate out of the explants (5–15 days), feed them by adding fresh medium.
9. Passage cells as described in Subheading 3.5 when they reach approximately 70 % confluence or become tightly packed around the explant. This can take about a month.

3.5 Long-Term Maintenance of Newt Cells

1. Do not allow cells to become overgrown as they will start to differentiate.
2. For culture expansion, passage cells when they are 70–90 % confluent and split them either 1:2 or 1:3. Do not passage cells that are less than 30 % confluent nor plate them at such a low density, as they will struggle to grow. Newt cells require some cell contact for optimal growth.
3. Rinse cells with Ca^{2+}/Mg^{2+}-free APBS prior to trypsinization to remove Ca^{2+}/Mg^{2+} ions as they affect trypsin activity.
4. Remove the Ca^{2+}/Mg^{2+}-free APBS.
5. Add a small volume of trypsin solution sufficient to cover evenly the surface of your dish/flask.
6. Incubate for 4–6 min in the amphibian tissue culture incubator.
7. Check your cells under the microscope and gently shake the dish/flask to aid cell detachment.
8. When the cells start to detach, neutralize the trypsin by adding AMEM or 60 % L-15 with 10 % FCS.
9. Remove all cells from the flask by gentle pipetting.
10. Transfer the suspension to a conical centrifuge tube.
11. Centrifuge the suspension at $176 \times g$ for 5 min in a standard tissue culture centrifuge.

12. Remove the supernatant and resuspend the cells in 200–1,000 μL of AMEM by gentle mixing; count them if necessary.
13. Dilute the cell suspension with AMEM as required and plate onto gelatine-coated plastic. Newt cells are usually plated at an approximate density of 6,000 cells/cm^2 (*see* **Note 8**). When grown under optimal conditions, they reach confluence in 7–10 days. After plating in fresh medium, cells can be fed, but the medium should not be changed to avoid removing essential autocrine factors essential for their growth. Frequent full changes of medium result in the arrest of cell proliferation.

3.6 Cryopreservation of Newt Cells

Newt cells can be frozen when they are 70 % confluent. A 75 cm^2 flask at 70 % confluence contains a sufficient number of cells for freezing. Freezing less than 500,000 cells should be avoided as it results in poor recovery after thawing.

1. Defrost the freezing medium and keep it on ice; it must never be warmed above room temperature.
2. Trypsinize the cells as described in Subheading 3.5, **steps 2–11**.
3. Resuspend the pellet in 200 μL of AMEM.
4. Transfer the cell suspension to 1.2 mL of cold freezing medium in a cryogenic vial (*see* **Note 9**).
5. Place the vial in a freezing box and keep it at −20 °C for 2–3 h.
6. Transfer the vial to a −70/−80 °C freezer.
7. After 16–24 h, move the vial into a liquid nitrogen storage container (*see* **Note 10**).

3.7 Thawing Newt Cells

The procedure for thawing newt cells is similar to that used for mammalian cells, with much care needed to rapidly dilute and eliminate the DMSO present in the freezing medium in order to avoid toxic effects and maximize cell recovery.

1. Dispense 8 mL of cold AMEM or L15 in a conical centrifuge tube.
2. Rapidly thaw the cells by immersing the vial in a water bath at 37 °C.
3. As soon as the cell suspension has thawed and is still cold, dilute it in the 8 mL of medium in the centrifuge tube.
4. Spin the cell suspension at 62 × *g* for 5 min. Decant the supernatant and resuspend the pellet in 5 mL of AMEM.
5. Transfer the cell suspension to a 75 cm^2 (or smaller) gelatine-coated flask and place it in the amphibian cell incubator.

3.8 Nucleofection of Newt Cells

1. Equilibrate 10 mL of complete AMEM in a tissue culture flask or a centrifuge tube in an amphibian cell culture incubator for 20 min. The cells must be transferred to this medium

immediately after nucleofection. The final volume of the medium added depends on the culture flask or Petri dish used.

2. Warm up the Nucleofector solution at room temperature.
3. Turn on the Nucleofector device.
4. Select the following program based on the cell type: (a) newt A1 cells—program number T-020; (b) newt B1H1 cells—program T-030.
5. Trypsinize the cells as described in Subheading 3.5, **steps 2–11**. A cell density of 70–80 % confluence is optimal for all amphibian lines (A1, B1H1).
6. Carefully remove all the supernatant with an aspirator, and any small volume of medium remaining should be removed using a pipette tip.
7. Resuspend the cells in 100 μL of Nucleofector solution and mix it gently without creating air bubbles (*see* **Note 11**).
8. Add 2–4 μg plasmid DNA, and mix the solution.
9. Transfer the cell suspension to 0.2 cm electroporation cuvette carefully through the sides of the cuvette without creating any air bubbles. Use about 5×10^5 cells per 0.2 cm cuvette. Due to the large size of amphibian cells, manufacturer's recommended number of cells is not required. Nucleofection can be successfully carried out also using a lower number of cells. However, under these conditions cell death is higher.
10. Replace the cap of the cuvette and place it in the cuvette holder of the Nucleofector Device.
11. Run the required program.
12. Immediately after completion of the program, take out the cuvette and transfer the cell suspension to the equilibrated complete medium.
13. Place the tube in the cell culture incubator for 30 min to allow the cells to recover before plating them.
14. Nucleofected cells from each tube can be plated on 2× 25 cm^2 flasks or 3–4 35 cm cultures dishes.
15. Transfer the cells to the amphibian incubator.
16. Check for cell attachment and transfection rate. Expression of transfected constructs should be detectable after 24–48 h. Cell survival rate after nucleofection is between 30 and 50 %. Cells at passage 30–45 show very good survival after nucleofection. High variability in cell survival has been noticed in cells at high passage number (>60). Expression of marker genes (eGFP, RFP; Fig. 2a–b) in nucleofected newt cells is 60–80 % (*see* **Note 12**).
17. Newt cells can be cultured for several weeks without loss of marker gene expression. We have also generated newt cells

with selectable markers (e.g., puromycin) to establish stable cell lines expressing marker genes. Heterogeneity in the level of gene expression is observed in these cells.

4 Notes

1. Sterile water bottles should be designated for amphibian cell culture only and prepared as follows: use new bottles, fill them with distilled water, and autoclave; discard the autoclaved water and then fill the bottles with HPLC grade water and autoclave. Do not wash the bottles with detergents.
2. Amphibian cells do not grow equally well in all fetal calf sera (FCS). Therefore, batch screening is essential. Usually testing 3–4 samples of serum from a supplier is sufficient for identifying a good growth promoting FCS. Once a suitable batch has been identified, an appropriate number of 500 mL bottles (e.g., 20–40 depending on the volume of work anticipated) of that FCS batch should be reserved. The FCS can be stored at –20 °C for several years.
3. The culture medium can be stored at 4 °C for a couple of weeks. The medium should be brought to room temperature before use. Do not warm it in a water bath at 37 °C.
4. As L-15 medium is amino acid rather sodium bicarbonate buffered, it effectively maintains a physiological pH when handling tissue and cells outside the CO_2 incubator. Cells can be cultured long-term in this medium if an incubator with 2 % CO_2 is not available.
5. Gentamicin is used at the dissection stage as it is rather effective in reducing contaminations, but it was found that its prolonged use affects cell growth. Hence, it should not be included in the growth medium.
6. All of the lines currently in use were started by plating cells or explants onto collagen-coated plastic, but as the newt limb cell lines attach well also on gelatine, it is expected this would be a suitable substrate for starting new lines.
7. Normal limb tissue is very difficult to dissociate, and only occasionally live cells are obtained following enzymatic digestion. In contrast, blastemas are more easily dissociated but more prone to contaminations.
8. If antibody staining and high-resolution imaging under oil objective is anticipated, culturing of newt cells on Nunc 35 mm Petri dishes is recommended. This will allow mounting of the dishes with a round cover slip and easy trimming of the edges of the plate with a pair of sharp scissors to accommodate the oil objectives. Newt cells usually do not grow optimally on glass

coverslips. If plating the cells on a coverslip, coating with polylysine before gelatine coating helps cell attachment.

9. The presence of antibiotics in the freezing medium does not appear to cause any adverse effect.
10. The cells can be kept at −70 °C for a few days but ideally should be transferred into liquid nitrogen within 24 h.
11. Efficiency of transfection using the Lonza kit and Mirus electroporation solution was assessed in parallel using various plasmid and comparable results obtained. Cuvettes with larger volumes can also be used for nucleofection of cells.
12. The nucleofection approach described here is also applicable to cells derived from the limb of axolotl, a paedomorphic salamander.

References

1. Boilly B, Albert P (1988) Blastema cell proliferation in vitro: effects of limb amputation on the mitogenic activity of spinal cord extracts. Biol Cell 62:183–187
2. Ferretti P, Brockes JP (1988) Culture of newt cells from different tissues and their expression of a regeneration-associated antigen. J Exp Zool 247:77–91
3. Corcoran JP, Ferretti P (1997) Keratin 8 and 18 expression in mesenchymal progenitor cells of regenerating limbs is associated with cell proliferation and differentiation. Dev Dyn 210:355–370
4. Corcoran JP, Ferretti P (1999) RA regulation of keratin expression and myogenesis suggests different ways of regenerating muscle in adult amphibian limbs. J Cell Sci 112:1385–1394
5. Sessions SK (2008) Evolutionary cytogenetics in salamanders. Chromosome Res 16: 183–201
6. Morrison JI, Borg P, Simon A (2010) Plasticity and recovery of skeletal muscle satellite cells during limb regeneration. FASEB J 24: 750–756
7. Ferretti P, Fekete DM, Patterson M, Lane EB (1989) Transient expression of simple epithelial keratins by mesenchymal cells of regenerating newt limb. Dev Biol 133:415–424
8. Casimir CM, Gates PB, Ross-Macdonald PB, Jackson JF, Patient RK, Brockes JP (1988) Structure and expression of a newt cardio-skeletal myosin gene. Implications for the C value paradox. J Mol Biol 202:287–296
9. Kosaka M, Kodama R, Eguchi G (1998) In vitro culture system for iris-pigmented epithelial cells for molecular analysis of transdifferentiation. Exp Cell Res 245:245–251
10. Simon A, Brockes JP (2002) Thrombin activation of S-phase reentry by cultured pigmented epithelial cells of adult newt iris. Exp Cell Res 281:101–106
11. Bettencourt-Dias M, Mittnacht S, Brockes JP (2003) Heterogeneous proliferative potential in regenerative adult newt cardiomyocytes. J Cell Sci 116:4001–4009
12. Lo DC, Allen F, Brockes JP (1993) Reversal of muscle differentiation during urodele limb regeneration. Proc Natl Acad Sci U S A 90: 7230–7234
13. Tanaka EM, Drechsel DN, Brockes JP (1999) Thrombin regulates S-phase re-entry by cultured newt myotubes. Curr Biol 9:792–799
14. Tanaka EM, Gann AA, Gates PB, Brockes JP (1997) Newt myotubes reenter the cell cycle by phosphorylation of the retinoblastoma protein. J Cell Biol 136:155–165
15. Beug S, Vascotto SG, Tsilfidis C (2006) Newt orthologue of Growth arrest-specific 6 (NvGas6) is implicated in stress response during newt forelimb regeneration. Dev Dyn 235:711–722
16. Giampaoli S, Bucci S, Ragghianti M, Mancino G, Zhang F, Ferretti P (2003) Expression of FGF2 in the limb blastema of two Salamandridae correlates with their regenerative capability. Proc Biol Sci 270:2197–2205
17. Kumar A, Velloso CP, Imokawa Y, Brockes JP (2000) Plasticity of retrovirus-labelled myotubes in the newt limb regeneration blastema. Dev Biol 218:125–136

18. Yun MH, Gates PB, Brockes JP (2014) Sustained ERK activation underlies reprogramming in regeneration-competent salamander cells and distinguishes them from their mammalian counterparts. Stem Cell Rep 3:15–23
19. Cash DE, Gates PB, Imokawa Y, Brockes JP (1998) Identification of newt connective tissue growth factor as a target of retinoid regulation in limb blastemal cells. Gene 222:119–124
20. Blassberg RA, Garza-Garcia A, Janmohamed A, Gates PB, Brockes JP (2011) Functional convergence of signalling by GPI-anchored and anchorless forms of a salamander protein implicated in limb regeneration. J Cell Sci 124:47–56
21. Velloso CP, Simon A, Brockes JP (2001) Mammalian postmitotic nuclei reenter the cell cycle after serum stimulation in newt/mouse hybrid myotubes. Curr Biol 11:855–858
22. Balls M, Worley RS (1973) Amphibian cells in vitro. II. Effects of variations in medium osmolarity on a permanent cells line isolated from Xenopus. Exp Cell Res 76:333–336
23. Roy S, Gardiner DM, Bryant SV (2000) Vaccinia as a tool for functional analysis in regenerating limbs: ectopic expression of Shh. Dev Biol 218:199–205
24. Yun MH, Gates PB, Brockes JP (2013) Regulation of p53 is critical for vertebrate limb regeneration. Proc Natl Acad Sci U S A 110:17392–17397

Chapter 15

Culture and Transfection of Axolotl Cells

Jean-François Denis, Fadi Sader, Patrizia Ferretti, and Stéphane Roy

Abstract

The use of cells grown in vitro has been instrumental for multiple aspects of biomedical research and especially molecular and cellular biology. The ability to grow cells from multicellular organisms like humans, squids, or salamanders is important to simplify the analyses and experimental designs to help understand the biology of these organisms. The advent of the first cell culture has allowed scientists to tease apart the cellular functions, and in many situations these experiments help understand what is happening in the whole organism. In this chapter, we describe techniques for the culture and genetic manipulation of an established cell line from axolotl, a species widely used for studying epimorphic regeneration.

Key words Axolotl, Cell culture, Electroporation, Limb regeneration, Lipofection, Salamander, Transfection

1 Introduction

The ability to grow cells from multicellular organisms in vitro was invented by R.G. Harrison at the beginning of the twentieth century. Over a century after this development, tissue culture has become an intrinsic part of the experimental arsenal used in every modern laboratory studying molecular and cellular biology, biochemistry, and genetics whether for biomedical or fundamental reasons. The ability to grow cells in culture is important to help our understanding of how different cell types proliferate or respond to specific signals. Tissue culture is therefore an important complement to studying processes that occur in vivo. We are particularly interested in the process of epimorphic regeneration in vertebrates and the animals of choice for this are the urodele amphibians (the most commonly used models are the axolotl and the newt). Although cultures from human and mouse cells have been established with tremendous success, it is often a challenge to set up cell cultures from lower vertebrates like salamanders.

The first, and presently the only, cell line available from an *A*xolotl *L*imb is the AL-1 cell line (Fig. 1). It was established in the

Anoop Kumar and András Simon (eds.), *Salamanders in Regeneration Research: Methods and Protocols*, Methods in Molecular Biology, vol. 1290, DOI 10.1007/978-1-4939-2495-0_15, © Springer Science+Business Media New York 2015

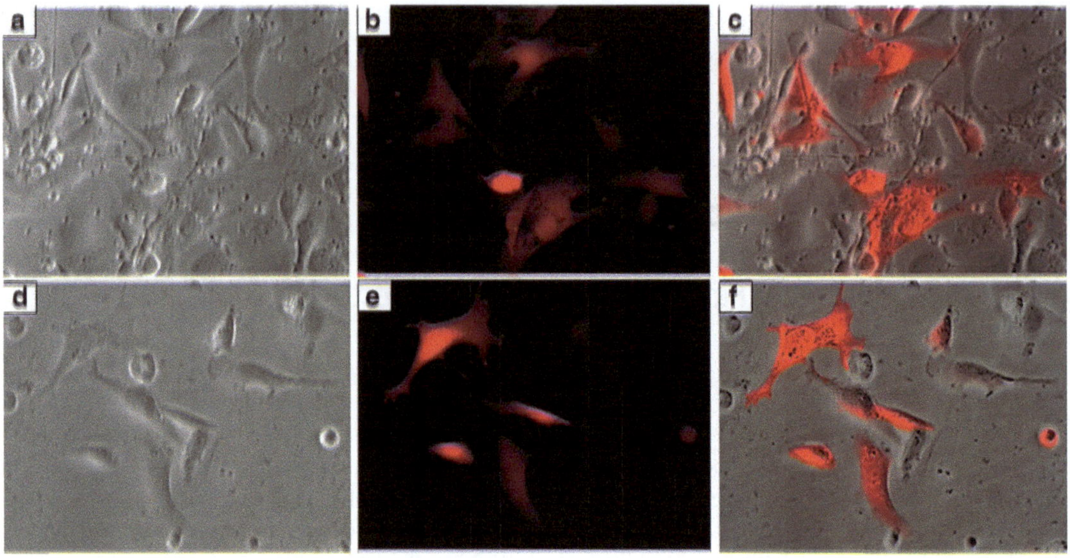

Fig. 1 AL-1 cells transfected with plasmid containing the CMV-mCherry expression cassette. (**a–c**) Transfection with 0.5 μg of plasmid using Lipofectamine 3000 at 37 °C for 4 h. (**a**) Bright-field view of transfected cells indicates that cells are attached and normal looking. (**b**) Same cells as (**a**) visualized with the mCherry filter. (**c**) Overlay of bright-field and mCherry from (**a**) and (**b**) indicates that about 40 % of cells are transfected. (**d–f**) Transfection using square wave electroporation of 5 μg of plasmid. (**d**) Bright-field view of transfected cells, less cells survive electroporation which explains the lower number of cells in the frame. (**e**) Same cells as (**d**) visualized with the mCherry filter. (**f**) Overlay of bright-field and mCherry from (**d**) and (**e**) shows that about 50 % of cells are transfected. The cells were photographed with a 20× objective

laboratory of David Gardiner and Susan Bryant at the University of California, Irvine, in the mid-1990s [1]. The AL-1 cells originate from migrated cells of a skin tissue explant from a normal axolotl limb. They display a fibroblastic morphology and attach and grow well on tissue culture plastic (Fig. 1). AL-1 cells have been used to test hormone response and pharmacological inhibitors [2] as well as viral vectors [1]. These cells have been important to help understanding the function of transcription factors and their downstream targets and to study tumor suppressor genes, such as p53 [3, 4]. In addition, the AL-1 cell line has been very valuable for testing plasmid constructs prior to electroporating them in vivo into regenerating limbs [5]. Recently, stable lines from AL-1 cells expressing fluorescent reporter genes have been developed by infecting them with pseudotyped retrovirus [6].

In this chapter we describe how to grow and maintain the AL-1 cell line as long-term cultures. Introduction of DNA into amphibian cells is difficult and we have screened a panel of commercially available transfection reagents to transfect AL-1 cells. Here we provide the details for the optimized transfection procedures for the transfection of plasmid DNA into AL-1 cells.

2 Materials

2.1 AL-1 Culture

1. Ultrapure water (e.g., we use Barnstead nanopure deionized water of 18 MΩ cm at room temperature).
2. Amphibian PBS (A-PBS): in 700 mL of ultrapure water, dissolve 8 g NaCl, 0.2 g KCl, 1.44 g Na_2HPO_4 anhydrous, and 0.24 g KH_2PO_4 anhydrous. Make up to 1 L with sterile water.
3. 100× insulin-transferrin-selenium (ITS; 1,000 mg/L insulin; transferrin 550 mg/L; sodium selenite 0.67 mg/L (Life Technologies, cat. no. 41400-045)). Keep stock solution at 4 °C.
4. 100× glutamine (200 mM stock): aliquot the stock solution and store at −20 °C.
5. 100× penicillin/streptomycin (PS; 10,000 units/mL of penicillin and 10,000 μg/mL of streptomycin): defrost the stock solution and aliquot as convenient (e.g., 2 mL aliquots if routinely preparing 200 mL of culture medium). Store at −20 °C.
6. Fetal bovine serum (FBS) or fetal calf serum (FCS).
7. Leibovitz's L-15 medium (L-15). Prepare a 64.8 % L-15 solution with sterile water and adjust the pH to 6.4 with 0.1 M HCl. Sterilize through a 0.22 μm filter.
8. L-15 complete medium: add 94 mL of L-15 prepared as in **item** 7, 5 mL FBS, 1× L-glutamine (1 mL of 100× stock), 0.2× ITS and P/S (*see* **Note 1**). Once the FBS, P/S, and ITS have been added, the final L-15 percentage is 60 %. One advantage of using L-15 medium is that it does not need to equilibrate with atmospheric CO_2 to maintain a constant pH. L-15 is buffered by its complement of salts, free-base amino acids, and galactose substituted for glucose to help maintain physiological pH control. Once the L-15 complete medium is prepared, it can be kept at 4 °C for 6 weeks without any adverse effects. Longer storage may result in adverse effect on cell growth and reattachment after passage.
9. Amphibian culture medium (AMEM; for growth in a CO_2 incubator): dispense 130 mL minimum essential medium (1× MEM), 20 mL of FCS, 2 mL of penicillin/streptomycin, 2 mL of glutamine and 2 mL of insulin in a sterile bottle (e.g., 75 cm^2 tissue culture flask; sterile glass bottle with a phenolic cap). Bring the volume up to 200 mL with sterile water. This dilution gives a medium of the appropriate osmolarity (approximately 225–250 mOsm) for amphibian cell culture.
10. Trypsin-EDTA working solution (0.25 % Life technologies™ cat. no. 25200-056): aliquot the stock solution and store at −20 °C. We keep a working aliquot at 4 °C for ease of use. Trypsin-EDTA will stay active for 3–4 weeks at 4 °C. Trypsin

is used at room temperature for no more than 3–4 min for complete cell detachment.

11. Freezing medium: mix 0.775 mL of L-15 medium with 75 μL of tissue culture grade dimethyl sulfoxide (DMSO, 7.5 %) and 150 μL of FBS (15 %).
12. Sterilizing filters (syringe filters and filter flask; 0.22 μm).
13. Amphibian cell culture incubator: set the incubator for amphibian cell cultures at 25–26 °C without CO_2. Use culture flasks which close tightly so that there is no media loss due to evaporation.

2.2 Equipment and Reagents for Transfection of AL-1 Cells

1. BTX Harvard apparatus ECM830 (capable of square wave electroporation).
2. Sterile 4 mm electroporation cuvettes (e.g., VWR cat. no. 89047-2010).
3. Lipofectamine® 3000 transfection reagent (Life technologies™, cat. no. L3000-015).
4. Opti-MEM® I reduced serum medium (Life technologies, cat. no. 310985-070).
5. Plasmid DNA.

2.3 Equipment and Reagents for Nucleofection of AL-1 Cells

1. Lonza nucleofector 2b device (basic model).
2. Ingenio 0.2 cm cuvettes (cat. no. 50121).
3. Cell line electroporation kit V (Lonza, cat. no. VCA-1003) or Ingenio electroporation solution (Mirus Bio LLC).
4. Fine-tip disposable pipettes.
5. Plasmid DNA.
6. 50 mL centrifuge tubes.

3 Methods

3.1 Thawing AL-1 Cells

1. Thaw cells rapidly by shaking vials in a 37 °C water bath until no ice is apparent.
2. When growing cells in the absence of CO_2, after **step 1**, add slowly 2 mL of complete L-15 to the freshly thawed cells.
3. Plate the cells in a small flask (35 mm or T-25) for optimal density.
4. Place the dish/flask in an incubator maintained at 25–26 °C without CO_2. Alternatively cells can be grown at room temperature if no incubator is available, but they will proliferate at a slower rate. Go to **step 8**.
5. When growing cells in a CO_2 incubator, after **step 1**, dilute the cell suspension with 8 mL of cold AMEM in a centrifuge tube.

6. Spin the cell suspension at 62 × *g* for 5 min. Decant the supernatant and resuspend the pellet in 5 mL of AMEM.
7. Transfer the cell suspension to a 75 cm^2 (or smaller) flask and place it in the amphibian cell incubator.
8. Change the medium on the following day.

3.2 Maintenance of AL-1 Cells

AL-1 cells can be grown either in L-15 complete medium at 26 °C in an incubator or in AMEM at 25 °C in a humidified CO_2 incubator. Split the cells 1:2 when they reach 80–85 % confluence, usually once a week (*see* **Note 2**).

1. Under sterile conditions, remove the medium within flask with a sterile pipette.
2. Rinse the AL-1 cells using A-PBS and remove PBS from flask.
3. Add 2 mL of 0.25 % trypsin.
4. Monitor the cells under an inverted microscope to see when they begin to detach.
5. When most cells have detached from the flask, neutralize the trypsin by adding four times the volume of L-15 complete medium (e.g., 8 mL of complete medium for 2 mL of trypsin) or AMEM.
6. Resuspend the cells by pipetting the medium up and down to break up clumps of cells and to make sure that no cells are left on the flask walls.
7. Transfer the cell suspension to a centrifuge tube (15 or 50 mL conical bottom tube).
8. Mix 100 μL of the cell suspension and 100 μL of 0.4 % trypan blue if cell counting is necessary.
9. Plate 10 μL in a hemocytometer and count the cells.
10. Centrifuge at 200 × *g* for 5 min at 4 °C to pellet the cells.
11. Remove the supernatant and discard it.
12. Add the desired volume of L-15 complete medium or AMEM (pipette up and down 4–5 times to break up clumps and disperse cells).
13. Plate the cell suspension.
14. If the cells are used for expansion, plate them in a flask; the target plating density should be 40 %.
15. If the cells are to be used for experimental procedures (e.g., effect of hormones or drug treatment on gene expression), plate them in the appropriate dish/well; the target plating density should be 60–75 %.
16. Check for cell attachment the following day.

3.3 Cryopreservation of AL-1 Cells

1. Harvest the cells when at 70–80 % density by trypsinization as described in Subheading 3.2, **steps 3–14**; one million cells should be used for freezing.
2. Resuspend the cell pellet in 1 mL of L-15 with 7.5 % DMSO and 15 % FBS.
3. Transfer to a 1 mL cryovial (any standard cryovial works fine).
4. Store the cryovials sequentially at −20 °C overnight inside a Styrofoam box to allow temperature to decrease slowly, then transfer the Styrofoam box containing the cells to a −80 °C freezer overnight. The cells can be stored at this temperature for 3–4 months without any adverse effects. Preferably, transfer the cells into liquid nitrogen for long-term storage.

3.4 Electroporation of AL-1 Cells

1. Place the microtubes and electroporation cuvettes on ice 30 min before the procedure.
2. L-15 complete medium, A-PBS, and the plasmids should be kept on ice during the entire procedure. Low temperature greatly helps cell survival.
3. Trypsinize the cells as described in Subheading 3.2, **steps 3–14**, and count.
4. Resuspend the cells using ice cold A-PBS.
5. Centrifuge at 200×*g* for 5 min at 4 °C to pellet the cells. Remove the supernatant and add A-PBS to obtain a concentration of 100,000 cells/100 μL.
6. Add to each tube containing the plasmid 100 μL of PBS containing the cells. The optimal plasmid amount is 1 μg but it can range from 0.5 to 5 μg. Addition of a fluorescent reporter expressing GFP or mCherry under the CMV promoter is useful to assess the transfection rate (typically addition of tenfold less of this plasmid is sufficient to assess transfection rate) (*see* **Note 3**).
7. Transfer the cell suspension to an electroporation cuvette.
8. Electroporate the cells at 200 V for 35 ms (*see* **Note 4**).
9. Quickly add 1 mL of L-15 complete medium to the transfected cells.
10. Plate the cells in a 12-well plate (one electroporation per well).
11. Repeat the **steps 5–10** for each transfection conditions.
12. Place the transfected cells in the incubator at 26 °C.
13. Change the medium the following day.
14. Check for cell attachment, survival, and transfection rate (Fig. 1d–f). Optimal expression should be seen after 24–48 h.

3.5 Transfection of AL-1 Cells with Lipofectamine® 3000

1. Use between 0.5 and 1 μg of plasmid DNA. This is the optimal amount, but a range between 0.5 and 2 μg works. A fluorescent marker can be used to assess transfection rate.
2. Trypsinize the cells as described in Subheading 3.2, **steps 3–14**, and count.
3. Dilute the cells in L-15 complete medium in order to obtain a concentration of 50,000 cells/0.5 mL.
4. Plate the cells: typically plate 50,000 cells in 0.5 mL per well in 24-well plates to obtain a confluence around 85 %; the cell density can range between 70 and 90 % but should not be lower or higher than these values to maintain good transfection rate.
5. Prepare the Lipofectamine® 3000 components for each transfection condition:
   (a) Add the desired amount of plasmid to 50 μL of Opti-MEM; 0.5–1 μg of plasmid DNA is the optimal amount but a range between 0.5 and 2 μg works. A fluorescent marker can be added to assess transfection rate.
   (b) Add 2 μL of P3000 reagent (part of Lipofectamine® 3000 kit) per 1 μg of plasmid.
   (c) In a second tube, mix 50 μL of Opti-MEM and 1.5 μL of Lipofectamine 3000.
   (d) Mix the content of the two tubes and incubate for 5 min at room temperature.
6. Remove the medium from the attached cells in the 24-well plate.
7. Rinse the cells once with A-PBS and then add 500 μL of L-15 medium containing 5 % FBS and 1× L-glutamine (without P/S).
8. Add the Lipofectamine 3000 solution from **step 5c**.
9. Incubate the cells at 37 °C for a maximum of 4 h (do not use a CO_2-enriched incubator). AL-1 will not survive well if kept longer in those conditions (*see* **Note 5**).
10. Transfer the cells to the 26 °C incubator.
11. Check for cell attachment and transfection rate. Expression of transfected constructs should be detectable after 24–48 h (*see* **Note 3**; Fig. 1a–c).

3.6 Transfection of AL-1 Cells by Nucleofection

1. Equilibrate 10 mL of complete AMEM in a tissue culture flask or a centrifuge tube in a CO_2 amphibian cell culture incubator for 20 min as cells must be transferred to this medium immediately after nucleofection. The final volume of the medium added depends on the culture flask or Petri dish used (*see* **Note 6**).

2. Warm up the nucleofector solution at room temperature.
3. Select the following program for AL-1 cells—program T-030.
4. Trypsinize the cells as described in Subheading 3.2, **steps 3–14**. A cell density of 70–80 % confluence is optimal.
5. Carefully remove all the supernatant with an aspirator and any small volume of medium remaining should be removed using a pipette tip.
6. Resuspend the cells in 100 μL of nucleofector solution and mix it gently without creating air bubbles.
7. Add 2–4 μg plasmid DNA and mix the solution.
8. Transfer the cell suspension to 0.2 cm electroporation cuvette carefully through the sides of the cuvette without creating any air bubbles (*see* **Note 7**).
9. Replace the cap of the cuvette and place it in the cuvette holder of the nucleofector device.
10. Run the program T-030.
11. Immediately after completion of the program, take out the cuvette and transfer the cell suspension to the equilibrated complete medium.
12. Place the tube in the cell culture incubator for 30 min to allow the cells to recover before plating them.
13. Nucleofected cells from each tube can be plated on 2× 25 cm^2 flasks or 3–4 35 mm cultures dishes.
14. Transfer the cells to the amphibian incubator.
15. Check for cell attachment and transfection rate. Expression of transfected constructs should be detectable after 24–48 h. Cell survival rate after nucleofection is between 30 and 50 %. Expression of marker genes (eGFP, RFP) in nucleofected cells is 60–80 % (*see* **Note 8**).
16. AL-1 cells can be cultured for several weeks without loss of marker gene expression. Heterogeneity in the level of gene expression is observed as for other methods of transfection (Fig. 1b, e).

4 Notes

1. Originally, AL-1 cells were grown in 10 % FBS with 1× ITS, but protocol optimization experiments showed that no difference in cell growth is observed by lowering FBS to 5 % and ITS concentrations to 0.2×.
2. AL-1 cells need to be maintained at 40–90 % confluence for optimal cell growth. If the density is too low, the cells will

proliferate very slowly. Over-confluent cells will stop growing and will not reattach after passage.

3. Optimal transfection is observed when plasmids are pure (we recommend using PureLink™ HiPure Plasmid Maxiprep kit) followed by extra purification with phenol/chloroform/isoamyl alcohol extraction (as described in Current Protocol) [7].
4. Use of other electroporation devices is feasible. However, the optimal conditions need to be determined based on the manufacturer's recommendation.
5. The method of transfection using the commercial transfection reagent, Lipofectamine® 3000, yields good transfection while allowing good cell survival when performed at 37 °C (Table 1). Optimal expression is observed around 72 h post-transfection.
6. Nucleofection has not been assessed in AL-1 cells grown in L-15 complete medium.

Table 1
Methods of transfection tested on AL-1 cells

Method of transfection	Result of transfection	Toxicity
Electroporation (square wave is necessary)	~50 % transfection rate	~50 % of cells don't reattach
Cytofectamine (Bio-Rad)	0 % (done at 26 °C)	None observed
Fugene-HD (Roche)	Less than 5 % (done at 26 °C)	None observed
Lipofectin (Invitrogen[a])	0 % (done at 26 °C)	Most cells died if left over 4 h
Lipofectamine 2000[a]	0 % (done at 26 °C)	Cells will die off if left over 4 h
Lipofectamine 2000[a]	~20 % if done at 37 °C 4 h	>50 % of cells will die within 72 h
Lipofectamine 3000[a]	0 % (done at 26 °C)	None observed
Lipofectamine 3000[a]	~50 % if done at 37 °C 4 h	Less than 5 % of cells detach
Endo-Porter (Gene Tools)	0 % (done at 26 °C) 4 and 16 h	Toxic if left on cells 16 h
Magnetofection PolyMag (Oz Biosc)	Less than 5 % at 26 °C	Depends on how much metal beads are used, the more beads are used, the more cells will die off
Nucleofection	60–80 %	30–50 % survival
Ca-Pi precipitation	0 % (done at 26 °C)	None observed

[a]N.B. for each method listed, various ratios of plasmid to reagents were tested and only the summary is presented in this table. The most efficient methods are the ones described in detail (i.e., electroporation, Lipofectamine 3000, and nucleofection)

7. Use about 5×10^5 cells per 0.2 cm cuvette. Due to the large size of amphibian cells, manufacturer's recommended number of cells is not required. Nucleofection can be successfully carried out also using a lower number of cells. However, under these conditions cell death is higher. Cuvettes with larger volumes can also be used for nucleofection of cells.
8. Efficiency of transfection using the Lonza kit and Mirus electroporation solution was assessed in parallel using various plasmids and comparable results were obtained.

References

1. Roy S, Gardiner DM, Bryant SV (2000) Vaccinia as a tool for functional analysis in regenerating limbs: ectopic expression of Shh. Dev Biol 218:199–205
2. Levesque M, Gatien S, Finnson K, Desmeules S, Villiard E, Pilote M, Philip A, Roy S (2007) Transforming growth factor: beta signaling is essential for limb regeneration in axolotls. PLoS One 2:e1227. doi:10.1371/journal.pone.0001227
3. Shaikh N, Gates PB, Brockes JP (2011) The Meis homeoprotein regulates the axolotl Prod 1 promoter during limb regeneration. Gene 484:69–74
4. Villiard E, Brinkmann H, Moiseeva O, Mallette FA, Ferbeyre G, Roy S (2007) Urodele p53 tolerates amino acid changes found in p53 variants linked to human cancer. BMC Evol Biol 7:180. doi:10.1186/1471-2148-7-180, 1471-2148-7-180 [pii]
5. Guimond JC, Levesque M, Michaud PL, Berdugo J, Finnson K, Philip A, Roy S (2010) BMP-2 functions independently of SHH signaling and triggers cell condensation and apoptosis in regenerating axolotl limbs. BMC Dev Biol 10:15. doi:10.1186/1471-213X-10-15
6. Whited JL, Tsai SL, Beier KT, White JN, Piekarski N, Hanken J, Cepko CL, Tabin CJ (2013) Pseudotyped retroviruses for infecting axolotl in vivo and in vitro. Development 140:1137–1146
7. Moore D, Dowhan D (2002) Purification and concentration of DNA from aqueous solutions. Curr Protoc Mol Biol, Chapter 2:Unit 2.1A. doi: 10.1002/0471142727.mb0201as59. John Wiley & Sons, Inc.

Chapter 16

Isolation and Culture of Neurospheres from the Adult Newt Brain

Liyakath Ali Shahul Hameed and András Simon

Abstract

Neural stem cells (NSCs) give rise to neurons in the adult brain and are possible targets in regenerative therapies. In vitro cultures of NSCs as neurospheres have been established from cells isolated from diverse species. Newts are exceptional regenerators among vertebrates. These animals are able to efficiently replace neurons following ablation of those by activation and subsequent differentiation of NSCs. Here we describe the method for isolating and culturing of NSCs from the newt brain both during self-renewing and differentiating conditions. Newt NSC culture provides a useful tool for functional studies of NSC fate with the potential of resulting in novel regenerative strategies.

Key words Primary culture, Neurosphere, Neural stem cells, Newt

1 Introduction

Newts have remarkable regeneration capacities among adult vertebrates [1], which extend also to the brain. In a series of experiments, it has been shown that the newt brain is able to activate sufficient cues to direct activated NSCs toward specific neuronal subtypes within an existing brain tissue [2–4]. As a complement to in vivo studies, in vitro cultures of NSCs offer additional opportunities to reveal regulatory mechanisms that control NSCs fate. A hallmark of NSCs is their potential to form neurospheres in culture. In cultures, NSC fate is easy to manipulate by over- and under-expression approaches and by treatment with small molecules. Such manipulations can be used to uncover critical regulatory mechanisms underlying proliferation and differentiation toward neurons. Recently, we used this approach to characterize the heterogeneity of newt NSCs isolated from defined brain regions at the molecular level and carry out functional manipulations of neurogenesis [5].

Here, we describe a detailed protocol for the formation and propagation of neurospheres from the adult newt brain (Fig. 1).

Anoop Kumar and András Simon (eds.), *Salamanders in Regeneration Research: Methods and Protocols*, Methods in Molecular Biology, vol. 1290, DOI 10.1007/978-1-4939-2495-0_16, © Springer Science+Business Media New York 2015

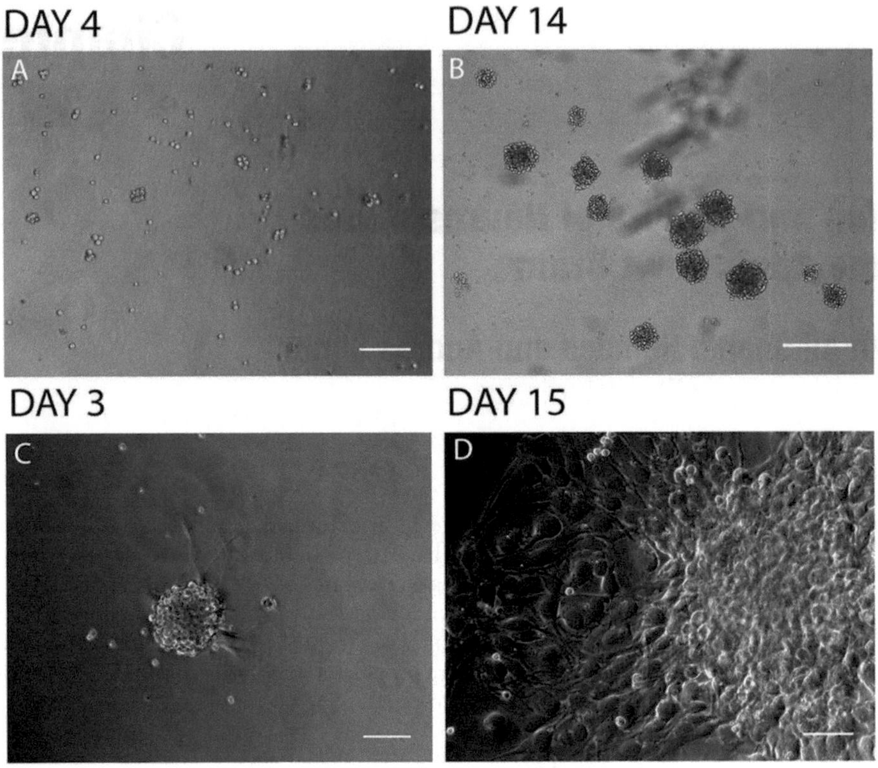

Fig. 1 Formation and differentiation of neurospheres. (**a**) Small spheres appearing 4 days after plating. (**b**) Spheres typically reach maximum size around day 14. (**c**) Plating neurospheres on PDL plates leads to attachment of the neurospheres to the substrate. (**d**) By day 15, differentiating neurospheres are apparent by bright-field microscopy

The basis of the procedure is the enzymatic digestion of newt brain into single cells. Among these single cells NSCs proliferate and form neurospheres as it has been described in other species [5]. We also outline how to promote adhesion of neurospheres to Poly-D-lysine-coated plates in order to induce differentiation of NSCs in defined culture medium (Fig. 2).

2 Materials

Prepare all solutions fresh before use. Dilute all culture medium to 66 % with Milli-Q water to adjust amphibian osmolarity (*see* **Note 1**), and filter using 0.22 μm syringe filter.

2.1 Dissection Tools and Culture Plates

1. Scalpel blade, forceps, and surgical scissors: Autoclave all instruments and sterilize with 70 % ethanol before usage (*see* **Note 2**).

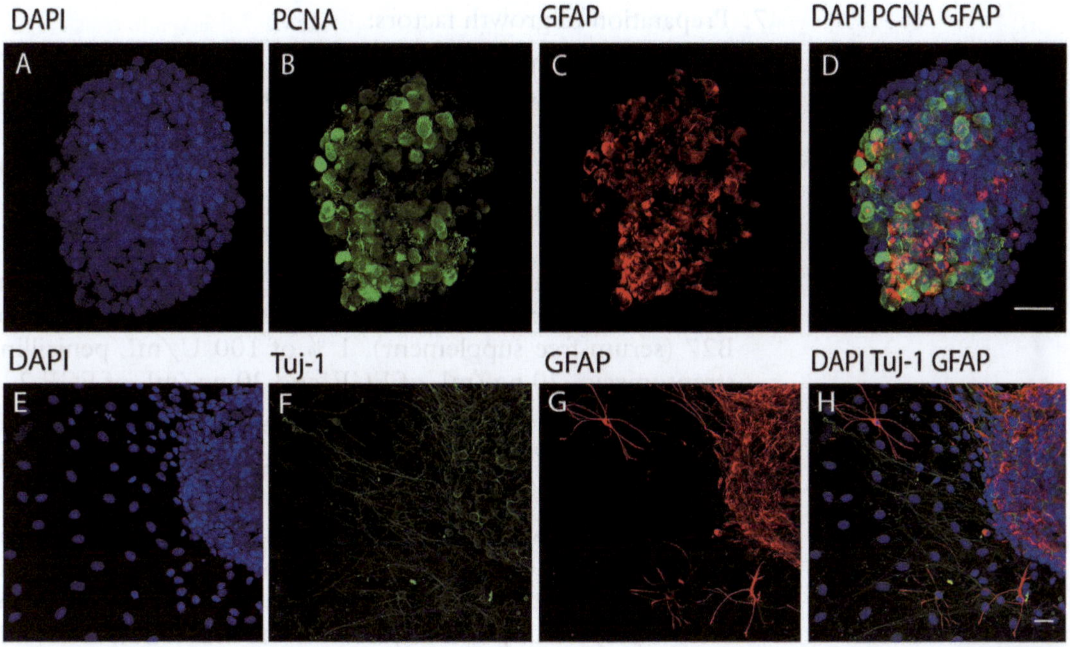

Fig. 2 Immunocytochemical characterization of neurospheres. (**a–d**) Growing neurospheres stained with stem cell marker GFAP and proliferation marker, PCNA. Most of the GFAP-expressing cells are marked by the proliferation marker, PCNA. (**e–h**) Differentiating neurospheres stained with the neuronal marker, Tuj-1, and the glial marker, GFAP. This image illustrates that most of the progeny are neurons and some progeny are glial cells

2. 8-well chamber slides, 24-well low-adherent plates, and 25 cm^2 cell culture flask.
3. 60 × 15 mm Petri dish, tissue strainer (40 μm).
4. Dissection microscope, light microscope, and fluorescence microscope.

2.2 Reagents for Neurosphere Culture

1. Anesthetic solution: Make 0.1 % of ethyl *p*-aminobenzoate (MS222) in tap water (*see* **Note 3**).
2. Dissection solution: L15 medium containing 1 % of 100 U/mL streptomycin-penicillin solution (*see* **Note 4**).
3. Enzyme solution A: Mix 30 U/mL papain and 40 μg/mL DNase in 1 mL of L15 medium.
4. Enzyme solution B: Dissolve 2 mg/mL of ovomucoid in 40 μg/mL DNase in containing L15 medium. Make 1 mL solution.
5. Enzyme digestion solution: Mix 250 μL of solution A and B, wait for 5 min (*see* **Note 5**).
6. Enzyme inhibitor solution: 2 mg/mL ovomucoid and 40 μg/mL DNase in L15.

7. Preparation of growth factors:
   (a) Dissolve 20 μg human FGF-2 in 1 mL of 5 mM Tris–HCl (pH 7.6) and sterilize through a 0.22 μm pore-size filter. Make 50 μL aliquots and store at −20 °C.
   (b) Dissolve 0.1 mg human EGF in 1 mL PBS containing 0.1 % bovine serum albumin. Make 50 μL aliquots and store at −20 °C.
8. Expansion medium: Dilute DMEM/F12/Glutamax medium to 66 % with sterile Milli-Q water and supplement with 2 % of B27 (serum-free supplement), 1 % of 100 U/mL penicillin-streptomycin, 20 ng/mL of EGF, and 20 ng/mL of FGF-2.
9. Differentiation medium: Neurobasal medium supplemented with 2 % of B27, 1 % of 100 U/mL penicillin-streptomycin, and 1 % of 2 mM Glutamax.
10. Poly-D-lysine coating (PDL): Prepare 1 mg/mL stock solution of PDL in Milli-Q water. The stock solution can be stored at −20 °C for up to 6 months. The working concentration of PDL is prepared by diluting 1 mL of stock solution in 50 mL of sterile PBS [6]. Coat the 8-well chamber slide by adding 250 μL of PDL on each well and incubate overnight at 25 °C. Wash the slides with Milli-Q water, three times for 5 min (*see* **Note 6**).

2.3 Reagents for Fixation and Staining

1. Sterile phosphate buffered saline (PBS), pH 7.4.
2. 4 % formaldehyde solution.
3. 0.5 % Tx-100 in PBS: Add 5.0 mL of Triton x-100 (Tx-100) to 995 mL of PBS.
4. Blocking solution: Add 5 mL of donkey serum to 95 mL of PBS containing 0.5 % Tx-100.
5. Primary antibodies such as anti-GFAP, PCNA, and Tuj-1. Dilute primary antibodies in blocking solution.
6. Alexa Fluor conjugated secondary antibodies. Dilute secondary antibodies in blocking solution.
7. Mounting medium: Dilute 5 mg/mL of DAPI (4′,6-diamino-2-phenylindole) by 1:1,000 in DAKO florescence mounting medium.

3 Methods

All the cell culture procedures are carried out under aseptic conditions.

3.1 Isolation of Brain Cells

1. Anesthetize newts in 0.1 % ethyl *p*-aminobenzoate (MS222). Typically it takes 20 min.

2. Sacrifice the animal and dissect out the brain under a dissection microscope.
3. Drop the extracted brain in to a 60 × 15 mm Petri dish containing dissection solution.
4. Remove carefully the meninges and blood vessels off the brain with fine forceps. Wash the tissue by transferring it to a new 60 × 15 mm Petri dish containing dissection solution (*see* **Note 7**).
5. Place the brain into fresh dissection medium and take the brain into culture hood. Use a scalpel to cut the brain into small (approximately 1 mm^3) pieces (*see* **Note 8**).
6. Collect the pieces of brain using a 1,000 μL pipette and transfer to the fresh 1.5 mL microtube. Allow the tissue to sink to the bottom by gravity, and remove the supernatant solution.
7. Add 500 μL freshly made enzyme digestion solution and incubate at room temperature for 1 h. During the incubation, stir the tube gently once every 15 min (*see* **Note 9**).
8. Add equal volume of enzyme inhibitor to stop the enzyme activity. Incubate 5 min at RT. Transfer the incubation mixture to 15 mL falcon tube and centrifuge for 5 min at 80 × *g* (*see* **Note 10**).
9. Remove supernatant and add 1 mL of fresh dissection medium to the precipitate.
10. Dissociate the newt brain cells by triturating with 1,000 μL pipette for 2 min slowly up and down (*see* **Note 11**).
11. Filter cell suspension through a 40 μm tissue strainer to remove any undissociated cells or tissue clumps.
12. Collect the filtered cells and centrifuge at 80 × *g* for 5 min.
13. Remove most of the supernatant, leaving 300 μL in the tube to make sure that cells are not lost.
14. Resuspend the cells in 4 mL of expansion medium and transfer them to a 25 cm^2 cell culture flask (*see* **Note 12**).

3.2 Neurosphere Formation and Differentiation

1. Plated isolated cells in the 25 cm^2 flask should be cultured in the incubator at 25 °C at 2 % CO_2 (*see* **Note 13**).
2. Change half of the medium once per week by centrifuging the suspension and then replacing it with half the volume of the medium (*see* **Note 14**).
3. After 14 days, collect the neurospheres by centrifuging at 80 × *g* for 5 min. These neurospheres can be either stained directly or induced to differentiate (*see* **Note 15**).
4. For differentiation, remove the supernatant, resuspend the spheres in the expansion medium, and then plate the spheres on PDL-coated slides.
5. After 24 h, when the spheres are attached, remove the expansion medium and change to neurobasal medium (*see* **Note 16**).

3.3 Immunostaining of Neurospheres and Differentiated Neurospheres

Between changing the solutions wait at least 3 min so that the neurospheres have time to resettle at the bottom of the well. Aspire the solutions by slightly tilting the plates and having the tip of the pipette touching the wall of the plate rather than the bottom [7]. Do not remove more than 75 % of the solutions.

1. Isolate the spheres by centrifugation and plate them into 24-well plates.
2. Add 4 % PFA and incubate for 15 min.
3. Remove PFA and wash three times with PBS for 5 min.
4. For proliferative cell nuclear antigen staining (PCNA), incubate neurospheres in 2 M HCl in 0.5 % Tx-100 in PBS for 20 min at 37 °C.
5. For GFAP and Tuj-1 staining, use 0.5 % Tx-100 in PBS for 15 min at RT.
6. Wash the neurospheres with PBS once for 5 min.
7. Change to blocking solution containing 5 % donkey serum in 0.5 % Tx-100 in PBS. Incubate for 30 min at RT.
8. Prepare primary antibodies in blocking buffer. Incubate with respective antibody overnight at 4 °C. Use anti-rabbit GFAP (1:500), anti-mouse PCNA (1:500), and anti-mouse Tuj-1 (1:500).
9. Wash the neurospheres with PBS once for 5 min.
10. Incubate neurospheres with Alexa Fluor conjugate secondary antibody, either Alexa Fluor 488 (1:500) or Alexa Fluor 594 (1:500) for 2 h at RT.
11. Wash the neurospheres with PBS once for 5 min.
12. Remove maximum PBS as possible, then collect the neurosphere with 1,000 μL pipette and place them on 76×26 mm glass slide. Add mounting medium containing DAPI on the slide and mount with a 24×50 mm coverslip.
13. For immunostaining differentiated neurospheres, follow all the **steps 1–12**, and omit the centrifugation steps.
14. Observe the samples under a fluorescence microscope with appropriate filters.

4 Notes

1. Amphibian osmolarity is 225±5 mOsmol. Therefore, the culture medium is diluted to 66 % with Milli-Q water to match the osmolarity of the newt cells.
2. It is important to remove any remaining ethanol from the surgical tools by rinsing them with sterile PBS before touching the tissue. Traces of ethanol will act as a fixative for the tissue.

3. MS222 should be dissolved in tap water, do not use Milli-Q water.
4. Dissection solution should be kept cold during preparation of the tissue explant; therefore, keep the solution on ice. The antibiotics, penicillin, and streptomycin prevent the bacterial contamination of the cell culture medium.
5. Prepare this solution just before use.
6. PDL coating can also be done at 37 °C for 2 h. PDL-coated slides can be kept maximum for a week before using.
7. Blood vessels are a source of contamination and should be removed from the culture.
8. Smaller pieces of tissue will dissociate better and yield higher number of neurospheres.
9. Check the enzymatic digestion every 15 min and stir gently to mix the solution. The solution becomes turbid due to digestion. Make sure that there are no clumps forming during digestion. Clumping usually occurs due to the release of DNA from dead cells.
10. Digestion time varies depending on the type of tissue and concentration of enzyme. Vigorous digestion leads to fewer viable cells. Digestion is one of the most critical steps, which, if not performed correctly, would substantially reduce the number of viable cells. Also, avoid formation of air bubbles; excessive air bubbles decrease cell viability.
11. NSCs are of ventricular origin. During the digestion of the tissue, cells from the ventricular zone are released first and later other cells start to dissociate. Therefore, it is not necessary to completely digest the entire tissue.
12. The optimal cell density for neurosphere formation is about 5,000–10,000 cells/mL but may be lowered for clonal analyses.
13. Optimal culturing temperature is 25 °C; however, the cultures tolerate temperature between 20 and 27 °C.
14. Instead of changing the medium completely, it is advised to change only half of the medium at the time because prosurvival factors produced by the cells may be present.
15. Small-sized neurospheres start to appear on day 4, grow bigger over time, and reach maximum size (100–300 μm) by 2–3 weeks (Fig. 1).
16. Just removing the growth factor can be sufficient to induce differentiation of neurospheres. Defined medium leads to more robust differentiation toward various lineages. Over time, neurospheres start to differentiate and produce progenies. These cells can be stained for differentiation markers (Fig. 2).

Acknowledgment

This work was supported by a grant from the Swedish Research Council to AS.

References

1. Brockes JP, Kumar A (2008) Comparative aspects of animal regeneration. Annu Rev Cell Dev Biol 24:525–549
2. Berg DA, Kirkham M, Beljajeva A, Knapp D, Habermann B, Ryge J, Tanaka EM, Simon A (2010) Efficient regeneration by activation of neurogenesis in homeostatically quiescent regions of the adult vertebrate brain. Development 137:4127–4134
3. Berg DA, Kirkham M, Wang H, Frisen J, Simon A (2011) Dopamine controls neurogenesis in the adult salamander midbrain in homeostasis and during regeneration of dopamine neurons. Cell Stem Cell 8:426–433
4. Parish CL, Beljajeva A, Arenas E, Simon A (2007) Midbrain dopaminergic neurogenesis and behavioural recovery in a salamander lesion-induced regeneration model. Development 134:2881–2887
5. Kirkham M, Hameed LS, Berg DA, Wang H, Simon A (2014) Progenitor cell dynamics in the newt telencephalon during homeostasis and neuronal regeneration. Stem Cell Reports 2:507–519
6. Ortega F, Costa MR, Simon-Ebert T, Schroeder T, Gotz M, Berninger B (2011) Using an adherent cell culture of the mouse subependymal zone to study the behavior of adult neural stem cells on a single-cell level. Nat Protoc 6:1847–1859
7. Sasaki R, Aoki S, Yamato M, Uchiyama H, Wada K, Ogiuchi H, Okano T, Ando T (2010) A protocol for immunofluorescence staining of floating neurospheres. Neurosci Lett 479:126–127

Chapter 17

Methods for Axolotl Blood Collection, Intravenous Injection, and Efficient Leukocyte Isolation from Peripheral Blood and the Regenerating Limb

Ryan J. Debuque and James W. Godwin

Abstract

The vertebrate immune system comprises both adaptive and innate immune cells with distinct functions during the resolution of inflammation and wound healing after tissue injury. Recent evidence implicates a requirement for innate immune cells from the myeloid lineage during the early stages of limb regeneration in the Mexican axolotl. Understanding the functions of innate and adaptive immune cells in the axolotl has been hampered by a lack of approaches to isolate and analyze these cells. Here we describe a protocol to isolate myeloid cells from the regenerating axolotl limb that incorporates intravenous delivery of physiological labels. In addition we provide a protocol to enrich for leukocytes in the peripheral blood. These protocols produce single-cell suspensions that can be analyzed using flow cytometry or sorted into specific subsets using fluorescent-activated cell sorting (FACS). FACS is a routine approach to sort cells based on their physical characteristics as well as their cell surface antigen repertoire. Isolated cell populations can then be analyzed in a wide range of downstream assays to facilitate a greater understanding of leukocyte biology in the axolotl.

Key words Salamander, Leukocyte biology, Myeloid cell, Flow cytometry and regeneration

1 Introduction

Urodele amphibians possess the remarkable capacity to resolve injury in a range of clinically relevant organs such as the heart, brain, and spinal cord as well as regenerate structures such as limbs and tails [1]. Resolution of tissue injury triggers an inflammatory response that is mediated by cells of the immune system, a mechanism conserved between mammals and vertebrates with robust regenerative abilities [2–6]. Cells of the immune system can be divided into two separate branches that include the evolutionarily ancient "innate" immune cell populations and the more recently evolved "adaptive" cell types. Adaptive immune cells comprise of T and B lymphocytes. Innate immune cells include phagocytic cells (monocyte, macrophage, and neutrophils), natural killer (NK) cells, basophils, eosinophils, mast cells, and platelets.

Anoop Kumar and András Simon (eds.), *Salamanders in Regeneration Research: Methods and Protocols*, Methods in Molecular Biology, vol. 1290, DOI 10.1007/978-1-4939-2495-0_17, © Springer Science+Business Media New York 2015

Macrophages, in addition to their role as professional phagocytes, also function as a source of both pro-inflammatory and anti-inflammatory signals during tissue homeostasis and wound healing in mammals [7–9]. Studies in the Mexican axolotl (*Ambystoma mexicanum*) have demonstrated that myeloid cells accumulate in the blastema during the early stages of limb regeneration and are most abundant at 4 days postamputation [10]. Systemic macrophage depletion by clodronate liposome delivery prior to amputation results in the failure of limb regeneration. This is characterized by deregulation of key regenerative genes and extensive fibrosis [10]. Interestingly this phenotype can be rescued after the endogenous macrophage population is replenished and the limb is reamputated, suggesting that there is a temporal requirement during the early stages of axolotl limb regeneration [10].

Delineating the roles of macrophages and other leukocyte subsets during axolotl limb regeneration is a critical step in understanding what molecular pathways permit efficient tissue injury resolution and regeneration. However, understanding the molecular basis for macrophage dependence and the functions of other leukocytes during limb regeneration has been impeded by the lack of established methods to isolate these cells. Here we describe a protocol to isolate myeloid cells, specifically monocytes and macrophages from the regenerating axolotl limb. This protocol incorporates intravenous delivery of reagents used to label macrophages based on their physiological function. The process of intravenous injection has yet to be formally described in the axolotl and we provide a step-by-step guide for this procedure. In addition we detail a protocol to magnetically deplete red blood cells and enrich for leukocytes within the circulating blood. Together these methods can be utilized in the axolotl to study leukocyte biology during homeostasis, inflammation, injury, and regeneration.

2 Materials

2.1 General Reagents, Equipment, and Consumables

1. Microsurgery scissors and forceps.
2. Disposable Pasteur pipettes, 35 mm Petri dishes.
3. 15 and 50 mL conical tubes and centrifuge tubes.
4. Cell strainers with 40 and 100 μm pore size.
5. 5 mL round-bottom FACS tubes with 40 μm cell strainer cap.
6. Poly-L-lysine microscope slides.
7. Mounting medium.
8. 0.5 M Ethylene diaminetetraacetic acid (EDTA), pH 8.0.
9. 10× Hank's balanced salt solution (HBSS).
10. HBSS-EDTA solution (0.7×): To prepare 200 mL solution, add 14 mL 10× HBSS to 184 mL sterile water. Add 2 mL 5 M EDTA to this solution.

11. Goat serum.
12. Blocking buffer solution: To prepare 50 mL add 1 mL goat serum to 49 mL HBSS-EDTA solution.
13. Holtfreter's stock solution: Dissolve 3.46 g of NaCl, 0.05 g of KCl, 0.1 g of $CaCl_2$, and 0.2 g of $NaHCO_3$ in 1 L of water.
14. Holtfreter's working solution (20 %): Dilute stock Holtfreter's solution 1:5 using sterile water.
15. Stereomicroscope with fluorescence attachment.
16. Microinjector with accessories.
17. Capillary glass filaments with an outer diameter of 0.94 mm.

2.2 Axolotl Surgery

1. 0.2 % ethyl-3aminobenzoate methanesulfonate salt (Tricaine). For 1 L solubilize 2 g in 1 L Holtfreter's working solution. Adjust pH to 7–7.5. Store at 4 °C away from light.
2. 2 MDa dextran-rhodamine: Solubilize 5 mg in 1 mL of 1× HBSS, aliquot, and store at −20 °C.

2.3 Axolotl Limb Blastema Isolation and Tissue Dissociation

1. 2 mg/mL collagenase II. For 10 mL solution, add 20 mg collagenase II to 10 mL HBSS-EDTA solution.
2. 100 mg/mL collagenase/dispase. Use at a 1:100 dilution.
3. gentleMACS™ C-Tubes (Miltenyi Biotec).
4. gentleMACS™ Dissociator (Miltenyi Biotec).

2.4 Antibody Staining and Flow Cytometry

1. 5 mg/mL 4′,6-diamidino-2-phenylindole (DAPI). Use at a 1:2,000 dilution.
2. 0.5 mg/mL Isolectin B4. Use at a 1:50 dilution.
3. Axolotl specific anti-Pan Ig. A kind gift from Françoise Salvadori. Use at a 1:400 dilution.
4. 2 mg/mL Streptavidin, Alexa Fluor® 488 Conjugate. Use at a 1:200 dilution.
5. 2 mg/mL Anti-mouse IgG (H+L), Alexa Fluor® 647 Conjugate. Use at a 1:1,000 dilution.
6. Standard flow cytometer and FACS machine.

2.5 Modified Giemsa-Wright Stain

1. Cytospin centrifuge (Thermo Scientific or similar).
2. 133 mM dibasic sodium phosphate anhydrous (Na_2HPO_4). Dissolve 18.89 g Na_2HPO_4 in 1 L of deionized water.
3. 133 mM monobasic potassium phosphate (KH_2PO_4). Dissolve 9.08 g KH_2PO_4 in 1 L of deionized water.
4. 133 mM Sorenson's buffer pH 6.8. To prepare 100 mL, add 43.75 mL of 133 mM Na_2HPO_4 to 56.25 mL 133 mM KH_2PO_4. Final pH should be assessed using a pH meter.

5. May Grünwald Stain. For 30 mL working dilution add 3.95 mL of May Grünwald Stain to 26.05 mL of 133 mM Sorenson's buffer pH 6.8.
6. Giemsa Stain. For 30 mL working dilution add 3.95 mL of Giemsa Stain to 26.05 mL of 133 mM Sorenson's buffer pH 6.8.

2.6 RNA Synthesis, cDNA Synthesis, and Quantitative Polymerase Chain Reaction

1. RNAqueous®-4 PCR Kit (Life Technologies).
2. SuperScript® VILO™ cDNA Synthesis Kit (Life Technologies).
3. SYBR green.
4. Standard thermal cycler.

2.7 Isolation of Circulating Blood in the Axolotl

1. 25G SURFLO® winged infusion set (Terumo Medical Products).
2. 5 mL syringe.

2.8 Magnetic Separation Reagents Using EasySep™

1. 10× sodium dithionite. Add 0.95 g sodium dithionite to 5 mL HBSS-EDTA solution.
2. EasySep™ magnet (STEMCELL Technologies).

3 Methods

It is recommended to prepare the necessary equipment and reagents prior to performing any of the methods described in this section. Working dilutions of any of the reagents described should be prepared in advance unless indicated otherwise.

3.1 Time Considerations

The protocols covered in Subheadings 3.4–3.7 (Fig. 1) span over a 5-day period with animal injections performed on the first (Subheading 3.4) and third (Subheading 3.5) days. Blastemas (the mass of progenitor cells which gives rise to the new limb) are harvested on the fifth day (4 days postamputation) (Subheading 3.6). Blastemas are immediately processed into single-cell suspensions after which cells can be analyzed on a flow cytometer or sorted using a FACS machine. The estimated time taken between Subheadings 3.6 and 3.7 is approximately 4–5 h. Blastema isolation, tissue mincing, and bone removal require approximately 30–90 min depending on the number of blastemas to be processed. The enzymatic digestion and cell suspension filtration require approximately 90 min. Antibody staining requires approximately 60 min. This estimation does not include the time required for data acquisition on the flow cytometer or any methods thereafter. In addition this estimation does not include the time taken to prepare any equipment, materials, and reagents.

The following red blood cell depletion protocol requires approximately 90–120 min to complete. The estimated time for

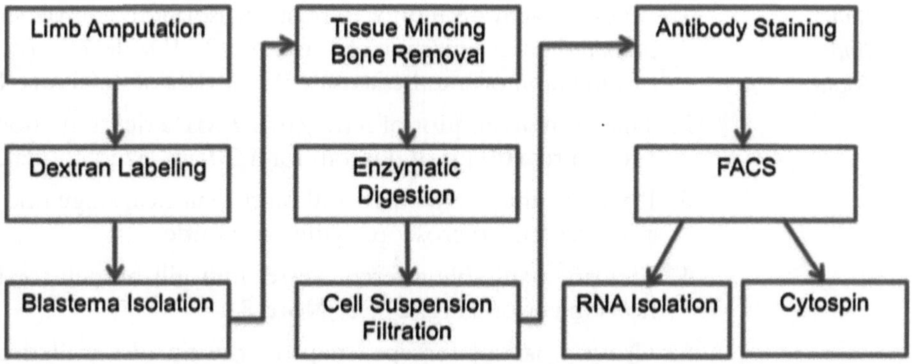

Fig. 1 Flowchart of the key steps performed to isolate and validate macrophages in the regenerating axolotl limb. The time taken between the limb amputation and FACS is 5 days with surgery occurring on the first and third days and tissue harvesting, processing, staining, and FACS all occurring on the fifth day

isolating blood using the winged infusion set is approximately 30–60 min depending on the number of animals used. Sodium dithionite treatment and magnetic depletion of red blood cells is estimated to take 30–60 min depending on the number of times each sample is put through the EasySep™ magnet.

3.2 Ethics Approval

Before proceeding with any animal experiments including the methods described below, please ensure that the necessary animal ethics approval is granted.

3.3 Animal Maintenance and Usage

Axolotls (*Ambystoma mexicanum*) are maintained in 20 % Holtfreter's solution at 19–22 °C on a 12-h light, 12-h dark cycle. To ensure enough cells are obtained for analysis, it is recommended that the animals used be at least 1 year of age and measure between 15 and 20 cm (snout to tail).

3.4 Axolotl Limb Amputation

1. Anesthetize axolotl in 0.2 % Tricaine. The time taken for an axolotl to become completely anesthetized is size dependent. This can take approximately 10–20 min. An anesthetized axolotl will not react to a skin pinch performed with a pair of surgical forceps (*see* **Note 1**).
2. Place the anesthetized axolotl onto the surgical stage lying on its side.
3. Secure the forelimb with a pair of surgical forceps. Using a pair of surgical scissors, amputate 1 mm below the wrist joint (for a distal amputation) or 1 mm below the elbow joint (for a proximal amputation) (*see* **Note 2**).
4. Place the axolotl into a fresh tank of 20 % Holtfreter's solution. It takes approximately 15–30 min for an anesthetized axolotl of this size to regain its consciousness.

3.5 Intravenous Injection of High Molecular Weight Dextran

1. Approximately 48 h prior to blastema isolation, animals should be prepared for i.v. injection of 5 mg/mL 2 MDa dextran-rhodamine by induction of anesthesia using 0.2 % Tricaine (*see* **Note 1**).
2. Thaw out an aliquot of 5 mg/mL 2 MDa dextran-rhodamine. Use up to 100 μg of dextran for a 20 cm-sized animal.
3. Place the anesthetized axolotl on the surgical stage underneath a fluorescent microscope lying on its side.
4. Identify a suitable afferent vessel in a gill branch for intravenous injection (Fig. 2) (*see* **Note 3**).
5. Using a pair of forceps, snap off the tip of a pulled capillary glass filament. The diameter of the capillary glass filament should closely match the diameter inside the afferent vessel chosen for injection.

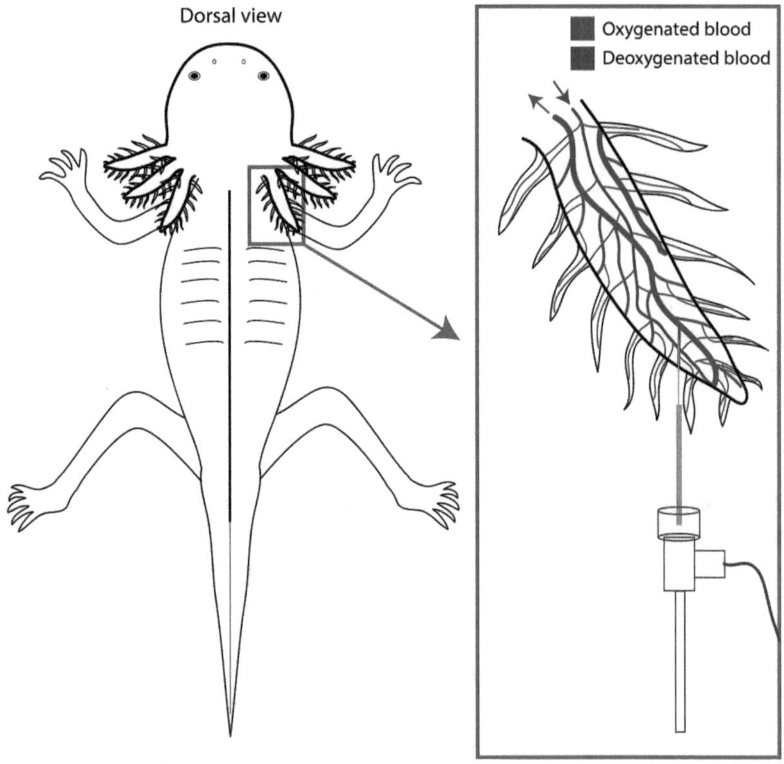

Fig. 2 A representative diagram depicting intravenous delivery of soluble reagents in the axolotl. The gills of the aquatic axolotl allow easy identification of the return flow of oxygenated blood flow back to the heart in afferent vessels before distribution to peripheral tissues. Under a dissection microscope, afferent vessels can be identified and capillary glass filaments (shown in *green*) can be used to deliver drugs or other compounds directly into the bloodstream (color figure online)

6. Load capillary glass filament with 5 mg/mL 2 MDa dextran-rhodamine using a 10 μL pipette tip. Ensure that no air bubbles are present within the injection volume (*see* **Note 4**).
7. Load the capillary glass filament into the microinjector.
8. Perform intravenous injection into the afferent vessel under a dissection microscope. Set injection pressure to 20 psi and slowly eject 5 mg/mL 2 MDa dextran-rhodamine from the glass needle into the target vessel using small pulses (5–10 ms) (*see* **Note 5**).
9. Slowly withdraw capillary glass filament from afferent vessel minimizing any damage to the vessel or surrounding tissue.
10. Slide a gloved finger over the injection site or hold it there for up to 30 s to accelerate clotting. This will aid in sealing the vessel at the entry point with skin mucus and minimize additional blood loss.
11. Using a fluorescent stereomicroscope, visualize a blood vessel in the tail to confirm delivery of the fluorescent 5 mg/mL 2 MDa dextran-rhodamine. Cells lining the vessels and within the bloodstream will be fluorescent for approximately 4–6 h postinjection.
12. Place the axolotl into a fresh tank of 20 % Holtfreter's solution for recovery in quiet individual housing for at least 24 h.

3.6 Blastema Isolation and Tissue Dissociation

1. 48 h after dextran labeling, anesthetize the amputated axolotl using 0.2 % Tricaine (*see* **Note 1**).
2. Cool C-tubes on ice. One C-tube is capable of processing two to four blastemas.
3. Place the anesthetized axolotl on the surgical stage lying on its side.
4. Use a pair of forceps to secure the forelimb. Cut 2 mm below the original line of amputation using a pair of surgical scissors.
5. Briefly wash any blood from the harvested blastema with HBSS-EDTA solution.
6. Place the isolated blastema onto a 35 mm tissue culture dish sitting on an elevated angle containing 3 mL of 2 mg/mL collagenase II in it. Harvest additional blastemas if required.
7. Use two forceps to carefully remove any pieces of bone or cartilage from the blastema. Grip the bone/cartilage with a pair of forceps and remove any tissues attached using a fresh pair of sterilized surgical scissors.
8. Finely mince the blastema tissue with surgical scissors into 0.2–0.5 mm-wide pieces.
9. Transfer the minced blastema pieces into a C-tube along with the 3 mL of 2 mg/mL collagenase II in the 35 mm tissue culture dish (*see* **Note 6**).

10. Rinse the tissue culture dish with an additional 3 mL of collagenase II. Transfer into the C-tube to ensure that all cells are collected.
11. Add enough collagenase II to the C-tube to a final volume of 10 mL.
12. Add 100 μL of collagenase/dispase (100 mg/mL) to the 10 mL 2 mg/mL collagenase II solution for a final concentration of 1 mg/mL (*see* **Note 7**).
13. Cap tube with lid, invert, and swirl gently to align the tissue to the plastic blades of the C-tube.
14. Place the inverted C-tube onto the gentleMacs™ apparatus ensuring that the C-tube is locked in place.
15. Run C-tube Heart 1 protocol. Maintain inverted position and place on ice. Process additional C-tubes if required (*see* **Note 8**).
16. Wrap C-tubes with foil and place on a roller at room temperature. Leave for 1 h to digest (*see* **Note 9**).
17. Run C-tube Heart 2 protocol and place tubes on ice with the lid facing upward. Process additional C-tubes if required.
18. Transfer the tissue suspension within each C-tube into a fresh 50 mL tube by passing through a 100 μm cell strainer (*see* **Note 10**).
19. Rinse the C-tube twice with 10 mL HBSS-EDTA solution and pass through a 100 μm cell strainer. The final volume should be 30 mL.
20. Centrifuge the cell suspension at 350 × *g* for 15 min at 4 °C.
21. Aspirate supernatant and resuspend the pelleted cells with blocking buffer solution at a final concentration of no more than 1×10^6 cells/100 μL.

3.7 Live-Cell Antibody Staining for Myeloid Cells from the Limb Blastema

1. All staining reactions are conducted in 1.7 mL microcentrifuge tubes (*see* **Note 11**).
2. Block cells on ice away from light for at least 20 min.
3. Prepare 1.7 mL microcentrifuge tubes with the correct dilution of primary antibody: 0.5 mg/mL biotinylated Isolectin B4 (use at a 1:50 dilution).
4. Add 100 μL to each individual 1.7 mL microcentrifuge tube. Mix by flicking or pipetting up and down.
5. Incubate on ice away from light for at least 20 min (*see* **Note 12**).
6. Wash any unbound primary antibodies with 900 μL of wash solution (HBSS-EDTA) and mix thoroughly and centrifuge at 350 × *g* at 4 °C for 5 min (*see* **Note 13**).
7. Prepare a working dilution of combined secondary antibodies: 2 mg/mL Streptavidin Alexa Fluor® 488 (1:200).

8. Aspirate supernatant and resuspend pellet in 100 μL of 1:200 Streptavidin Alexa Flour® 488. Incubate on ice, away from light for a minimum of 10 min (*see* **Note 14**).
9. Wash any unbound secondary antibodies with 900 μL of wash solution, mix thoroughly, and centrifuge at 4 °C for 5 min (*see* **Note 13**).
10. Prepare a working dilution of DAPI (1:2,000).
11. Aspirate supernatant and resuspend pellet in 1:2,000 DAPI at a final concentration of 1×10^6 cells/100 μL. Transfer stained cell suspension to a 5 mL FACS tube by passing cells through a 40 μm filter.
12. Proceed to analyze the single-cell suspensions using a flow cytometer. Alternatively cells can be sorted for applications in downstream assays using FACS. A gating strategy and the expected results using this method are shown in Figs. 3 and 4, respectively.

3.8 Cytospin

1. Using a FACS machine, isolate 10–50,000 cells per individual population into 1.7 mL microcentrifuge tubes.

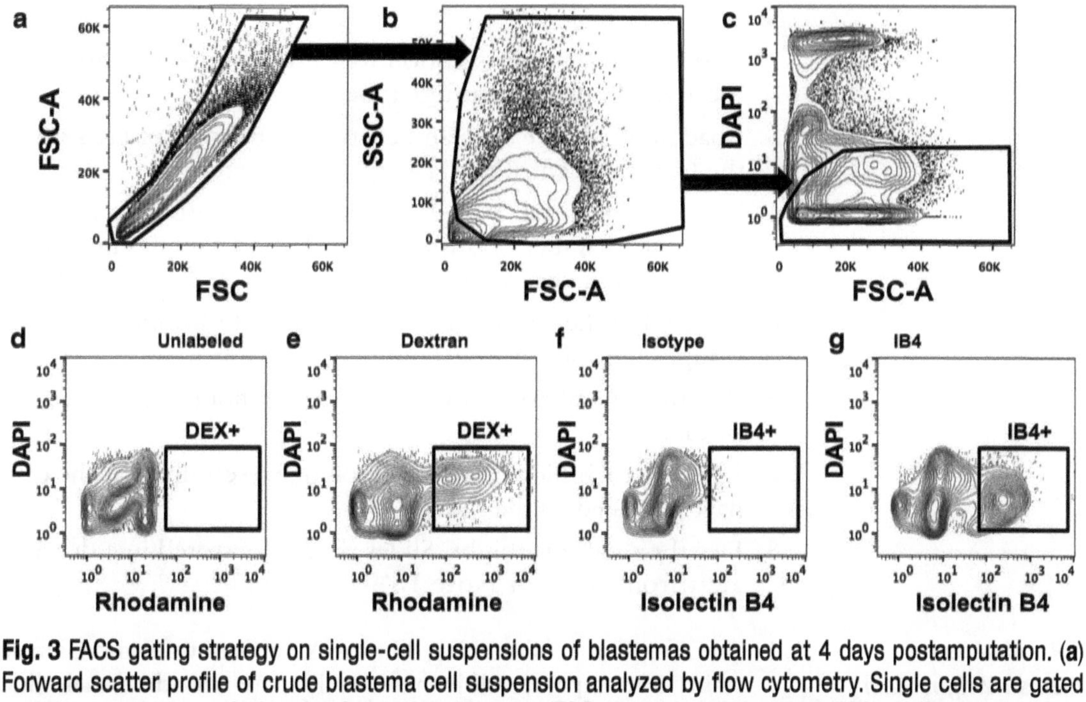

Fig. 3 FACS gating strategy on single-cell suspensions of blastemas obtained at 4 days postamputation. (**a**) Forward scatter profile of crude blastema cell suspension analyzed by flow cytometry. Single cells are gated on for downstream analysis using forward scatter area (FSC-A) and forward scatter (FSC). (**b**) Scatter profile of single cells and tissue debris. Tissue debris are gated out using side scatter area (SSC-A) and forward scatter area (FSC-A). (**c**) Scatter profile of nonviable cells and non-debris. Nonviable cells are excluded by DAPI staining as dead cells become permeable allowing DAPI to bind to the cell's nucleus. (**d–g**) Single-stain controls for dextran-rhodamine (DEX) and Isolectin B4 (IB4). (**d**) Detection of DEX+ cells in unlabeled (negative control). (**e**) Detection of DEX+ cells in positive control. (**f**) Detection of IB4+ cells in isotype (negative control). (**g**) Detection of IB4+ cells in positive control

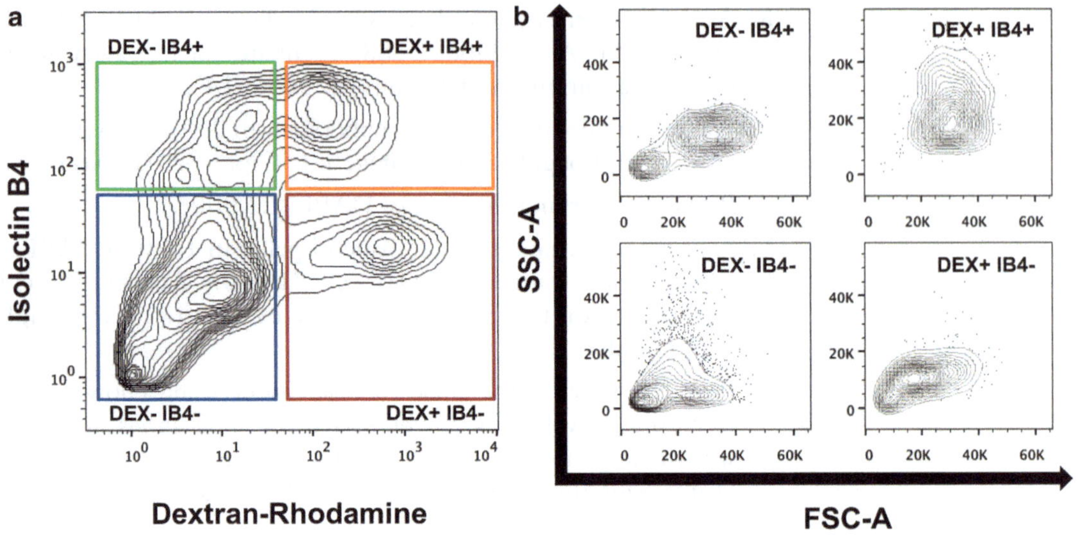

Fig. 4 Detection of myeloid cells in the regenerating axolotl limb at 4 days postamputation by flow cytometry. (**a**) Four distinct populations can be detected when cells are co-labeled with DEX and IB4: DEX− IB4−, DEX− IB4+, DEX+ IB4+, and DEX+ IB4+. (**b**) Side scatter area (SSC-A) and forward scatter area (FSC-A) profiles of each of the four individual populations are shown individually as indicated

2. Add blocking buffer solution to a final volume of 350 μL to the 1.7 mL microcentrifuge tubes.
3. Load poly-L-lysine slide and chamber into appropriate slots of the Cytospin centrifuge.
4. Transfer the total 300 μL of single-cell suspension into a Cytospin chamber. Transfer any additional samples into separate Cytospin chambers if required.
5. Place lid of the Cytospin centrifuge over the samples and centrifuge at 1,600 × *g* for 10 min.
6. Remove chambers from their slides without contacting the smears.
7. Examine each slide under a microscope to ensure that the cells have annealed to the slide.
8. Dry the slides overnight. Slides should be stored in a dry area away from light (*see* **Note 15**).

3.9 Modified Giemsa-Wright Stain

1. Transfer 30 mL of working dilution May-Grunwald and Giemsa stains into separate slide mailers. Prepare an additional slide mailer containing distilled water for the subsequent wash steps.
2. Transfer the poly-L-lysine slides into the slide mailer containing the May-Grunwald staining solution. Leave for 5 min at room temperature.
3. Wash by transferring the slides into a slide mailer containing distilled water. Ensure that any precipitation is removed.

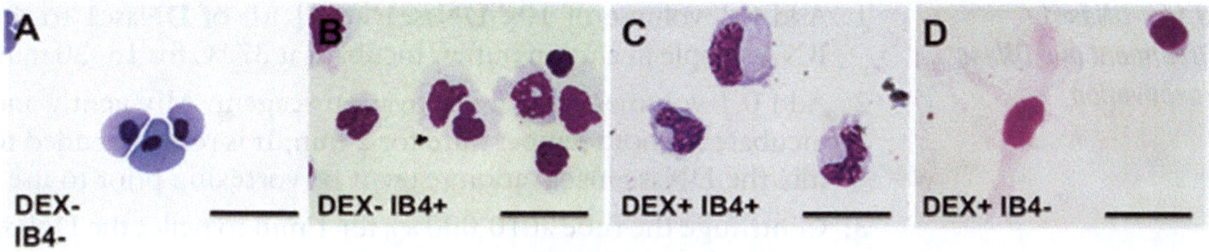

Fig. 5 Modified Giemsa-Wright stains on sorted cell populations. (**a**) IB4– Dex– cells consist predominantly of red blood cells and nonmyeloid cells. (**b**) IB4+ Dex– cells consist of a heterogeneous population of myeloid cells including cells with mononuclear phagocyte morphology (monocytes and macrophages) as well as cells with polymorphonuclear morphology (neutrophils). (**c**) IB4+ Dex+ cells are a homogenous population of cells displaying clear mononuclear phagocyte morphology. (**d**) IB4– Dex+ cells comprise of a heterogeneous population of cells displaying nonmyeloid cell morphology. This includes cells with dark round nuclei with small spherical cytoplasm and cells with dendritic/fibroblast-like morphology displaying light colored large elliptical nuclei with large elongated multipolarized cytoplasm. Scale bars indicate 50 μm

4. Transfer slides into the slide mailer containing a working dilution of Giemsa staining solution. Leave for 15 min at room temperature.
5. Wash by transferring slides into a slide mailer containing fresh distilled water. Ensure that any precipitation is removed.
6. Air dry slides (*see* **Note 16**).
7. Mount slides using mounting media and coverslip.
8. Slides can then be stored or imaged using a color microscope. Examples of the expected results for each of the FACS-isolated cell populations in the regenerating limb are shown in Fig. 5.

3.10 Isolation of RNA from FACS-Sorted Cells

1. Centrifuge isolated cell populations at 350×*g* at 4 °C for 5 min (*see* **Note 17**).
2. Aspirate and resuspend pellet in 350 μL of lysis/binding solution. Lyse cells thoroughly by pipetting or vortexing (*see* **Note 18**).
3. Add equal volume of 64 % ethanol (final volume of 700 μL) to the lysate and mix gently by pipetting.
4. Transfer lysate/ethanol mixture into filter cartridge assembled in a collection tube.
5. Centrifuge at 15,000×*g* for 1 min. Discard flow through.
6. Add 700 μL of wash solution 1. Centrifuge at 15,000×*g* for 1 min and discard flow through and reuse collection tube for washing steps.
7. Add 500 μL of wash solution 2/3. Centrifuge at 15,000×*g* for 1 min and discard flow through. Repeat twice.
8. Transfer filter cartridge into a fresh collection tube. Add 30–50 μL of elution buffer and centrifuge at 15,000×*g* for 1 min.

3.11 DNase1 Treatment and DNase Inactivation

1. Add 0.1 volume of 10× DNase1 and 1 μL of DNase1 to the RNA sample in elution buffer. Incubate at 37 °C for 15–30 min.
2. Add 0.1 volumes DNase inactivation reagent. Mix gently and incubate at room temperature for 2 min. It is recommended to mix the DNase inactivation reagent by vortexing prior to use.
3. Centrifuge the tube at 10,000 × *g* for 1 min to pellet the DNase inactivation reagent.
4. Transfer RNA solution into fresh microcentrifuge tube leaving behind the DNase inactivation reagent pellet.
5. Measure RNA quality and concentration via a spectrophotometer.
6. Store at −20 °C until further use.

3.12 cDNA Synthesis

1. For a single cDNA synthesis reaction, combine the following components in a 0.2 mL PCR tube on ice. If multiple reactions are being performed, it is recommended to prepare a master mix without the RNA samples (*see* **Note 19**).

Component	Volume
5× VILO™ Reaction Mix	4 μL
10× SuperScript® Enzyme Mix	2 μL
RNA (up to 2.5 μg)	*X* μL
DEPC-treated water	To 20 μL

2. Gently mix and incubate at 25 °C for 10 min.
3. Incubate at 42 °C for 60 min.
4. Terminate reaction at 85 °C for 5 min.
5. Store at −20 °C until further use.

3.13 Quantitative Real-Time PCR Gene Expression Analysis

1. Quantitative polymerase chain reaction (qPCR) assays are performed using SYBR green and analyzed using a thermal cycler instrument.
2. cDNA samples synthesized from the RNA of cell populations isolated by FACS at 4 days postamputation can be analyzed for gene expression relative to whole blastema tissue at 4 days postamputation.
3. Gene expression levels are calculated using the Pfaffl method and is normalized using the geometric mean of at least two housekeeping (normalizer) genes.
4. Repeat experiments at least three times from two biological replicates containing cDNA pooled from eight individual blastemas.
5. Primer sequences used in qPCR gene expression analysis are listed in Table 1.

Table 1
Primer sequences used for qPCR

Primer	Forward	Reverse
RBL27	CATCAGATCAAGCAAGCAGTA	CCAATGCAGCAGTTTAGATG
B-actin	TCCATGAAGGCTGCCCAACT	TGGCGCCACATCTGATTGAT
CD11b	CTCGCCCTACTGTGGATTGT	AAGGCCACCTTCTCCAGATT
Neutrophil elastase	CGCCTCCCTACAGTTCAGAG	ATCGGACTCATTCCATCCAA
Myeloperoxidase	TCAACAGCTGGAGAATCGTG	ATGTTGATGGCGGCTAAATC
CSF1R	CTCCAGGATGGGACTGTCAT	CCGCTTGGAGGTAGAGTCTG
MPEG1	CTTGCAAGACAGTTCGTCCA	TGGTTCGGTTTCCCAGATAG
MMP9	GCATCGTAGGATTCTCCATCA	ACCAGTGAAGGCCGTTCCGAT
VEGFR2	ACTGAAAACGCTGGGAAATG	AGAAGGTGGTGGTGTCCAAG
VCAM1	AAGACGACCGAAGTGTACGT	CCTGGTGAGTTTCAAGGTGC
ICAM1	AGCATAACGGCAAGATGGTC	GGATTTTCCCTCAATCAGCA
EpCAM	TAATGGCACAGACACCTGCT	TCTGCAACGTCATTCTCAGG
Collagen A2	ACCTGGTGGAAAGGGAGACT	GAAGATCCAGGTTCCCCATT
Fibronectin	ATTGCGTACTCCCAACTTCG	ACATGCTTCTCCCACGAGTC

3.14 Isolation of Peripheral Blood in the Axolotl

1. Anesthetize the axolotl in 0.2 % Tricaine (*see* **Note 1**).
2. Place anesthetized axolotl onto surgical stage on its back with the ventral side facing up (Fig. 6).
3. Place a pad of lint-free tissue under the loose skin of the ventral side of the head to expose the gill branches and efferent vessels.
4. Prepare cold HBSS-EDTA solution in a 50 mL tube.
5. Connect a 25G winged infusion set to a 5 mL syringe. Wash the tubing of the winged infusion set by drawing 3 mL of HBSS-EDTA solution into the 5 mL syringe. Flush 2 mL HBSS-EDTA solution out of the 5 mL syringe (*see* **Note 20**).
6. Using one hand to hold the wings of the infusion set, gently insert the needle into an efferent vessel. Take care to insert the needle into the vessel and not through to the other side (*see* **Note 21**).
7. Using the remaining free hand, slowly draw blood from the efferent vessel through the connecting tube and into the syringe. A maximum of 10 % of total blood volume can be drawn from one animal (*see* **Note 22**).
8. Slowly withdraw the needle minimizing any damage to the vessel or surrounding tissue.

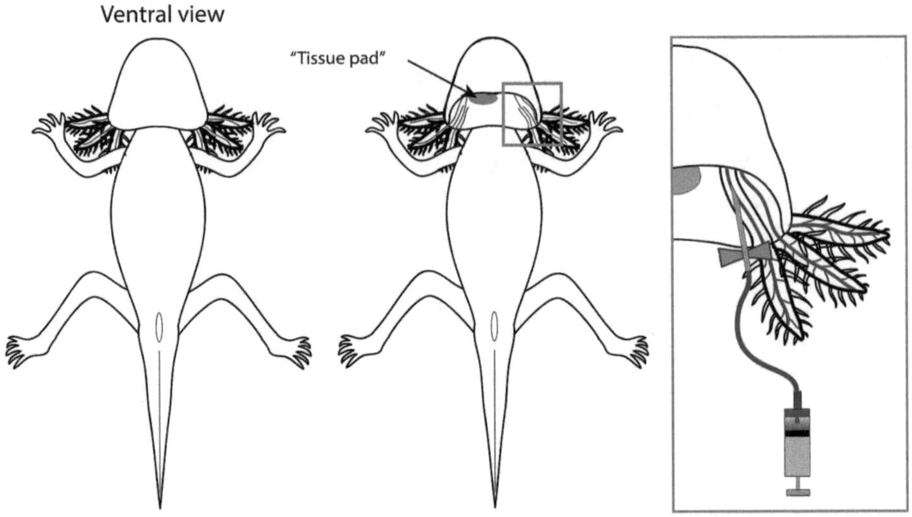

Fig. 6 Isolation of peripheral blood in the adult axolotl. Deoxygenated blood vessels in the gills can be exposed after a pad of lint-free tissue is placed under the ventral side of the head. This allows easy access to insert the needle of the SURFLO® winged infusion set into the vessels shown in *blue* (color figure online)

9. Transfer blood into a 15 mL tube containing cold HBSS-EDTA solution (*see* **Note 23**).
10. Slide a gloved finger over the injection site or hold it there for up to 30 s to accelerate clotting. This will aid in sealing the vessel at the entry point with skin mucus and minimize additional blood loss.
11. Place the axolotl into a fresh tank of 20 % Holtfreter's solution, for recovery in quiet individual housing for at least 24 h.

3.15 Magnetic Depletion of Red Blood Cells Using EasySep™ Magnets

1. Chill EasySep™ magnets on ice prior to use.
2. Prepare 1× concentration of sodium dithionite from 10× stock solution (*see* **Note 24**).
3. Centrifuge peripheral blood for 10 min at 350×*g* at 4 °C. Remove the supernatant.
4. Resuspend cell pellet in 1× sodium dithionite solution (solution should turn dark red-purple color).
5. Pour 3 mL of resuspended cell solution into 5 mL polystyrene FACS tube and slot into EasySep™ magnet (ensure that tube does not go all the way through).
6. Pour the remaining 3 mL of cell suspension into separate 5 mL polystyrene FACS tube and slot into a separate magnet. (Alternatively if two magnets are not available, cells can be resuspended in 4 mL and performed in one tube.)
7. Incubate on ice for 5 min. During this time red blood cells will become magnetically attracted toward the walls of the 5 mL polystyrene FACS tube and EasySep™ magnet.

8. Without removing the tube from the EasySep™ magnet, pour out cell solution into a new 5 mL polystyrene FACS tube. Any cells that are not bound to the magnet will pour out into the next tube (*see* **Note 25**).
9. Place the new 5 mL polystyrene FACS tube with the first round of depleted red blood cells into EasySep™ magnet and incubate on ice for 5 min.
10. Repeat **steps 8** and **9** in Subheading 3.14 and at least four times (*see* **Note 26**).
11. Transfer 1× solution of sodium dithionite cell suspension into a 50 mL Falcon tube and add blocking buffer solution to a final volume of 30 mL.
12. Wash out sodium dithionite by centrifuging cells for 15 min at 350 × *g* at 4 °C.
13. Remove supernatant. The cell pellet should have a visible layer of white blood cells.

Proceed to downstream experiments such as direct culturing of cells or flow cytometric analysis. The following steps provide an example of how to perform live-cell staining prior to flow cytometric analysis. This method follows the same live-cell staining method described in Subheading 3.7.

1. Resuspend the cell pellet in blocking buffer solution at a final concentration of no more than 1×10^6 cells/100 μL.
2. Block on ice for a minimum of 20 min.
3. Prepare primary antibodies: 0.5 mg/mL Isolectin B4 (1:50) and axolotl monoclonal antibody Pan Ig (1:400). Incubate the cells on ice away from light.
4. Add 100 μL to each individual 1.7 mL microcentrifuge tube and mix by flicking or pipetting.
5. Incubate on ice away from light for at least 20 min (*see* **Note 12**).
6. Wash any unbound primary antibodies with 900 μL of wash solution (HBSS-EDTA), mix thoroughly, and centrifuge at 350 × *g* at 4 °C for 5 min (*see* **Note 13**).
7. Prepare a working dilution of combined secondary antibodies: 2 mg/mL Streptavidin, Alexa Fluor® 488 (use at a 1:200 dilution) and 2 mg/mL Anti-mouse IgG (H + L), Alexa Fluor® 647 Conjugate (use at a 1:1,000 dilution).
8. Aspirate supernatant and resuspend pellet in 100 μL of combined secondary antibody solution. Incubate on ice, away from light for a minimum of 10 min (*see* **Note 14**).
9. Wash any unbound secondary antibodies with 900 μL of wash solution, mix thoroughly, and centrifuge at 4 °C for 5 min (*see* **Note 13**).
10. Prepare a working dilution of DAPI (1:2,000).

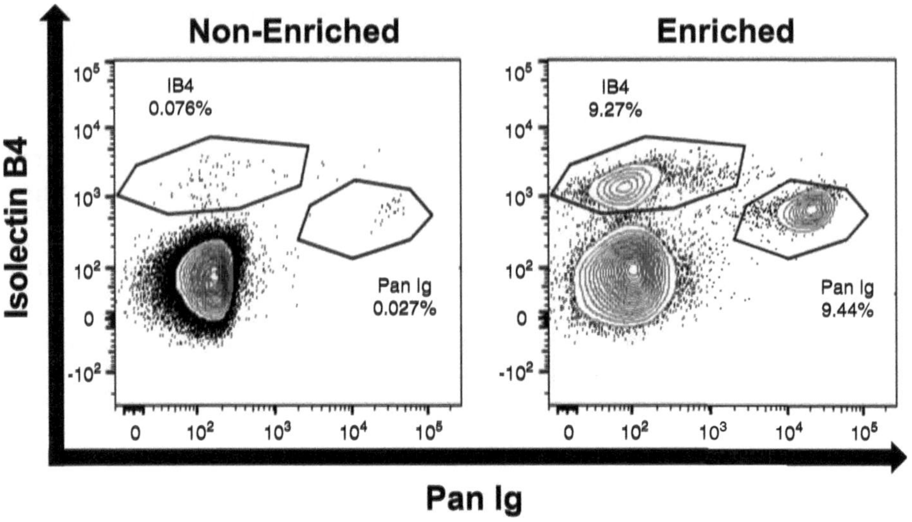

Fig. 7 Leukocytes can be enriched from total peripheral blood following magnetic depletion of red blood cells using the EasySep™ system. A comparison between normal peripheral blood (*left*) and leukocyte-enriched (*right*) samples using flow cytometric analysis. Treatment with sodium dithionite allows for red blood cells to be depleted. This results in an over 100-fold increase in the detection of innate (IB4) and adaptive (Pan Ig) immune cells

11. Aspirate supernatant and resuspend pellet in 1:2,000 DAPI at a final concentration of 1×10^6 cells/100 μL. Transfer stained cell suspension to a 5 mL flow cytometry tube by passing cells through a 40 μm filter.
12. Proceed to analyze cell populations using a flow cytometer. An example of the expected result is shown in (Fig. 7).

3.16 Expected Results and Conclusions

Here we describe a protocol that enables the isolation of myeloid cells at the 4-day postamputation time point. This method is designed to isolate cells using commercially available reagents and specialized laboratory equipment. This method incorporates intravenous delivery reagents into the axolotl. For our experiments we have used 2 MDa dextran-rhodamine. Alternative soluble reagents may also be delivered using this method. The 2 MDa dextran is taken up by macrophages via macropinocytosis, the endocytic function by which large antigens are captured [11]. Cells are liberated through a combination of mechanical and enzymatic digestion where the extracellular matrix components are digested. This allows for a single-cell suspension to be obtained. These cells are viable in abundance and maintain their surface antigens. Preservation of the cell surface integrity enables live-cell surface labeling using antibodies which identify activated myeloid cells that can be later analyzed using flow cytometry [12].

We demonstrate that a distinct population is labeled by both dextran-rhodamine and Isolectin B4 a marker of activated myeloid cells in the axolotl: DEX+ IB4+ (Fig. 4a) [13]. The DEX+ IB4+

population is shown to be homogenous in morphology displaying typical mononuclear phagocyte characteristics are large in size (Fig. 5c). The large cell size observed is consistent with side scatter area (SSC-A) and forward scatter area (FSC-A) profile observed on the flow cytometer (Fig. 4b). In addition a heterogeneous population of myeloid cells can be detected within the DEX– IB4+ population (Fig. 4). This includes the presence of polymorphonuclear cells (neutrophils) and mononuclear cells (monocytes and macrophages) (Fig. 5b). Gene expression analysis on these cell population confirms that both the DEX– IB4+ and DEX+ IB4+ are highly enriched for genes expressed by myeloid cells (CD11b), neutrophils (neutrophil elastase and myeloperoxidase), and monocytes and macrophages (CSF1R and MPEG1) (Fig. 8). Conversely the DEX– IB4+ and DEX+ IB4+ populations are not enriched for genes expressed by other mesenchymal cell types found in the regenerating limb such as fibroblasts and endothelial cells (Fig. 9). Together these results demonstrate a validated approach to isolate an abundant population of myeloid cells with specific enrichment of monocytes and macrophages in the regenerating axolotl limb.

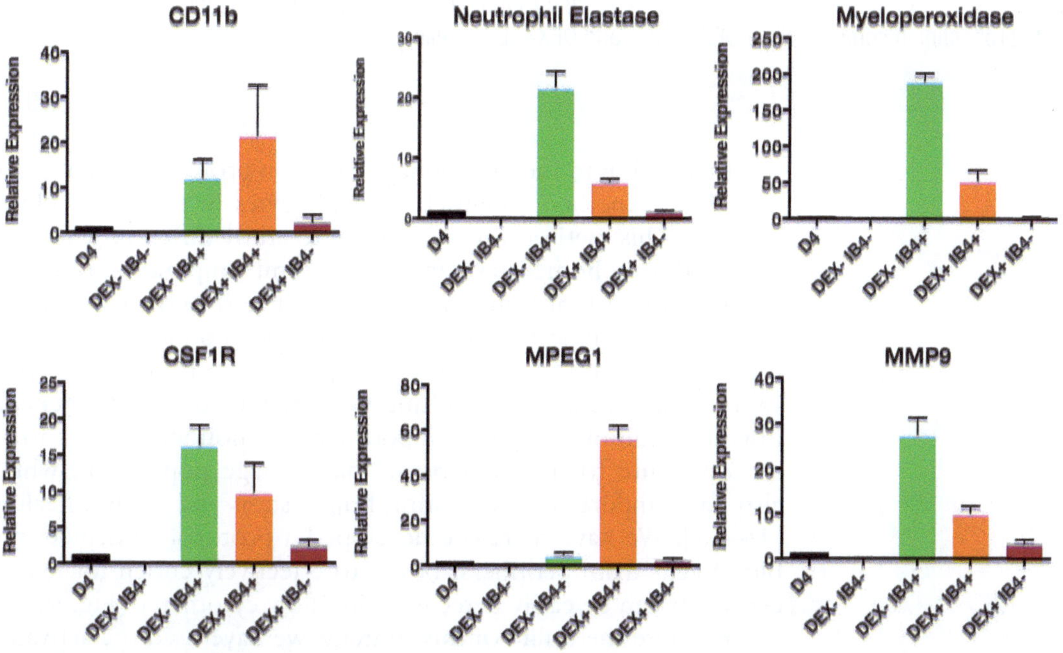

Fig. 8 Gene expression analysis on sorted cell populations using qPCR. Gene expression is relative to a whole blastema at 4 days postamputation. DEX– IB4+ and DEX+ IB4+ cells are enriched for genes restricted to the myeloid lineage in comparison to DEX– IB4– and DEX+ IB4– cells. DEX– IB4+ cells are a heterogeneous population of myeloid cells indicated by high enrichment of myeloid cells (CD11b), neutrophils (neutrophil elastase and myeloperoxidase), and monocytes/macrophages (CSF1R). In contrast DEX+ IB4+ cells are predominantly enriched for monocyte and macrophage genes (CSF1R, MPEG1, and MRC-1)

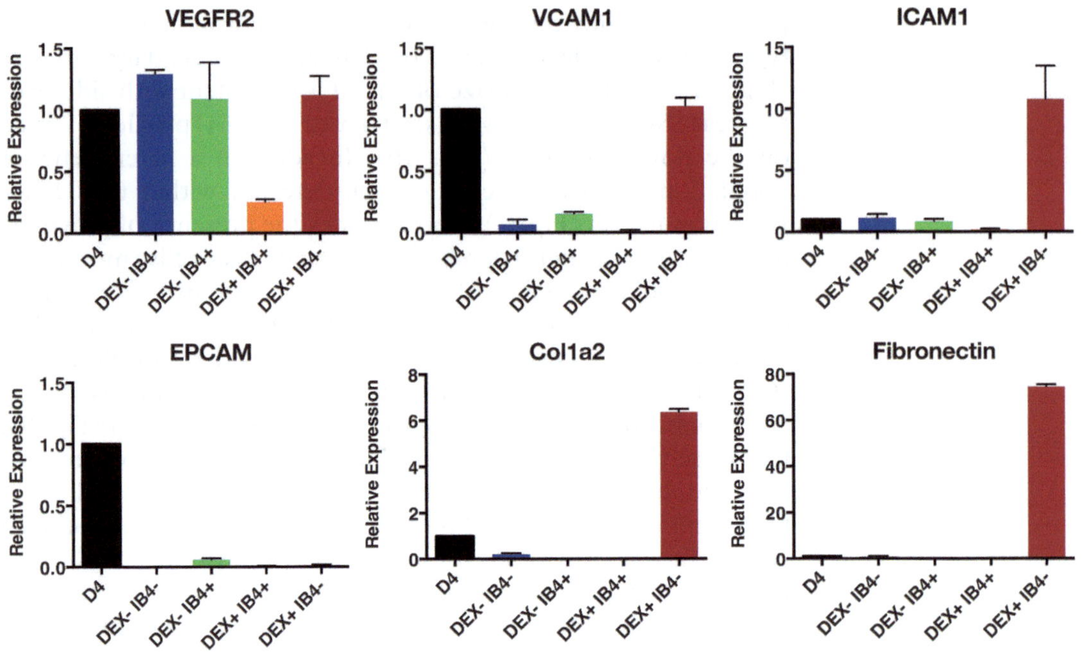

Fig. 9 Gene expression analysis on sorted cell populations using qPCR. Gene expression is relative to a whole blastema at 4 days postamputation. DEX– IB4+ and DEX+ IB4+ cells are not enriched for genes commonly expressed by endothelial cells (VEGFR2, VCAM1, and ICAM), epithelial cells (EpCAM), and fibroblasts (Col1a2 and fibronectin) in comparison to DEX– IB4– and DEX+ IB4– cells

In addition to the macrophage isolation protocol, we have also described a method to enrich for leukocytes in peripheral blood (Fig. 7). This method can be utilized to study different leukocyte subsets during homeostasis and injury. Limb amputation results in the induction of inflammatory signals that lead to the recruitment of different leukocyte subsets that include monocytes from the peripheral blood [6]. Infiltrating peripheral blood monocytes play various roles during the resolution of inflammation and regeneration. In addition peripheral blood-derived monocytes are known to contribute to the local tissue macrophage population, which also have important roles during homeostasis and wound healing [14–18]. We have therefore developed a protocol to deplete red blood cells from peripheral blood to effectively enrich for leukocytes that can be easily detected using flow cytometry (Fig. 9). To demonstrate the utility of this strategy, we have used the activated myeloid cell marker Isolectin B4 and axolotl specific pan-immunoglobulin marker Pan Ig [12, 13, 19]. We demonstrate that this strategy allows for over a 100-fold enrichment of leukocytes by flow cytometric analysis. This enables cells to be concentrated in a smaller volume, significantly decreasing the time required to analyze and isolate cells on the flow cytometer.

The protocols described in this chapter provide a detailed guide to isolate cells from the myeloid lineage including macrophages from the axolotl blastema as well as enrich for leukocytes found in the circulating blood. We have shown that a combination of physiological and cell surface labeling techniques is capable of identifying distinct myeloid cell populations in the regenerating limb. Leukocytes from the circulating blood can be enriched by magnetically depleting nucleated red blood cells after treatment with sodium dithionite. Both protocols are ideal for isolating specific cell populations that can be utilized in other downstream applications such as cytochemistry staining, gene expression analyses, and in vitro assays. More significantly these protocols provide methods that can be used to aid in the study of macrophages and additional leukocyte subsets in the context of the regenerating axolotl limb.

4 Notes

1. The percentage of Tricaine used to anesthetize the axolotl is dependent on the size of the animal. It is necessary to adjust this percentage to avoid any potential lethal doses. For example, animals that are 3 cm (snout to tail) in size are routinely anesthetized using 0.05 % Tricaine, whereas 0.2 % Tricaine is used in animals that are greater than 15 cm (snout to tail) in size.
2. It is recommended that surgeries are performed under a dissection microscope. Following amputation of the limb (proximal or distal), it is common to find that the bone/s have not been completely removed. To prevent any bone/cartilage protruding into the regenerating wound epithelium, it is recommended that bones be trimmed to fall in line with the plane of amputation.
3. Choosing an injection point in an afferent vessel upstream of branch points will assist in maintaining a continuous mixture of the reagent into the oxygenated blood flow back to the heart. This is especially important for fine vessels where blood flow from other vessels downstream will help pull the blood flow toward the heart.
4. Capillary glass filaments of 0.94 mm diameter are most suitable in our hands but other diameters can also be used. Optimizing the parameters for pulling glass filaments should be determined empirically and long fine needles can have their tips snapped to an appropriate size.
5. Injection of air into the bloodstream should be avoided in all instances.

6. Alternatively, isolated blastemas may be placed into microcentrifuge tubes containing 500 μL of 2 mg/mL collagenase II. Using a pair of surgical scissors, users can take advantage of the microcentrifuge tube's curved bottom to mince the blastema into fine pieces. It is important that all tissue pieces are cut very finely and there are no large chunks remaining when using this method to avoid any potential jamming of the C-tube during tissue processing stage using the gentleMacs™.
7. In addition, it is recommended that 2 μL of DNase1 (1 mg/mL) is added to the final solution to prevent any clumping of cells.
8. Additional protocols on the gentleMacs™ can be utilized; however, we have found that the two-step heart isolation protocol gives the best ratio of total viable cells.
9. The minimum time C-tubes can be placed on a roller at room temperature is 40 min; lower incubation time decreases yield of cells. Incubation times spanning greater than 2 h results in a decrease in cell viability.
10. This protocol does not yield cells from the epidermal layer (skin). Therefore, it is expected that there will be tissue that is not digested. This will appear as small white pieces and will not pass through the 100 μm cell strainer.
11. In order to maximize cell viability, it is recommended that all steps in this section be carried out on ice.
12. Normal time for primary antibody incubation is approximately 40–60 min. We have found that this can be decreased to a minimum of 20 min without any decrease in efficiency of antibody binding capacity.
13. More than 900 μL of wash solution can be used. Additional washes can be performed but are not necessary.
14. Normal time for secondary antibody incubation is approximately 30 min. We have found that this can be decreased to a minimum of 10 min without any decrease in secondary antibody binding capacity to the primary antibody.
15. Cells can be fixed onto the side using 4 % paraformaldehyde.
16. Slides can be left to dry overnight and mounted the next day.
17. For this protocol we have chosen to use the commercially available RNA isolation kit from Life Technologies. Other RNA isolation kits may be used in this method.
18. The exact volume of lysis/binding solution used is not critical. A general tip is that low-end volumes (100 μL) be used for fewer cells (<1,000 cells) and high-end volumes (500 μL) used for large amounts of cells (1×10^7 cells).

19. For this protocol we have chosen to use the commercially available cDNA synthesis kit available from Invitrogen. Other cDNA synthesis kits may be used in this method.
20. Flushing the winged infusion set with ice-cold HBSS-EDTA solution will minimize clotting within the tubing.
21. The pressure required to draw the blood is low when correctly positioned within a vessel. Do not apply too much pressure as vessels may collapse and cause clotting.
22. Blood collection of more than 10 % total blood volume in any 3-week period is not recommended. For a point of reference, approximately 1 mL of blood can be drawn from animals 15–20 cm in size (snout to tail). Collection of blood exceeding 15 % of total blood volume may require special permission from local ethics committees.
23. The collected blood can be used directly for some applications; however, spinning for 15 min at $350 \times g$ will separate blood cells from platelets, serum, and clotting factors and allow other downstream applications.
24. Sodium dithionate is an oxidizing agent. In order to ensure optimal leukocyte enrichment and viability, it is recommended that the 10× stock solution is prepared fresh, shortly before making the final 1× dilution.
25. When pouring out cells dark red/purple droplet will remain at the edge of the tube. Do not transfer this into the subsequent tube.
26. Subsequent rounds of red blood cell depletion can be performed by repeating **steps 8** and **9** in Subheading 3.6.

Acknowledgments

We thank Nadia Rosenthal and the members of her laboratory for the advice and helpful discussions. We thank members of the FlowCore facility at Monash University for performing FACS experiments. This work was supported by grants from Stem Cells Australia (Australian Research Council). The Australian Regenerative Medicine Institute is supported by grants from the State Government of Victoria and the Australian Government.

References

1. Tanaka EM, Reddien PW (2011) The cellular basis for animal regeneration. Dev Cell 21: 172–185
2. Harty M, Neff AW, King MW, Mescher AL (2003) Regeneration or scarring: an immunologic perspective. Dev Dyn 226:268–279
3. Mescher AL, Neff AW (2005) Regenerative capacity and the developing immune system. Adv Biochem Eng Biotechnol 93:39–66
4. Godwin JW, Brockes JP (2006) Regeneration, tissue injury and the immune response. J Anat 209:423–432

5. Mescher AL, Neff AW, King MW (2013) Changes in the inflammatory response to injury and its resolution during the loss of regenerative capacity in developing Xenopus limbs. PLoS One 8:e80477
6. Godwin JW, Rosenthal N (2014) Differentiation. Differentiation 87:66–75
7. Delavary BM, van der Veer WM, van Egmond M (2011) Macrophages in skin injury and repair. Immunobiology 216:753–762
8. Newman AC, Hughes CCW (2012) Macrophages and angiogenesis: a role for Wnt signaling. Vasc Cell 4:13
9. Tidball JG, Dorshkind K, Wehling-Henricks M (2014) Shared signaling systems in myeloid cell-mediated muscle regeneration. Development 141:1184–1196
10. Godwin JW, Pinto AR, Rosenthal N (2013) Macrophages are required for adult salamander limb regeneration. Proc Natl Acad Sci U S A 110:9415–9420
11. Lim JP, Gleeson PA (2011) Macropinocytosis: an endocytic pathway for internalising large gulps. Immunol Cell Biol 89:836–843
12. Miron VE, Boyd A, Zhao J-W, Yuen TJ, Ruckh JM, Shadrach JL, van Wijngaarden P, Wagers AJ, Williams A, Franklin RJM, ffrench-Constant C (2013) M2 microglia and macrophages drive oligodendrocyte differentiation during CNS remyelination. Nat Neurosci 16:1211–1218
13. Zammit PS, Clarke J, Golding JP, Goodbrand IA, Tonge DA (1993) Macrophage response during axonal regeneration in the axolotl central and peripheral nervous system. Neuroscience 54:718–719
14. Gordon S, Taylor PR (2005) Monocyte and macrophage heterogeneity. Nat Rev Immunol 5:953–964
15. Shi C, Pamer EG (2011) Monocyte recruitment during infection and inflammation. Nat Rev Immunol 11:762–774
16. Davies LC, Jenkins SJ, Allen JE, Taylor PR (2013) Tissue-resident macrophages. Nat Immunol 14:986–995
17. Pinto AR, Paolicelli R, Salimova E, Gospocic J (2012) An abundant tissue macrophage population in the adult murine heart with a distinct alternatively-activated macrophage profile. PLoS One 7:e36814
18. Epelman S, Lavine KJ, Beaudin AE, Sojka DK, Carrero JA, Calderon B, Brija T, Gautier EL, Ivanov S, Satpathy AT, Schilling JD, Schwendener R, Sergin I, Razani B, Forsberg EC, Yokoyama WM, Unanue ER, Colonna M, Randolph GJ, Mann DL (2014) Embryonic and adult-derived resident cardiac macrophages are maintained through distinct mechanisms at steady state and during inflammation. Immunity 40:91–104
19. Tournefier A, Guillet F, Ardavin C, Charlemagne J (1988) Surface markers of axolotl lymphocytes as defined by monoclonal antibodies. Immunology 63:269–276

Chapter 18

Assessing Cardiomyocyte Proliferative Capacity in the Newt Heart and Primary Culture

Hans-Georg Simon and Shannon Odelberg

Abstract

Unlike humans, adult newts possess extraordinary abilities to functionally regenerate lost and injured organs, including cardiac muscle. The most remarkable feature of mature newt cardiomyocytes is their ability to reenter the cell cycle, undergo cell division, and serve as a reservoir for progenitor cells. There are, however, a number of unsolved questions concerning the cellular and molecular mechanisms that underlie this plasticity; for example, we still lack a deeper understanding of the cell-inherent properties of newt cardiomyocytes and to what degree they differ from their mammalian counterparts. Along with considerable morphological changes at the wound site, a striking feature shared by different regenerating tissues in the newt is an extensive and dynamic remodeling of the extracellular environment. The dynamic signaling between cardiomyocytes and extracellular environment is of eminent importance in the control of the differentiated state of the cell, but the molecular details remain elusive. In this chapter, we describe methods to assess cardiomyocyte proliferation in vivo and enrich primary cardiomyocytes from newt hearts to study their behavior, taking extracellular matrix components into consideration.

Key words Heart regeneration, Cardiomyocyte proliferation, EdU, Cardiomyocyte culture, Extracellular environment, ECM

1 Introduction

1.1 In Vivo Cardiac Regenerative Activities

The regenerative ability of the adult mammalian heart is very limited. Adult mammalian cardiomyocytes withdraw from the cell cycle soon after birth and the default response to damage is characterized by formation of a collagen-rich scar. In contrast, after injury newt cardiomyocytes can proliferate and contribute to the functional restoration of the tissue architecture without scarring. Three different injury models have been described: (1) surgical resection of approximately 20 % of the ventricular apex [1, 2], (2) needle puncture of the ventricle wall and removal of pierced tissue [3], and (3) internal cardiomyocyte crushing using mechanical squeezing of the heart [4, 5]. In each case, bleeding is arrested by the formation of a blood clot and minor myocardial contraction.

Anoop Kumar and András Simon (eds.), *Salamanders in Regeneration Research: Methods and Protocols*, Methods in Molecular Biology, vol. 1290, DOI 10.1007/978-1-4939-2495-0_18, © Springer Science+Business Media New York 2015

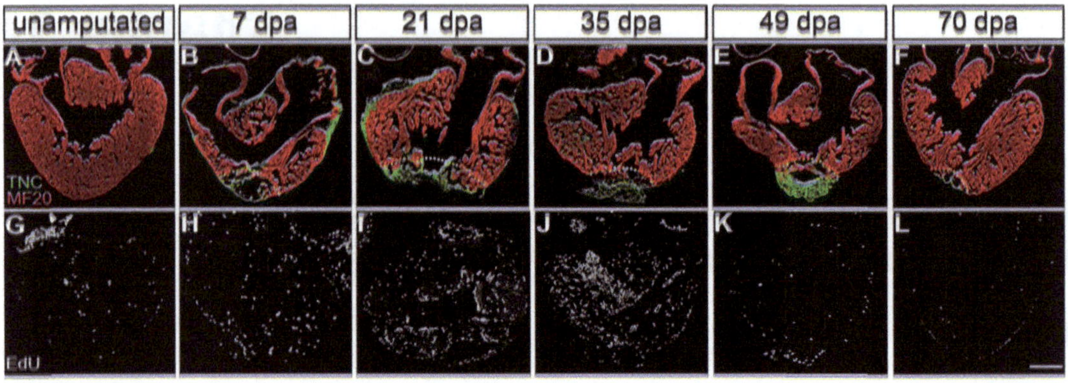

Fig. 1 Time course of cardiac regeneration. Cardiac injury induces the deposition of a tenascin-C (TNC)-rich transitional extracellular matrix (*green*). While the unamputated heart contains only a few cells actively synthesizing DNA (EdU = *white*), there is a substantial proliferative response throughout the myocardium between 7 and 35 dpa. Cardiac muscle is labeled with MF20 (*red*). *White dotted lines* demarcate the amputation site. Scale bar: 400 μm; *dpa* days postamputation

Within 2–3 days postamputation (dpa), a dense fibrin matrix replaces the clot, and dedifferentiation of a subset of cardiomyocytes is thought to begin during this early regenerative period, as emphasized by downregulation of cardiac markers, such as α-myosin heavy chain and cardiac troponins [4, 5]. Along with these changing intracellular activities, we identified a dramatic reorganization of the ECM (Fig. 1). A transitional extracellular matrix (ECM) that is shared by different regenerating tissues and bears astounding similarities to matrix components critical in embryonic development [1, 6–10] apparently provides cues to the embedded cells and controls the temporal and spatial regulation of cardiac regeneration [1]. Mounting evidence suggests that this cellular dedifferentiation facilitates cardiomyocyte proliferation with significant cell cycle reentry beginning by 7 dpa (Fig. 1). There is some controversy as to the location of the proliferative activity [1, 3, 11, 12]; however, recent work using pulse-chase labeling techniques in our laboratory revealed a dynamic redistribution of proliferative cells from cell cycle reentry to the final destination at the site of tissue replacement [1]. Cell proliferation gradually subsides and the expression of mature muscle markers returns, with functional restoration of the wound by 60–70 dpa (Fig. 1) [3, 4].

1.2 In Vitro Cardiomyocyte Activities

Long-term cultures revealed that the majority of newt cardiomyocytes can enter into S phase. Surprisingly, while all cells were exposed to serum-containing medium, only about a third of newt cardiomyocytes went through more than one complete cycle of cell division [2]. Interestingly, the picture from in vivo labeling studies after injury of the newt ventricle also indicates that only a subset of differentiated cardiomyocytes seem to be responsible for the increase in cell number observed upon regeneration [1]. It is noteworthy that S phase entry seems apparently driven by serum-induced pathways that are dependent on Rb inactivation and shared in cardiomyocytes and skeletal myotubes [2]. We found that ECM components that reflect the cellular environment during regeneration promote similar behaviors of cultured cardiomyocytes and skeletal myotubes, including proliferation [1, 6, 13].

Thus, experimental evidence suggests that newt primary cardiomyocyte cultures reflect properties of the population of cardiomyocytes after heart injury and can account for these properties at the single-cell level [1, 2]. However, the heterogeneity in proliferative potential remains an important question. In this regard, it is of interest to test the functional properties of various ECM components in isolation and combination on cardiomyocyte explants. A molecular comparison of the cells in culture should help elucidate the regulation of the differentiated state and cell cycle progression in adult cardiomyocytes. This chapter provides an experimental toolbox ranging from in vivo to in vitro assessment of single cardiomyocytes, which we believe is important for an integrated perspective of their functional properties.

2 Materials

Prepare all solutions using ultrapure water and analytical-grade reagents. Prepare and store all reagents at room temperature (unless indicated otherwise). Diligently follow all waste disposal regulations when disposing waste materials. The use of animals must follow the guidelines of and be approved by the Institutional Animal Care and Use Committee.

2.1 Surgical Procedures

1. Six to eight adult newts per preparation.
2. Newt water: Deionized and dechlorinated water supplemented with 0.0375 % Instant Ocean (available from suppliers of aquarium supplies). We use Instant Ocean salt at a very low concentration, which is not comparable to the concentrations typically used in saltwater aquariums.
3. Anesthesia solution: 0.1 % w/v ethyl 3-aminobenzoate methanesulfonate salt (Tricaine) in 0.0375 % Instant Ocean.
4. IC-spacing perfboard (approximately 13 cm × 8 cm) (available from suppliers of electronic components).
5. Stretch Magic elastic cord.
6. Alcohol-free povidone-iodine (Betadine®).
7. Forceps, No. 5 forceps, fine spring scissors, and microscissors. Sterilize in autoclave.
8. C-13 Sofsilk™ sutures.
9. Recovery tanks: Clear plastic containers (30 cm × 18 cm × 13 cm) with cover, filled with approximately 1.0 L newt water.

2.2 Collection of Regenerating Tissue

1. Tissue-Tek™ OCT compound.
2. Dry ice slurry in isopentane.

2.3 Monitoring Proliferative Progenitor Cells

1. Phosphate buffered saline (PBS).
2. 1 mM 5-ethynyl-2'-deoxyuridine (EdU, Invitrogen) in 70 % PBS.
3. 100 mM thymidine in 70 % PBS.
4. 4 % paraformaldehyde (PFA) in PBS.
5. 0.5 % Triton X-100 in PBS.
6. Blocking buffer: 20 % goat serum, 0.2 % bovine serum albumin (BSA), 50 mM ammonium chloride, 25 mM glycine, 25 mM lysine, and 0.02 % sodium azide in PBS.
7. EdU detection buffer: 1 ng/mL Alexa Fluor (AF) 594 conjugated azide (Life Technologies) diluted in 100 mM Tris–HCl (pH 8.5), 1 mM copper (II) sulfate, and 100 mM ascorbic acid.
8. 0.1 % Tween-20 in PBS.
9. Nonfluorescent mounting medium (e.g., FLUORO-GEL).

2.4 Imaging Cellular and Extracellular Proteins and Matrices

1. Superfrost Plus slides.
2. 0.5 % Triton X-100 in PBS.
3. Blocking buffer: 20 % goat serum, 0.2 % bovine serum albumin (BSA), 50 mM ammonium chloride, 25 mM glycine, 25 mM lysine, and 0.02 % sodium azide in PBS.
4. Antibodies: Primary antibody information to detect cellular and extracellular matrix proteins is summarized in Table 1. All primary antibodies were tested for species specificity with appropriate controls. ECM antibodies have been verified in the newt model [1, 6, 13]. For immunofluorescence studies, primary antibodies were detected with appropriate species-specific Alexa Fluor (AF)-conjugated secondary antibodies, Table 2.
5. 0.1 % Tween-20 in PBS.
6. Biotinylated hyaluronic acid binding protein (HABP, Calbiochem), 1 μg/μL in PBS with 0.2 % BSA.
7. Avidin/biotin blocking kit (Zymed Labs).
8. Hyaluronic acid sodium salt (Calbiochem) 1 mg/mL.

2.5 Collection of Myocardial Tissue for Primary Cultures

1. Sterile 60 mm plastic cell culture dish.
2. 70 % L15 media. Use sterile ddH_2O to dilute the medium to 70 %.
3. 70 % L15 media supplemented with 1× pen/strep/fungizone.

Table 1
Primary detection reagents used for immunohistochemistry

Probe	Company	Product no.	Host	Dilution	Antigen/source
Tenascin-C	Millipore	AB19013	Rabbit	1:500	Embryonic chick brain
Fibronectin	Sigma	F0791	Mouse—IgG1	1:200	IST-3 hybridoma
HABP-biotinylated	Calbiochem	385911	Bovine	1:300	Bovine nasal cartilage
MF20	DSHB	MF20	Mouse—IgG2b	1:2,000 ascites; 1:300 Bioreactor supernatant	Light meromyosin
Wilms Tumor 1	Calbiochem	CA1026	Rabbit	1:150	Human WT1
Phospho-histone H3	Cell Signaling Technology	3377	Rabbit	1:200	Ser10 of human histone H3

See **Note 6**

Table 2
Secondary detection reagents used for immunohistochemistry

Fluorophore	Company	Product no.	Secondary Ab/target	Dilution or concentration	Imaging channel
AF488	Invitrogen	A11034	Anti-rabbit	1:500	Green
AF488	Invitrogen	A11029	Anti-mouse	1:500	Green
AF546	Invitrogen	A21123	Anti-rabbit	1:500	Near red
AF594	Invitrogen	A11005	Anti-mouse	1:500	Near red
AF647	Invitrogen	A21242	Anti-mouse IgG2b	1:300	Far red
AF546	Invitrogen	S11225	Streptavidin	1:300	Near red
AF594 conjugated azide	Invitrogen	A10270	EdU	1 ng/mL	Near red
DAPI	Roche	10236276001	Nuclear DNA	1:1,000	Blue

4. 100 % ethanol.
5. Forceps and iridectomy scissors. Sterilize in autoclave and keep stored in 70 % ethanol during procedure.
6. 20 mL glass scintillation vials and caps. To sterilize expose to UV light for at least 30 min.

7. 70 % calcium- and magnesium-free PBS (nPBS).
8. Enzyme cocktail: Under sterile conditions, make fresh every time, 4× 3 mL 70 % calcium- and magnesium-free nPBS with 0.5 % trypsin (Sigma), 400 U/mL type II collagenase (Sigma), 0.15 % BSA, 0.3 % glucose, and 1× pen/strep/fungizone. The dissociation enzymes are critical; other suppliers are possible, but concentrations and conditions need to be optimized.
9. Shaking water bath with variable speed.
10. Sterile fire-polished glass Pasteur pipette.
11. 50 mL conical tubes.
12. Newt minimum essential medium (newt MEM): 70 % MEM with 10 % FBS and 1× pen/strep/fungizone.
13. Centrifuge with swinging bucket rotor (e.g., Beckman Allegra 6KR).
14. Uncoated 60 mm plastic tissue culture dishes.
15. Tissue culture incubator set to 25 °C with 2.0 % CO_2.
16. ECM-coated 60 mm plastic tissue culture dishes: tenascin-C (Millipore, catalog number CC065), fibronectin (Millipore, Catalog no. FC010), and laminin (Millipore, Catalog no. CC095). The source of the matrices is critical; other suppliers are possible, but concentrations and conditions need to be optimized.
17. Inverted microscope with phase and fluorescence optics (e.g., Leica DMI6000).
18. 12-well tissue culture polystyrene dishes.

3 Methods

Carry out all procedures at room temperature unless otherwise specified. The use of animals must follow the guidelines of and be approved by the Institutional Animal Care and Use Committee.

3.1 Surgical Procedures for Ventricular Resection

1. Newts are anesthetized by submersion in room temperature 0.1 % Tricaine in newt water for 10–15 min and then placed on ice covered with plastic wrap for 15–20 min. Prior to heart ventricle amputation, verify that newts are nonresponsive to touch and pain through both the tail pinch and the supine positioning test (*see* **Note 1**).
2. Newts are subsequently turned on their back, gently strapped into correct position on an IC-spacing perfboard with Stretch Magic elastic cord loosely wrapped around the neck and tail (*see* Fig. 2a). To minimize infection apply approximately 50 μL alcohol-free povidone-iodine to the incision site.

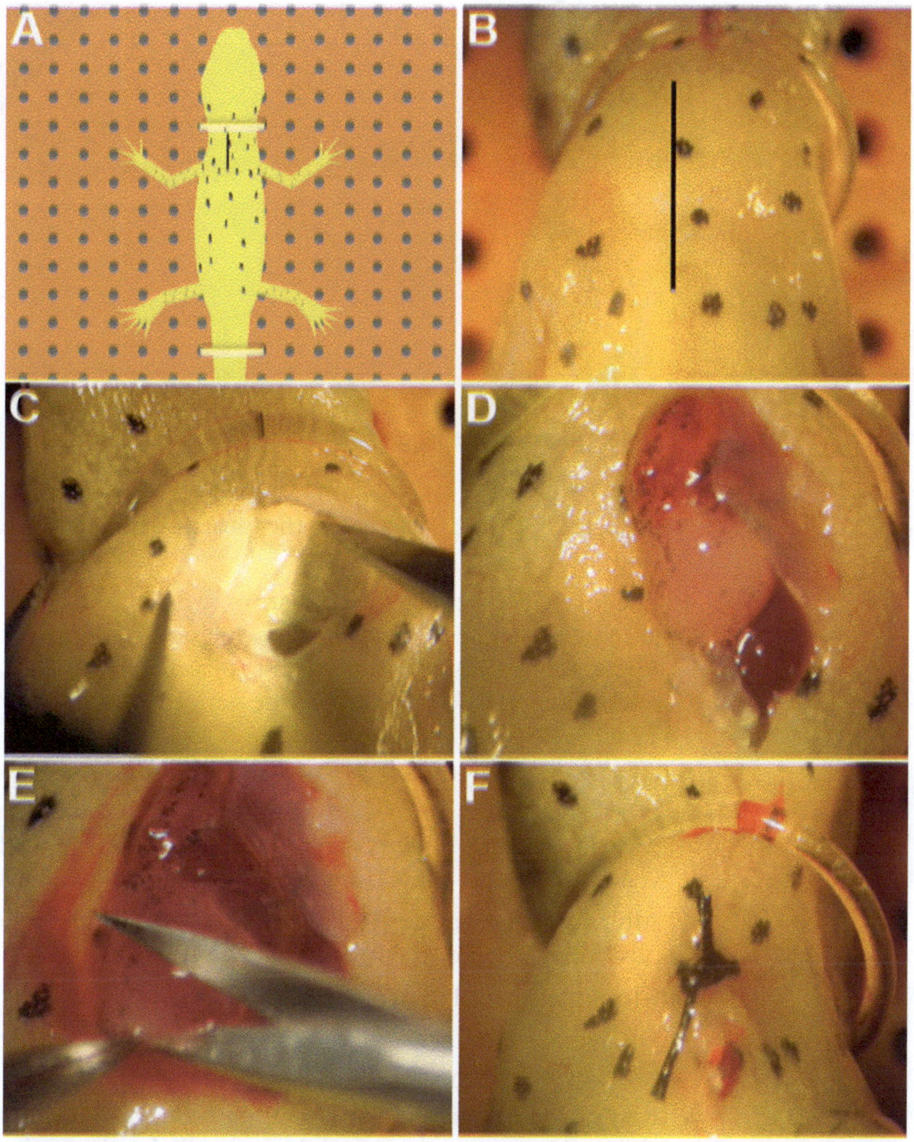

Fig. 2 Newt ventricular resection surgery. (**a**) Cartoon demonstrating how anesthetized newts are held in place on a perfboard with elastic cord. (**b**) *Vertical line* indicates position of ventral near-midline incision. (**c**) Resection of cartilaginous sternum and exposure of pericardial sac. (**d**) Gentle external pressure releases the heart from the chest cavity. (Note the two darker red atria on *top* and the single pale red ventricle at the *bottom*.) (**e**) Resection of approximately 20 % of the ventricle. (**f**) A single suture closes the external chest wound following ventricular resection

3. For ventricular amputation surgeries (Fig. 2b–f), cut the outer layer of the skin and tissue using No. 5 forceps and microscissors. Make the initial incision approximately 2 mm from the midline, just below the neck and above the posterior side of the forelimbs (Fig. 2b). Beneath the incision, the sternum is easily visualized as an opaque, cartilaginous material, which is

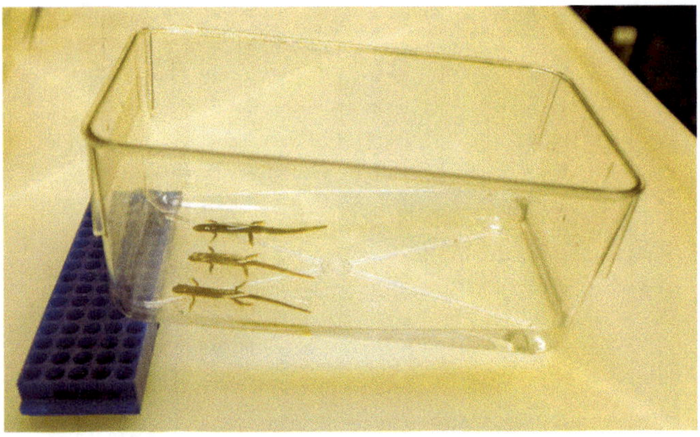

Fig. 3 After surgery newts recover immersed in shallow water

dissected through to reveal the pearly white pericardial sac (Fig. 2c).

4. Use forceps to expose the pericardial sac in the chest cavity and use fine spring scissors to make an incision in the pericardium. Apply gentle pressure to the chest (near the shoulders) to fully release the exposed heart from the pericardial sac and the chest cavity (Fig. 2d). The newt has a three-chambered heart, visualized as two dark red atria superior to a lighter pink ventricle.
5. Hold the lower tip of the ventricle with No. 5 forceps, and with fine spring scissors carefully resect 15–20 % of the total ventricular tissue mass from the apex (Fig. 2e) (*see* **Note 2**).
6. After resection, place the injured heart back into its original position in the pericardial sac and chest cavity.
7. Close the wound with a single C-13 Sofsilk™ suture (Fig. 2f).
8. For recovery, keep the newts on ice covered with plastic wrap for approximately 30 min.
9. Return the newts to a tank containing standard newt water and elevated at one end so that the animal's head remains out of the water for 1–3 h to avoid drowning during recovery (Fig. 3). Keep the animals in these small plastic containers filled partway with newt water until regenerating tissues are collected. Cover with a lid and change water daily.

3.2 Collection of Regenerating Tissue

1. At the time of tissue collection, newts will be sacrificed. Newts are re-anesthetized as described in Subheading 3.1, **step 1**.
2. For immunohistochemistry studies, harvest regenerating hearts at designated time points, embed in Tissue-Tek™ OCT compound, freeze in an isopentane-dry ice slurry, and store at −80 °C.
3. For RNA-based investigations, remove from the apex of the heart regenerating tissue with fine spring scissors and immediately flash freeze in liquid nitrogen.

3.3 Monitoring Proliferative Progenitor Cells

1. For cardiac regeneration studies, 3 days before tissue harvest, newts are injected intraperitoneally with 100 μL of 1 mM EdU in 70 % PBS daily to allow for the identification of cells that had entered the cell cycle (*see* **Note 3**).
2. For pulse-chase studies, after the 3-day labeling period, newts are injected again with 100 μL of 100 mM thymidine in 70 % PBS to dilute residual EdU and greatly reduce further incorporation of the EdU label.
3. At the time of sacrifice, harvest the tissue, embed in Tissue-Tek™ OCT as described in Subheading 3.2, **step 2**, and collect serial sections of 14 μm on Superfrost Plus slides using a cryostat (e.g. Leica CM3050S).
4. Fix the sections in 4 % PFA for 5 min, rinse in PBS, permeabilize with 0.5 % Triton X-100 in PBS for 1 min, and then rinse with PBS.
5. Block sections for 30 min in blocking buffer.
6. To identify cells that incorporated EdU, incubate sections for 30 min with AF594 conjugated azide (final concentration 1 ng/mL) diluted in 100 mM Tris–HCl (pH 8.5), 1 mM copper (II) sulfate, and 100 mM ascorbic acid.
7. Apply primary antibodies for 1 h and rinse sections with 0.1 % Tween-20 in PBS.
8. Block slides again for 5 min before staining with the appropriate secondary detection reagents for 30 min.
9. Rinse slides with 0.1 % Tween-20 in PBS and mount with FLUORO-GEL.

3.4 Imaging Cellular and Extracellular Proteins and Matrices

1. For immunohistochemistry studies (*see* **Notes 4** and **5**), collect serial cryosections of 14 μm on Superfrost Plus slides.
2. Fix sections in 4 % PFA for 5 min, rinse in PBS, permeabilize with 0.5 % Triton X-100 in PBS for 1 min, and then rinse with PBS.
3. Treat sections for 30 min with blocking buffer (20 % goat serum, 0.2 % BSA, 50 mM ammonium chloride, 25 mM glycine, 25 mM lysine, and 0.02 % sodium azide in PBS).
4. Apply primary antibodies for 1 h at room temperature and rinse sections with 0.1 % Tween-20 in PBS.
5. Block slides again for 5 min before staining with the appropriate secondary detection reagents for 30 min.
6. Rinse slides with 0.1 % Tween-20 in PBS and mount with FLUORO-GEL.
7. Image sections on a fluorescence or confocal microscope (e.g., Leica DMI6000, Zeiss LSM 510 Meta).

8. To assemble individual sections into a single composite image we employ Image-Pro Analyzer.
9. To quantify cell staining, we use Adobe Photoshop and subsequently process cell count data in Microsoft Excel.

3.5 Collection of Myocardial Tissue for Primary Cultures

We prepare primary cultures of newt cardiomyocytes using a modified version of previously described methods [2].

1. For each preparation, six to eight adult newts are anesthetized as described in Subheading 3.1, **step 1**.
2. To minimize possible contamination, generously cover the newt chest and forelimbs in alcohol-free povidone-iodine.
3. Expose hearts as described in Subheading 3.1, **steps 3** and **4**, and with iridectomy scissors, cut and remove the single ventricle free of atrial tissue.
4. Place the ventricles quickly and carefully into a sterile 60 mm cell culture dish with room temperature 70 % L15 media.
5. Store ventricles overnight (12–18 h) at 25 °C in fresh 70 % L-15 media supplemented with 1× pen/strep/fungizone.
6. In a designated tissue culture hood, rinse sterile forceps and iridectomy scissors in 70 % calcium- and magnesium-free PBS (nPBS) and dissect each ventricle into four to six pieces.
7. Using the forceps, transfer dissected ventricles to a sterile scintillation vial containing 3 mL of digestion mixture [nPBS with 0.5 % trypsin, 400 U/mL type II collagenase, 0.15 % BSA, 0.3 % glucose, 1× pen/strep/fungizone].
8. Place the standing scintillation vial into a shaking water bath at 27 °C and digest the heart tissue for 8 h, with a change of enzyme cocktail every 2 h. Rotate the scintillation vial containing the dissected heart muscle at 90 rpm for the first 2 h and 60 rpm for the final 6 h. To protect cells from light during the incubation, cover the scintillation vial with aluminum foil.
9. Following the 8-h digestion, triturate the cell suspension with a sterile fire-polished glass Pasteur pipette for 2–5 min.
10. To neutralize the digestion mixture, add the cell suspension to 50 mL conical tubes containing 30 mL newt minimum essential media (70 % MEM) with 10 % FBS and 1× pen/strep/fungizone.
11. In a low-speed centrifuge with swinging bucket rotor (e.g., Beckman Allegra 6KR), centrifuge the suspension for 10 min at 200 ×*g* to collect the cells.
12. Remove the supernatant and resuspend the sedimented cells with 4 mL newt MEM with 10 % FBS and 1× pen/strep/fungizone.

3.6 Enrichment of Cardiomyocytes and Culture of Primary Cells

1. Following the enzymatic digestion of cardiac tissue and removal of debris by centrifugation, plate the resuspended cell mixture in 60 mm uncoated tissue culture dishes and store the dishes undisturbed for 2–3 days at 25 °C in a humidified 2.0 % CO_2 incubator. During this time, blood cells, connective tissue cells, etc., will attach to the plastic surface, while cardiomyocytes remain nonadherent.
2. Carefully collect the enriched floating cardiomyocytes with a glass Pasteur pipette, count them on a hemocytometer, and finally plate at a density of 3,000 cells/cm^2 onto ECM-coated dishes (see below) in newt MEM with 10 % FBS and 1× pen/strep/fungizone. To support cardiomyocyte adherence, leave this culture dish undisturbed for 2–3 days at 25 °C in a humidified 2.0 % CO_2 incubator before the first media change.
3. During culture, monitor the cardiomyocytes regularly and image them for analysis using an inverted tissue culture microscope with 10× or 20× objectives. In culture, the newt cardiomyocytes maintain their differentiated status and display an elongated shape with the typical branched morphology (Fig. 4a) and positive reaction with the MF20 antibody (Fig. 4c). Over 10–12 days in culture, the cardiomyocytes will spread, form intricate junctions, and display spontaneous beating (Fig. 4b).

3.7 Assessing DNA Synthesis in Cultured Cells

1. For in vitro proliferation studies, we employ commercially available ECMs, including tenascin-C, fibronectin, and laminin [1, 6, 13]. Dilute the ECMs in serum-free MEM and allow them to passively adsorb to 12-well tissue culture polystyrene dishes at 1 μg/cm^2 for a minimum of 1 h but typically overnight at 25 °C in a sterile incubator, before rinsing the plates with nPBS. For DNA synthesis assays, coat wells in triplicate. Confirm successful coating in 1 well per ECM via IHC with respective antibodies. Importantly, do not allow ECM-coated dishes to dry at any time before plating with cells.
2. Culture cardiomyocytes in 70 % MEM with 10 % FBS and 1× pen/strep/fungizone. Approximately 7 days after plating, the cells reach a density suitable to assess their proliferative activity (Fig. 4c). By 10–14 days post-plating, the cardiomyocytes begin to form complex networks that prevent proper observation and counting (Fig. 4b).
3. To the media of each well, add EdU at a final concentration of 1 μM and incubate for 12 h.
4. After the overnight incubation with EdU, fix, permeabilize, and stain the cells with AF594 azide and other antibodies as described above for immunohistochemical staining procedures (Fig. 4d).

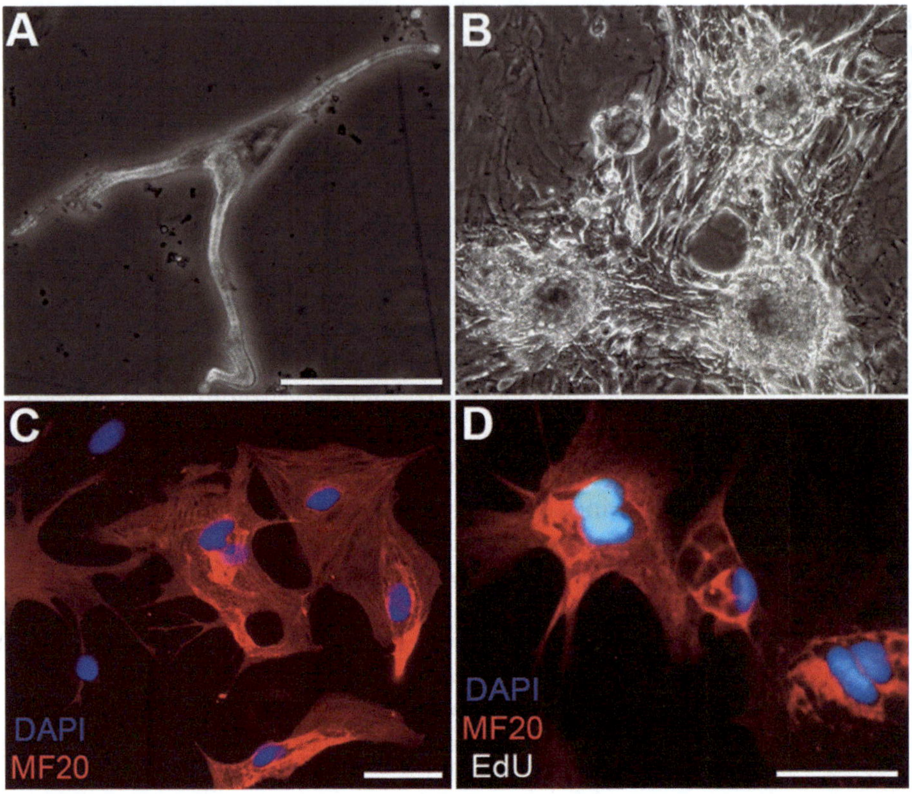

Fig. 4 Explanted newt cardiomyocytes in culture. (**a**) Early on in diluted cultures, cardiomyocytes display the typical bifurcated structure. (**b**) Picture frame of movie displaying beating cardiomyocytes. Within 2 weeks the cardiomyocytes reach a considerable density, interconnect, and beat in synchrony. (**c** and **d**) Throughout in vitro culture, the cardiomyocytes maintain their contractile machinery as verified by MF20 staining. (**d**) Cardiomyocytes actively synthesize DNA as determined by EdU incorporation. Scale bar for **a**, **c**, **d**: 50 μm

5. Use an inverted microscope with phase contrast and fluorescence illumination at 10× to acquire ≥10 non-overlapping images for each well.
6. Count cells in Adobe Photoshop and process cell count data in Microsoft Excel. To statistically assess the effects of different matrices on in vitro DNA synthesis, we typically use one-way ANOVA in PAST software.

4 Notes

1. The tail pinch test involves pinching the tail with enough force to normally cause a newt to respond with a jerk. If the newt fails to respond, this suggests that the newt is sufficiently anesthetized. The supine positioning test involves rolling the newt onto its back in the supine position. A newt does not appreciate being rolled onto its back and will immediately flip its body to the prone position if not sufficiently anesthetized.

2. When resecting 15–20 % of the ventricular tissue mass from the apex, be careful not to cut too deeply into the ventricular lumen as this will cause massive bleeding resulting in the animal's death.
3. Both BrdU and EdU can be used to identify cardiomyocytes that are actively synthesizing DNA (i.e., have entered the cell cycle). However, we prefer using EdU as the mild click chemistry protocol does not destroy cellular epitopes (unlike BrdU protocols) and is compatible with immunofluorescent techniques aimed at identifying proteins or other markers present in the cardiomyocytes or ECM. Either AF594 azide or AF488 azide can be used for the EdU detection.
4. The fluorescent staining procedures described in this chapter represent a method that we have found to produce reliable and reproducible results and clear images for subsequent analysis. However, as with any immunohistologic studies, there are alternative methods that can be employed that will also produce reliable and meaningful results. Each investigator should examine and validate the methods that best suit their particular study.
5. To visualize hyaluronic acid distribution, we use biotinylated hyaluronic acid binding protein (HABP) as an indirect probe [6]. After fixation and permeabilization, and prior to labeling sections with HABP, inactivate endogenous avidin and biotin activity with an avidin/biotin blocking kit. After these blocking steps, wash sections two times for 5 min each in PBS and then follow with the HABP incubation as described for primary antibodies. Confirm the staining specificity by preincubating HABP with a solution of 1 mg/mL hyaluronic acid sodium salt, and then incubate a tissue section with this absorbed reagent. A marked reduction in fluorescent staining when compared to the unabsorbed HABP labeled indicates that the HABP labeling is specific.
6. If using antibodies not listed in this table, investigators should validate their use for newts either through a literature search or through conducting their own studies to ensure that the antibodies react with the appropriate antigen and are specific for that antigen. This is especially true when using antibodies that have been generated for use in other species.

Acknowledgements

We would like to acknowledge both current and former research laboratory personnel who helped develop the protocols described in this chapter. These individuals include Claudia Guzman and Sarah Mercer. We thank Drs. Jeremy Brockes and Anoop Kumar for sharing their laboratory protocol for establishing newt primary cardiomyocyte cultures. We would also like to acknowledge Dr. Ken Poss who initially helped us establish the heart surgery procedure we describe in this chapter.

References

1. Mercer SE, Odelberg SJ, Simon H-G (2013) A dynamic spatiotemporal extracellular matrix facilitates epicardial-mediated vertebrate heart regeneration. Dev Biol 382:457–469
2. Bettencourt-Dias M, Mittnacht S, Brockes JP (2003) Heterogeneous proliferative potential in regenerative adult newt cardiomyocytes. J Cell Sci 116:4001–4009
3. Witman N, Murtuza B, Davis B, Arner A, Morrison JI (2011) Recapitulation of developmental cardiogenesis governs the morphological and functional regeneration of adult newt hearts following injury. Dev Biol 354:67–76
4. Laube F, Heister M, Scholz C, Borchardt T, Braun T (2006) Re-programming of newt cardiomyocytes is induced by tissue regeneration. J Cell Sci 119:4719–4729
5. Piatkowski T, Mühlfeld C, Borchardt T, Braun T (2013) Reconstitution of the myocardium in regenerating newt hearts is preceded by transient deposition of extracellular matrix components. Stem Cells Dev 22:1921–1931
6. Calve S, Odelberg SJ, Simon H-G (2010) A transitional extracellular matrix instructs cell behavior during muscle regeneration. Dev Biol 344:259–271
7. Mercer S, Cheng C-H, Atkinson DL, Krcmery J, Guzman CE, Kent DT, Zuckor K, Marx KA, Odelberg SJ, Simon H-G (2012) Multi-tissue microarray analysis identifies a molecular signature of regeneration. PLoS One 7(12):e52375
8. Tassava RA, Nace JD, Wei Y (1996) Extracellular matrix protein turnover during salamander limb regeneration. Wound Repair Regen 4:75–81
9. Toole BP, Gross J (1971) The extracellular matrix of the regenerating newt limb: synthesis and removal of hyaluronate prior to differentiation. Dev Biol 25:57–77
10. Vinarsky V, Atkinson DL, Stevenson TJ, Keating MT, Odelberg SJ (2005) Normal newt limb regeneration requires matrix metalloproteinase function. Dev Biol 279:86–98
11. Bader D, Oberpriller J (1979) Autoradiographic and electron microscopic studies of minced cardiac muscle regeneration in the adult newt, Notophthalmus viridescens. J Exp Zool 208: 177–193
12. Oberpriller JO, Oberpriller JC (1974) Response of the adult newt ventricle to injury. J Exp Zool 187:249–253
13. Calve S, Simon H-G (2012) Biochemical and mechanical environment cooperatively regulate skeletal muscle regeneration. FASEB J 26:2538–2545

Chapter 19

Long-Term Organ Cultures of Newt Hearts

Tanja Piatkowski and Thomas Braun

Abstract

Adult newts regenerate their hearts after injury by initiating proliferation of cardiac muscle and non-muscle cells. Mechanistic studies in vivo to analyze heart regeneration are challenging due to the long reproduction cycle of newts and the complexity of the genome. Culture of primary newt cells might offer alternative experimental approaches, but monolayers of newt cardiomyocytes and slice cultures of newt hearts show extensive morphological changes during cultivation. Hence, we developed a protocol to culture intact newt hearts in vitro, avoiding major morphological changes of explanted organs during a 5-week cultivation. The model provides improved accessibility and allows manipulation of cultured organs by small molecules and viral vectors. We found that dedifferentiation and S-phase entry of cardiomyocytes, which are hallmarks of cardiac regeneration in vivo, can be recapitulated in cultured hearts in vitro. We reason that long-term organ cultures of newts are a versatile tool for mechanistic studies on organ regeneration.

Key words Organ culture, Newt heart, 3D culture, S-phase labeling, Dead cell stain

1 Introduction

The spatial and temporal organization of a tissue is crucial for many cellular processes such as proliferation, migration, and apoptosis [1]. Cell-cell and cell-environment interactions play an important role during tissue morphogenesis, homeostasis, and remodeling [2]. Tissue formation and remodeling depend on expression of receptors and cell adhesion molecules, intracellular signaling cascades, and mechanical junctions [3, 4], which are easily disrupted when cells are separated from their native environment. In addition, the extracellular matrix (ECM) affects cellular behavior and gene expression by providing a scaffold with a tissue-specific 3D ultrastructure [5] and allowing transmission of mechanical forces formed by interactions of cells. The ECM consists of fibrous and noncollagen components, including collagens, elastin, fibronectin, laminin, and tenascin. Glycosamine and proteoglycans form the hydrous part of the ECM, in which different cell types such as fibroblasts, nerve endings, pericytes, vessels, and macrophages are located [6].

Anoop Kumar and András Simon (eds.), *Salamanders in Regeneration Research: Methods and Protocols*, Methods in Molecular Biology, vol. 1290, DOI 10.1007/978-1-4939-2495-0_19, © Springer Science+Business Media New York 2015

Monolayer cultures of cells fail to form the three-dimensional bioscaffold and do not provide the characteristic interaction with adjacent cells typical of organs in vivo. As a result, cells in monolayer cultures show a deviant gene expression pattern and behave differently compared to their in vivo counterparts. 3D cultures of primary cells are better suited to recapitulate the tissue microenvironment and physiology. Therefore, several different 3D culture systems such as organotypic explants, spheroids consisting of aggregated single cells, polarized cell cultures by cultivation on porous membranes, in collagen or Matrigel, microcarrier culture, or tissue-engineered models have been developed [1, 2]. Such models are mainly used for drug screening [4] or toxicity tests, but also for cancer, cell adhesion, migration, and gene expression studies [2].

Unlike mammals, newts are able to regenerate heart tissue after injury [7, 8]. Dedifferentiation of cardiomyocytes occurs by changes in myofibrillar morphology including division of myofibrils, cross-links between adjacent myofibrils, or radial spreading of myofilaments into sarcoplasm [7, 9–14]. Adult newt cardiomyocytes reenter the cell cycle and proliferate in vivo after cardiac damage [8, 9, 11–15]. Similar processes occur during cultivation of larval newt hearts [10], slice cultures [9], or isolated cardiomyocytes [16–19] presumably induced by damages inflicted by cutting hearts into slices or as a result of disrupting the 3D tissue organization. So far, attempts to genetically manipulate newts to study newt heart regeneration in vivo mechanistically showed only limited success. Therefore, we developed an ex vivo cultivation system of newt hearts, allowing us to avoid excessive cell death or major morphological changes during a 5-week cultivation period (Figs. 1 and 2). Similar to the in vivo situation, S-phase entry of cardiac cells in explanted hearts increased after experimental damage of the ventricle (Fig. 3). Application of the long-term heart organ culture system uncovered major differences in the ability of cultured newt hearts to rebuild trabeculated ventricular structures depending on the timepoint of the injury. We found that hearts explanted 24 h after initial injury showed a much better ability to regenerate compared to hearts explanted immediately after damage. Improved regenerative capacity was reflected by migration of dedifferentiated cardiomyocytes into damaged areas, a process that is a characteristic for newt heart regeneration in vivo. Hearts, cultivated immediately after injury, showed significant lower levels of cardiomyocyte dedifferentiation and migration and failed to rebuild ventricular trabeculations. We are currently studying the molecular basis of the signals, which trigger cardiac regeneration in newts using transcriptomic and proteomic analysis in combination with inhibition of signaling pathways with small molecules [20–25].

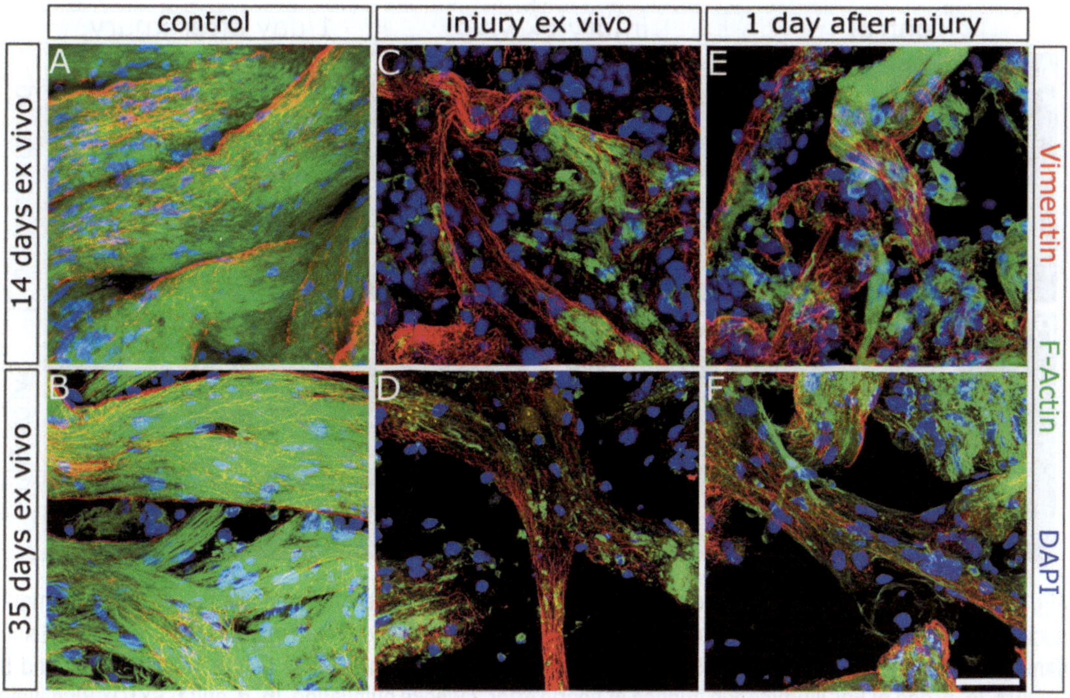

Fig. 1 Immunofluorescence staining of tissue sections from hearts after 14 and 35 days of ex vivo culture. (**a**) Control hearts show a normal morphology after 14 days in culture as indicated by staining for F-actin, Vimentin, and nuclei (DAPI). (**b**) Only slight changes in the trabeculated network of the ventricle are visible after 5 weeks of cultivation. (**c** and **d**) Thin trabeculae with scattered cardiomyocytes are visible in the damaged region of hearts cultivated directly after injury. (**e** and **f**) Hearts placed in culture 1 day after experimental injury show trabeculae that contain multiple dedifferentiated cardiomyocytes. *Red*, Vimentin; *green*, F-actin; *blue*, DAPI. Scale bar: 75 μm

2 Materials

2.1 Animal Handling and Heart Injury

1. Sulfamerazine solution: Dissolve 5 g sulfamerazine (Sigma) in 1 L tap water (*see* **Note 1**).
2. Alvegesic solution: Alvegesic (butorphanol 10 mg/mL) is diluted in tap water (0.5 mg/L) [26] (*see* **Note 2**).
3. Tricaine solution: 1 g/L tricaine (ethyl 3-aminobenzoate methanesulfonate or MS 222) is dissolved in tap water, and the pH is adjusted to 7.4 with $NaHCO_3$.
4. Tissue glue Histoacryl (B. Braun) (*see* **Note 3**).
5. Sterile microscissors and fine forceps with curved and straight-shaped tips (*see* **Note 4**).
6. Stereomicroscope.

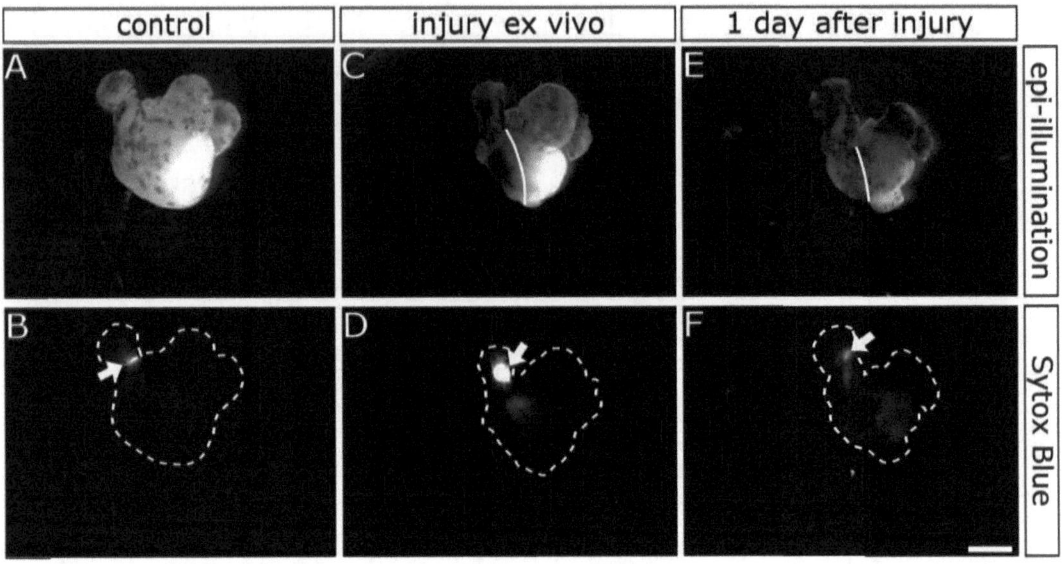

Fig. 2 SYTOX Blue staining of cultured newt hearts. The cell membrane-impermeable dye SYTOX Blue marks nuclei of cells with damaged membranes undergoing cell death. To generate an internal positive control, tissue damage was induced by squeezing the aortic trunk with forceps immediately before staining (indicated by *arrows*). (**a**, **c**, and **e**) Epi-illumination images of hearts after 2 weeks cultivation. (**b**, **d**, and **f**) SYTOX Blue staining of the hearts is shown in the *upper panel*. Cultured hearts do not show significant signs of cell death after 14 days in culture irrespective of previous experimental damage. *Solid lines* indicate injured regions of hearts (**c** and **e**). *Dashed lines* indicate the outer borders of cultured hearts (**b**, **d**, and **f**). Scale bar: 500 μm

2.2 Organ Removal

1. 70 % ethanol.
2. L15 Leibovitz medium: To 35 mL L15 Leibovitz medium with GlutaMAX, add 1 mL penicillin/streptomycin solution, and make up to 50 mL with sterile cell culture water (*see* **Note 5**).
3. Sterile scissors for decapitation, microscissors, and fine forceps with curved and straight tips.
4. Stereomicroscope.

2.3 Cultivation

1. Cultivation medium: To 30 mL MEM (minimum essential medium) with GlutaMAX, add 5 mL FCS (fetal calf serum) and 1 mL penicillin/streptomycin solution. Additionally, add ciprofloxacin (CiproHEXAL 100 mg/50 mL) at a final concentration of 10 μg/mL, and fill up to 50 mL with sterile cell culture water.
2. 3.5 cm plastic dishes and 24-well plates.
3. EdU (5-ethynyl-2′-deoxyuridine/component A) from a Click-iT EdU Alexa Fluor 488 Imaging Kit (Life Technologies) for S-phase labeling.

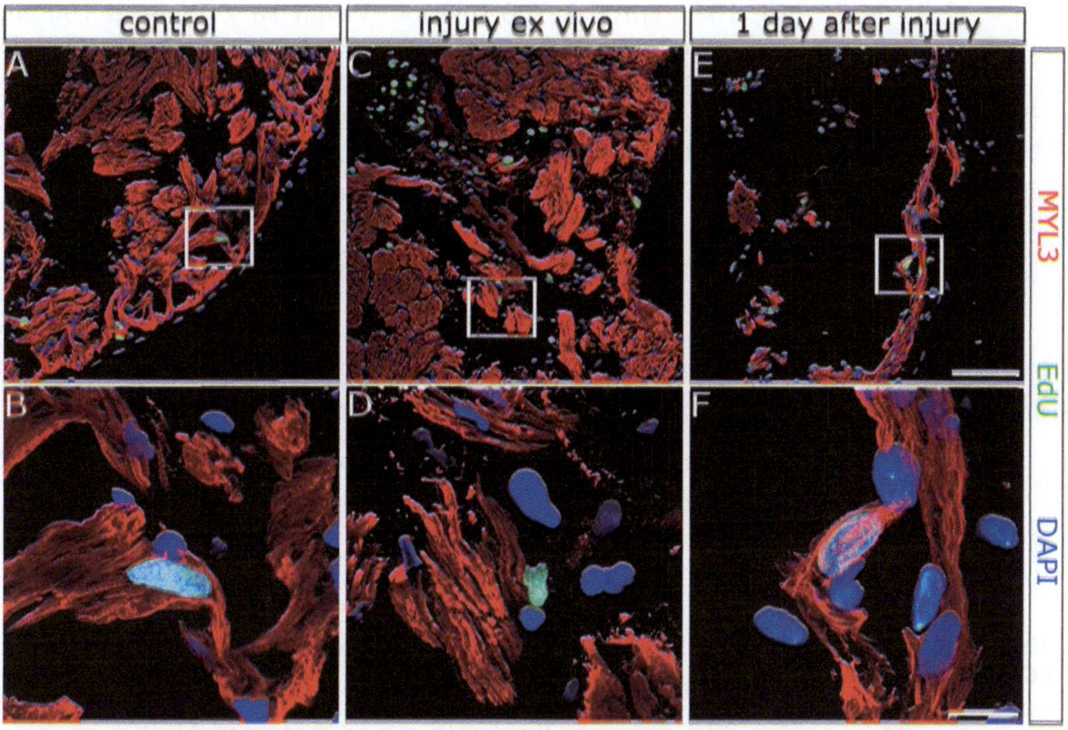

Fig. 3 EdU labeling of hearts after 14 days in culture. Cardiac cells enter S-phase and proliferate in vivo after injury. Low-level EdU labeling is also present in uninjured hearts [8]. (**a–f**) EdU incorporation of cells in cultured hearts indicating S-phase entry. (**a**) Control hearts showing low numbers of EdU-positive cells. (**b**) EdU-labeled cardiomyocytes in uninjured control hearts. (**c–f**) After heart injury, EdU-positive cardiac cells increase in number. *Red*, myosin light chain 3 (MYL3); *green*, EdU; *blue*, DAPI. *White borders* in the *upper row* indicate magnified areas. Scale bar (**a**, **c**, and **e**), 150 μm; (**b**, **d**, and **f**), 25 μm

2.4 Tissue Staining Against Vimentin and F-Actin

1. 10× APBS (amphibian phosphate buffered saline): Dissolve 80 g NaCl, 14.4 g $NaHPO_4 \times 2H_2O$, 2 g KH_2PO_4, and 2 g KCl; make up to 1 L with deionized water; and stir well.
2. 1× APBS: Dilute 75 mL 10× APBS to 1 L with deionized water and adjust the pH to 7.4 with NaOH.
3. 4 % paraformaldehyde (PFA) solution: Add 4 g paraformaldehyde to 80 mL of APBS. For better solubility, add a pellet of NaOH and stir the solution. Next, pass the solution through a pleated filter, and fill up to 100 mL with APBS. Adjust the pH to 7.4 with NaOH (*see* **Note 6**).
4. 2 % agarose gel: Dissolve 0.2 g agarose (low-melting agarose is preferred) in 1× APBS by using a microwave. To embed the hearts, the gel should be cooled down.
5. Antibody against Vimentin, conjugated with Cy3 (Sigma, Clone V9).
6. Phalloidin, conjugated with FITC (Sigma).

7. 10 % Triton X-100 solution: Weigh 10 g Triton X-100 and make up to 100 mL with APBS, and stir well to dissolve the solution. The final concentration of Triton X-100 will be 0.1 %, by adding 1 μL 10 % Triton X-100 solution to 99 μL of staining solution (*see* **Note** 7).
8. DAPI (4′,6-diamidino-2-phenylindole) nuclei staining solution: Dilute DAPI stock solution (1:1,000) in APBS.
9. Razor blades.
10. Glass bottom dishes.

2.5 SYTOX Blue Staining of Dead Cells

1. SYTOX Blue nuclei dye (Life Technologies).
2. Forceps.
3. Stereomicroscope with an appropriated filter (e.g., Leica M205 FA).
4. Cultivation medium (*see* Subheading 2.3, **item 1**).

2.6 Sample Preparation for Cryosections

1. 1× APBS (*see* Subheading 2.4, **item 2**).
2. 4 % PFA solution (*see* Subheading 2.4, **item 3**).
3. 20 % sucrose: Dissolve 10 g D-saccharose in APBS.
4. 30 % sucrose: Dissolve 15 g D-saccharose in APBS.
5. OCT Tissue-Tek embedding media (Sakura).
6. Cryomolds.
7. SuperFrost Plus slides (Menzel).

2.7 Reagents for S-phase Labeling and Antibody Staining Against Myosin Light Chain 3 (MYL3)

1. Click-iT EdU Alexa Fluor 488 Imaging Kit (Life Technologies).
2. 1× APBS (*see* Subheading 2.4, **item 2**).
3. 0.5 % Triton X-100 in 1× APBS (*see* **Note** 7).
4. 3 % bovine serum albumin (BSA) in APBS.
5. Deionized water.
6. Antibody against MYL3 (Sigma HPA 016564).
7. 5 % BSA in 1× APBS.
8. 10 % Triton X-100 solution (*see* Subheading 2.4, **item 7**).
9. Cy3-conjugated secondary antibody against rabbit (Dianova).
10. DAPI nuclei stain solution (*see* Subheading 2.4, **item 8**).
11. Mowiol (mounting medium).

3 Methods

Two different injury models are possible in newts: (1) an immediate ex vivo injury model, where the hearts are damaged directly before cultivation (injury ex vivo) and (2) a delayed ex vivo injury model, where the injury is applied 24 h before cultivation (1 day after injury).

3.1 Animal Pretreatment, Heart Injury, and Organ Removal

1. For "1 day after injury" heart cultures, keep animals 6 h before operation in Alvegesic solution.
2. Afterward, deeply anesthetize the newts by incubating them for 15–20 min in tricaine (*see* **Note 8**).
3. To induce heart injury, keep the newt facing ventral side and make a small skin incision with microscissors. Cut the exposed pericardium and fix the heart at the aortic trunk using curved fine forceps.
4. For heart injury, squeeze the right half of the ventricle 15 up to 20 times with a straight fine forceps in orthogonal direction. Afterward, carefully return the heart into the thoracic cavity, and close the wound with Histoacryl.
5. For disinfection, incubate the animals in sulfamerazine solution until recovery from anesthesia, and transfer them into husbandry aquaria with Alvegesic solution [7, 27].
6. To avoid contaminations during cultivation, immerse the animals for 2 h in sulfamerazine solution before organ removal.
7. After deep anesthesia in tricaine solution, decapitate the newts and wash them with 70 % ethanol.
8. Remove the hearts, transfer them to L15 Leibovitz media, and wash three times to remove the blood (*see* **Note 9**).
9. Prepare the hearts for the group "direct injury" by fixing them at the aortic trunk, and damage one half of the ventricle by squeezing with forceps, following the procedure described in in vivo operated hearts for "1 day after injury" cultures (*see* **step 4**).

3.2 Preparation and Cultivation of Ex Vivo Hearts

1. After the transfer of the removed organs to a laminar flow hood, wash the hearts three times in cultivation medium, and culture each heart in 3.5 cm cell culture dishes in an incubator maintained at 25 °C with 5 % CO_2 (*see* **Note 10**). Perform first medium change after 24 h and subsequently twice a week. Transfer hearts into 24-well plates at the second medium change.
2. For S-phase labeling, add EdU (200 μM) to the cultivation medium 3 days before checkpoint (*see* **Note 11**).

3.3 Vimentin and F-Actin Tissue Staining

1. Wash the hearts twice in APBS, embed them in 2 % agarose gel, and cut them with a razor blade in two halves.
2. Wash them again in APBS.
3. Fix the samples for 30 min with 4 % PFA solution.
4. Wash them again three times with APBS.
5. Label endothelia cells by incubating the samples in a 1:300 dilution of the Vimentin-Cy3 antibody in APBS for 2 h at room

temperature (*see* **Note 12**). Use an FITC-coupled phalloidin in a 1:100 dilution in APBS with 0.1 % Triton X-100 to detect F-actin of cardiomyocytes. Incubate the samples therein overnight at 4 °C. Thereafter, incubate the samples for 30 min at room temperature in DAPI solution (1:1,000) for counterstaining the nuclei.

6. Wash the samples extensively for the next 2 h with APBS by changing the APBS every 15 min.
7. Embed the tissue in 2 % agarose gel in glass bottom dishes and take images immediately with a confocal microscope.

3.4 SYTOX Blue Dead Cell Stain

1. The cell membrane-impermeable dye SYTOX Blue is used to stain cells with damaged membranes, a sign of pending cell death. Add SYTOX Blue to the cultivation media to a final concentration of 5 μM (*see* **Note 12**). To control staining intensity, squeeze the aortic trunk with sterile forceps to induce necrosis and incubate the hearts for 2 h in an incubator at 25 °C with 5 % CO_2.
2. Wash the hearts twice with cultivation medium.
3. Observe the hearts immediately under a stereofluorescence microscope with appropriated filter and record the images (*see* **Note 13**).

3.5 EdU Detection, Antibody Staining Against MYL3 (Myosin Light Chain 3) in Cardiomyocytes, and Nuclei Labeling with DAPI

1. Wash the hearts twice in APBS and fix them for 1 h with 4 % PFA solution.
2. Wash the tissues 3× for 10 min each in APBS.
3. Incubate the samples in 20 % sucrose overnight at 4 °C.
4. Thereafter, transfer the hearts to 30 % sucrose, and incubate them for the next 48 h at 4 °C.
5. For embedding, transfer the heart to a cryomold containing OCT Tissue-Tek and incubate for 10 min at room temperature.
6. Freeze the cryomold in liquid nitrogen and store the samples at −80 °C (*see* **Note 14**).
7. Cut 6 μm sections by using a cryostat microtome and mount the sections on SuperFrost Plus slides.
8. Perform EdU labeling as described for cell culture by using the protocol provided by the supplier, but without fixation.
9. Afterward, wash the slides 3×for 10 min in 5 % BSA solution.
10. Label the cardiomyocytes by using an antibody against MYL3. Dilute the antibody 1:500 in 5 % BSA and 0.1 % Triton X-100 in APBS and incubate the slides therein overnight at 4 °C.
11. Wash the samples 3×for 10 min each with APBS.

12. To detect the primary antibody, incubate with a Cy3-conjugated secondary anti-rabbit antibody at 1:300 dilution in APBS for 45 min.
13. Counterstain the nuclei with DAPI (1:1,000) for 20 min at room temperature.
14. Finally, wash the samples 3× for 10 min with APBS, and mount the samples with a cover slip by using Mowiol.
15. Let the slides dry overnight at 4 °C. For imaging, use a confocal microscope.

4 Notes

1. Sulfamerazine inhibits folic acid synthesis and is antibacterial. It is a fine white powder. Protect yourself and work under a hood.
2. Alvegesic (butorphanol) solution is given as a pre- and postoperative analgesic to the newts.
3. Histoacryl is a tissue glue, which harden within seconds. Always wear gloves and avoid skin contact.
4. Dissection instruments are sterilized by autoclaving or by incubation in 70 % ethanol for 15 min.
5. L15 Leibovitz medium is buffered by a combination of salt, sugar, and amino acids and can be used for cultivation without CO_2/sodium bicarbonate.
6. When using PFA powder protect yourself by working under the hood. Wear nitrile gloves, a lab coat and protective goggles. Avoid skin contact. To enhance solubility of PFA in APBS, you might increase the temperature by using a temperature-regulated magnetic mixer, but do not exceed 50 °C.
7. Triton X-100 is highly viscous solution and pipetting an exact volume is difficult. Therefore, weigh out Triton X-100 using a laboratory balance.
8. After 15–20 min, newts should not show any swim or tail movement, not even when the animals are lying on their backs. Prolong the incubation time in tricaine solution, if animals show any movements.
9. The newt skin is contaminated by bacteria, which is only reduced after disinfection with sulfamerazine. Therefore, first remove control hearts and hearts subjected to damage after organ removal. Hearts that are removed 1 day after injury should be dissected last. Separate the hearts in single tubes during transfer to cell culture and always disinfect the dissecting stools between the procedures.

10. Contaminations by bacteria will be visible during the first 24 h after setting up the cultures. Therefore, use single 3.5 cm culture dishes, so you can simply remove contaminated dishes, without taking a risk for the whole culture.
11. For the preparation of solutions and EdU detection, follow the protocol provided by the supplier.
12. Fluorescence dyes are light sensitive and should be protected from light to avoid photo bleaching. Samples should be stored in the dark, for example, by covering with aluminum foil.
13. SYTOX Blue is a cell membrane-impermeable dye, which only stains nuclei after cell membrane damage. Fixation with PFA affects membrane integrity of cells. If PFA fixation is done after SYTOX Blue staining, the remaining dye will stain the nuclei of previously negative cells, resulting in false-positive signals.
14. Freezing of hearts in OCT Tissue-Tek without fixation and incubation in sucrose results in the formation of ice crystals inside the hearts and impairs tissue morphology when placing samples in liquid nitrogen.

Acknowledgments

This work was supported by the Max-Planck-Society, the Excellence Cluster Cardiopulmonary System (ECCPS), the University of Giessen-Marburg Lung Center (UGMLC), the Cell and Gene Therapy Center (CGT) of the University of Frankfurt, and the DFG (SFB TRR81).

References

1. Haycock JW (2011) 3D cell culture: a review of current approaches and techniques. Methods Mol Biol 695:1–15
2. Pampaloni F, Reynaud EG, Stelzer EH (2007) The third dimension bridges the gap between cell culture and live tissue. Nat Rev Mol Cell Biol 8:839–845
3. Green JB (2008) Sophistications of cell sorting. Nat Cell Biol 10:375–377
4. Griffith LG, Swartz MA (2006) Capturing complex 3D tissue physiology in vitro. Nat Rev Mol Cell Biol 7:211–224
5. Badylak SF (2002) The extracellular matrix as a scaffold for tissue reconstruction. Semin Cell Dev Biol 13:377–383
6. Hein S, Schaper J (2001) The extracellular matrix in normal and diseased myocardium. J Nucl Cardiol 8:188–196
7. Piatkowski T, Muhlfeld C, Borchardt T, Braun T (2013) Reconstitution of the myocardium in regenerating newt hearts is preceded by transient deposition of extracellular matrix components. Stem Cells Dev 22:1921–1931
8. Witman N, Murtuza B, Davis B, Arner A, Morrison JI (2011) Recapitulation of developmental cardiogenesis governs the morphological and functional regeneration of adult newt hearts following injury. Dev Biol 354:67–76
9. Nag AC, Healy CJ, Cheng M (1979) Organ culture of adult amphibian heart: a fine structural analysis. Tissue Cell 11:231–248
10. Millhouse EW Jr, Chiakulas JJ, Scheving LE (1971) Long-term organ culture of the salamander heart. J Cell Biol 48:1–14
11. Bader D, Oberpriller J (1979) Autoradiographic and electron microscopic studies of minced

cardiac muscle regeneration in the adult newt, Notophthalmus viridescens. J Exp Zool 208: 177–193

12. Bader D, Oberpriller JO (1978) Repair and reorganization of minced cardiac muscle in the adult newt (Notophthalmus viridescens). J Morphol 155:349–357
13. Oberpriller J, Oberpriller JC (1971) Mitosis in adult newt ventricle. J Cell Biol 49:560–563
14. Oberpriller JO, Oberpriller JC (1974) Response of the adult newt ventricle to injury. J Exp Zool 187:249–253
15. Oberpriller JO, Oberpriller JC, Arefyeva AM, Mitashov VI, Carlson BM (1988) Nuclear characteristics of cardiac myocytes following the proliferative response to mincing of the myocardium in the adult newt, Notophthalmus viridescens. Cell Tissue Res 253:619–624
16. Bettencourt-Dias M, Mittnacht S, Brockes JP (2003) Heterogeneous proliferative potential in regenerative adult newt cardiomyocytes. J Cell Sci 116:4001–4009
17. Tate JM, McDonnell TJ, Oberpriller JC, Oberpriller JO (1987) Isolation of cardiac myocytes from the adult newt, Notophthalmus viridescens. An electron microscopic and quantitative light microscopic analysis. Tissue Cell 19:577–585
18. Tate JM, Oberpriller JO (1989) Primary cell culture and morphological characterization of ventricular myocytes from the adult newt, Notophthalmus viridescens. Anat Rec 224: 29–42
19. Tate JM, Oberpriller JO, Oberpriller JC (1989) Analysis of DNA synthesis in cell cultures of the adult newt cardiac myocyte. Tissue Cell 21:335–342
20. Looso M (2014) Opening the genetic toolbox of niche model organisms with high throughput techniques: novel proteins in regeneration as a case study. Bioessays 36:407–418
21. Looso M, Preussner J, Sousounis K, Bruckskotten M, Michel CS, Lignelli E, Reinhardt R, Hoffner S, Kruger M, Tsonis PA, Borchardt T, Braun T (2013) A de novo assembly of the newt transcriptome combined with proteomic validation identifies new protein families expressed during tissue regeneration. Genome Biol 14:R16. doi:10.1186/gb-2013-14-2-r16
22. Looso M, Michel CS, Konzer A, Bruckskotten M, Borchardt T, Kruger M, Braun T (2012) Spiked-in pulsed in vivo labeling identifies a new member of the CCN family in regenerating newt hearts. J Proteome Res 11:4693–4704
23. Bruckskotten M, Looso M, Reinhardt R, Braun T, Borchardt T (2012) Newt-omics: a comprehensive repository for omics data from the newt Notophthalmus viridescens. Nucleic Acids Res 40(Database issue):D895–D900
24. Looso M, Borchardt T, Kruger M, Braun T (2010) Advanced identification of proteins in uncharacterized proteomes by pulsed in vivo stable isotope labeling-based mass spectrometry. Mol Cell Proteomics 9:1157–1166
25. Borchardt T, Looso M, Bruckskotten M, Weis P, Kruse J, Braun T (2010) Analysis of newly established EST databases reveals similarities between heart regeneration in newt and fish. BMC Genomics 11:4. doi:10.1186/1471-2164-11-4
26. Koeller CA (2009) Comparison of buprenorphine and butorphanol analgesia in the eastern red-spotted newt (Notophthalmus viridescens). J Am Assoc Lab Anim Sci 48:171–175
27. Laube F, Heister M, Scholz C, Borchardt T, Braun T (2006) Re-programming of newt cardiomyocytes is induced by tissue regeneration. J Cell Sci 119:4719–4729

Chapter 20

In Vitro Preparation of Newt Inner Ear Sensory Epithelia as a Model for Repair and Regeneration

Ruth R. Taylor

Abstract

The sensory "hair" cells of the inner ear transform sound energy into electrical signals, but are readily lost through aging, excessive noise, and ototoxic agents. The newt provides an excellent model in which to explore regeneration and whilst loss of hair cells from inner ear epithelia does not require whole organ regeneration, new hair cells are generated from differentiated supporting cells that transdifferentiate without an intervening mitotic event. Here we describe the methods for maintaining the sensory epithelia in long term culture; for the use of the aminoglycoside, gentamicin, to kill the hair cells; and for the examination of the tissue by electron microscopy or fluorescence microscopy. Demembranation of the epithelium reveals the underlying ultrastructure of the tissue for examination by scanning electron microscopy (SEM) and is a technique that can be utilized with immunogold labelling.

Key words Hair cell, Salamander, *Notophthalmus viridescens*, Stereocilia, Utricle, Saccule

1 Introduction

The inner ear contains the mechanosensitive sensory epithelia that detect and transduce sound and those in the vestibular portion that detect changes in head position and are involved in maintaining balance and posture. The inner ear of the adult newt, *Notophthalmus viridescens* comprises seven sensory epithelia: the three cristae located within the ampullae of the semicircular canals that detect angular motion of the head; the utricle which detects gravity and acceleration; the amphibian papilla and saccule which are auditory end organs and the lagena which performs mostly a vestibular function. In some other newt species there is also a basilar papilla but this has been lost from *Notophthalmus viridescens*.

The mechanosensory "hair" cells of the inner ear epithelia transform sound waves and vibrations into electrical signals that ultimately are detected and interpreted by the cortex of the brain [1]. In mammals, the loss of these cells from the auditory system results in a progressive and permanent deafness. Ototoxic agents,

Anoop Kumar and András Simon (eds.), *Salamanders in Regeneration Research: Methods and Protocols*, Methods in Molecular Biology, vol. 1290, DOI 10.1007/978-1-4939-2495-0_20, © Springer Science+Business Media New York 2015

excessive noise, and aging can trigger this loss and new hair cells are not generated. Following damage to the mammalian vestibular system there is a very limited level of hair cell replacement [2]. These new hair cells are thought to arise from the non-sensory supporting cells that surround each hair cell via direct phenotypic conversion (transdifferentiation) without a mitotic event [3]. The capacity of urodele amphibians, such as *Notophthalmus viridescens*, to regenerate various body parts and tissues such as limb and the lens has lead us to investigate the generation of new hair cells arising spontaneously following their ablation in the inner ear of this vertebrate [4].

It has long been known that aminoglycosides, used to treat tuberculosis, gram-negative sepsis, and many other infections, are ototoxic and can cause deafness through the loss of hair cells and consequent loss of spiral ganglion cells [5]. They have been used in many inner ear studies both in lower vertebrates and in mammals. In mammals administration of aminoglycoside such as gentamicin by a series of systemic injections will cause permanent damage to the cochleae and some aminoglycosides will affect the vestibular sensory epithelia. In the newt, attempts to systemically deliver aminoglycosides to ablate hair cells have proved to be difficult as sufficiently high dosage results in death of the animal due to kidney damage. However, in vitro organotypic long term cultures provide a suitable preparation to examine recovery and regeneration of the inner ear epithelia.

2 Materials

Adult newts are available from animal suppliers from the USA (e.g., Charles D. Sullivan & Co.). Maintain the animals according to the regulations of the institution and/or the guidelines of the authorities.

2.1 Culture of Sensory Epithelia

1. Anesthetic: Prepare 0.1 % 3-aminobenzoic acid ethyl ester (MS-222) in distilled water.
2. 70 % Ethanol.
3. Amphibian PBS (APBS): To prepare 100 mL APBS, mix 80 mL Dulbecco's phosphate buffered saline with 20 mL sterile distilled water.
4. Amphibian minimal essential medium (AMEM): Prepare 80 % Minimal Essential Medium with GlutaMAX (MEM) + 20 % sterile water. Then add 1 % Hepes buffer (4-(2-hydroxyethyl)-1-piperazineethanesulfonic acid) and 10 % horse serum (donor herd collected).
5. Hair cell culture medium (H-AMEM): To the AMEM add 1 % amphotericin B (antimycotic) and 100 µg/mL ciprofloxacin (a non-ototoxic broad spectrum antibiotic).

6. Gentamicin sulfate (aminoglycoside antibiotic): Prepare a stock solution of 10 mM in sterile water and store at –20 °C for up to 6 months.
7. Hair cell ablation medium: To the H-AMEM solution (**item 5**), add 2 mM gentamicin solution.
8. Surgical tools: Fine forceps (no. 3 and 5 fine tip medical grade tweezers suitable for dry heat sterilizing/autoclaving), spatula and curettes and assorted small dental surgical tools.
9. Sterile plasticware: 35 mm petri dishes, 24-well plates, 15 mL falcon tubes.
10. Stereo dissecting microscope with incident light illumination.

2.2 Scanning Electron Microscopy (SEM)

1. Cacodylate buffer: Prepare a stock solution of 0.2 M sodium cacodylate with 6 mM $CaCl_2$, adjust to pH 7.3 with 0.2 M HCl. Store at 4 °C.
2. Cacodylate buffer working solution: Buffer diluted 1:1 with deionized water for working concentration of 0.1 M cacodylate.
3. Primary fixative: 2.5 % glutaraldehyde in 0.1 M cacodylate buffer. Prepare from 10 mL individual vials of 25 % glutaraldehyde (electron microscopy grade) in water. Prepare a working solution by mixing one part 25 % glutaraldehyde, five parts 0.2 M cacodylate buffer, and four parts sterile water. Store at 4 °C.
4. Osmium tetroxide (OsO_4) post-fixative. Break open a 1 g vial of solid OsO_4 in a fume hood and add to 50 mL of distilled or deionized water to provide a 2 % solution. The OsO_4 requires some time to dissolve. Leave overnight before first use. Store at room temperature in dark-glass bottle to reduce exposure to light (*see* **Note 1**).
5. 1 % OsO_4 solution: Mix equal volumes of 2 % OsO_4 and 0.2 M cacodylate to give a 1 % solution in 0.1 M cacodylate buffer as a post-fixative; dilute 1:1 with water during processing with thiocarbohydrazide.
6. Thiocarbohydrazide (TCH): Make a filtered saturated solution in water. 0.5 g/100 mL provides a saturated solution. Filter before use. Use fresh solution each time.
7. Ethanol series: Dilute 100 % ethanol with deionized water to give 30, 50, 70, 85 and 90 %. Store at room temperature.
8. 100 % Ethanol (pure, anhydrous).
9. Small sample holders, e.g., specimen containers for critical point dryer.
10. Small glass bottles suitable size for the sample holders.
11. Critical point dry device: from liquid CO_2.

12. Specimen support stubs for SEM. Use appropriate size for the instrument according to manufacturer's recommendations for their particular holders.
13. Silver Paint: Electrolube Silver Conductive paint or Electrodag Fast drying conductive paint (provides a conductive adhesive for attaching specimens to support stub).
14. Sputter coater: platinum foil.

2.3 Demembranation of Tissue

1. Cytoskeletal buffer with phallacidin (CPP buffer): Prepare a fresh solution of 5 mM KCl, 137 mM NaCl, 4 mM $NaHCO_3$, 0.4 mM KH_2PO_4, 1.1 mM Na_2HPO_4, 2 mM $MgCl_2{\cdot}6H_2O$, 5 mM PIPES, 2 mM EGTA, and 5.5 mM Glucose. Adjust the pH 6.0–6.1 if necessary using NaOH (*see* **Note 2**). Add 5 μM Phallacidin 1mg/mL in methanol stock solution and store at −20 °C. To this solution, add Protease inhibitor tablets (EDTA free): one tablet dissolved just before use as per supplier's instruction.
2. 0.25 % Triton X-100: dilute Triton X-100 stock solution in CPP buffer.
3. 0.25 % saponin: dilute saponin in CPP buffer.
4. Fixative: 2.5 % glutaraldehyde in CPP buffer (*see* **item 1**).
5. 0.2 M cacodylate buffer (*see* Subheading 2.2, **item 1**).
6. 2 % OsO_4 (*see* Subheading 2.2, **item 4**).
7. Specimen holders and small glass bottles.

2.4 Immunogold Labelling

1. Fixative: 4 % paraformaldehyde and 0.05 % glutaraldehyde in CPP buffer (Subheading 2.3, **item 1**).
2. Blocking solution: Prepare a solution of 1 % bovine serum albumin, 0.15 % Triton X-100 in CPP buffer (Subheading 2.3, **item 1**).
3. Primary antibody: Dilute anti Myosin VI 1:50 in blocking solution (**item 2**).
4. Secondary antibody: Gold-conjugated goat anti-rabbit IgG—Dilute 1:20 in blocking solution (**item 2**).
5. PBS wash solution: PBS with 0.15 % Triton X-100 for rinsing.

2.5 Transmission Electron Microscopy (TEM)

1. Prepare 2.5 % glutaraldehyde fixative, osmium tetroxide, sodium cacodylate buffer as described in Subheading 2.2, **items 1–3**.
2. Decalcification buffer: 4.13 % EDTA in 0.1 M cacodylate buffer.
3. Ethanol series: 30–90 % in deionized water.

4. Ethanolic uranyl acetate: saturated solution of uranyl acetate in 70 % ethanol.
5. Aqueous uranyl acetate: 1 g uranyl acetate per 100 ml deionized water.
6. Lead citrate.
7. Plastic resin for embedding.
8. Silicon-coated rubber molds for embedding.
9. Embedding oven.

2.6 Immunofluorescence

1. Phosphate buffer solution (PBS): Prepare as a 1× solution by dissolving one phosphate buffered saline tablet in 200 mL in deionized water (yields 0.1 M phosphate buffer, 0.0027 M potassium chloride and 0.137 M sodium chloride, pH 7.4 at 25 °C). Alternatively, dissolve 137.0 mM NaCl, 2.7 mM KCl, 8.0 mM Na_2HPO_4, 1.5 mM KH_2PO_4, in deionized water.
2. Paraformaldehyde (PFA) fixative. Prepare a 4 % (w/v) solution in PBS, warming carefully on a stirring hotplate in a fume hood. Solution is cooled to room temperature before use. For storage at −20 °C, for up to 6 months, prepare 8 % (w/v) solution in ×2 PBS. Dilute to 4 % prior to use.
3. Permeabilization: Prepare 0.5 % (v/v) Triton X-100 in sterile water.
4. Blocking solution: Prepare a solution of 0.15 % Triton X-100 and 10 % (v/v) normal goat serum in PBS. Store for 2–3 days at 4 °C.
5. PBS wash solution: 0.15 % Triton X-100 in PBS.
6. Antibody dilution buffer: 0.1 % (v/v) Triton X-100 and 100 mM L-lysine in PBS. Store for up to 7 days at 4 °C.
7. Primary antibodies: antibodies to calretinin, otoferlin (HCS-1), neurofilament.
8. Secondary antibodies: Goat anti-mouse IgG conjugated to Alexa 488 or similar, Goat anti-rabbit IgG conjugated to Alexa 555 or similar.
9. Phalloidin conjugated to TRITC or FITC.
10. Antifade mountant with DAPI .
11. Glass slides: PTFE multi-spot glass slides.

3 Methods

3.1 Dissection and Culture of the Inner Ear (See Note 3)

1. Anesthetize the animals by placing them in a 0.1 % solution of 3-aminobenzoic acid ethyl ester for approximately 15 min.
2. Remove the animal from the water and kill in accordance with local regulations. Following decapitation, remove the lower jaw and bisect the head.

3. Place bisected head in 70 % ethanol for 5 min to reduce the chance of bacterial infection (*see* **Note 4**).
4. Rinse and continue dissection in APBS. The otic capsule lies within close proximity of the eye and careful separation of tissue will expose the bright, hard bony capsule. Gently remove the connective tissue and muscle surrounding the otic capsule using fine surgical tools (dental probe) to reveal the bony outer wall of the capsule.
5. Retract the columellar bone to one side and excise the otic capsule (Fig. 1a) from the head.
6. Use a curette to open the thick membranous covering of the oval window and expand the window opening by scraping away some of the edge. This will allow medium to infiltrate the inner ear epithelia more readily during culturing.
7. The utricular and saccular maculae are normally covered with an "otoconial membrane" that consists of a fibrous extracellular matrix covered with calcium carbonate crystalline particles (otoconia). Gently aspirate the otoconia from the otic capsule with APBS using a syringe with a fine needle and incubate in 500 μL H-MEM medium in a sterile 24-well plate. Otoconia remaining during fixation tend to attach to the epithelium obscuring the surface and will show autofluorescence.
8. To ablate hair cells, incubate as above in hair cell ablation medium. For maximal hair cell ablation, incubate for 48 h and rinse with fresh medium.
9. Maintain the cultures by removing and supplementing 50 % of the medium on alternate days.
10. Cultures can be maintained for up to 28 days (*see* **Note 5**).

3.2 Scanning Electron Microscopy (SEM)

1. Fix the inner ear tissues in situ in the otic capsule. For this, place the otic capsule in a 35 mm Petri dish and immobilize using tweezers. Gently perfuse 2.5 % glutaraldehyde using a syringe at the edge of the oval window taking care not to damage the labyrinth.
2. Immerse the entire otic capsule in glutaraldehyde solution for at least 90 min at room temperature on a mechanical rotator, or overnight at 4 °C.
3. Rinse the samples three times for 10 min in 0.1 M cacodylate buffer,
4. Post-fix the tissues in OsO_4 for 2 h at room temperature.
5. To dissect the inner ear epithelial tissues from the otic capsule, place the tissue in cacodylate buffer under a stereo microscope.
6. Gently crack open the otic capsule using a limited amount of force at the edge of the oval window.

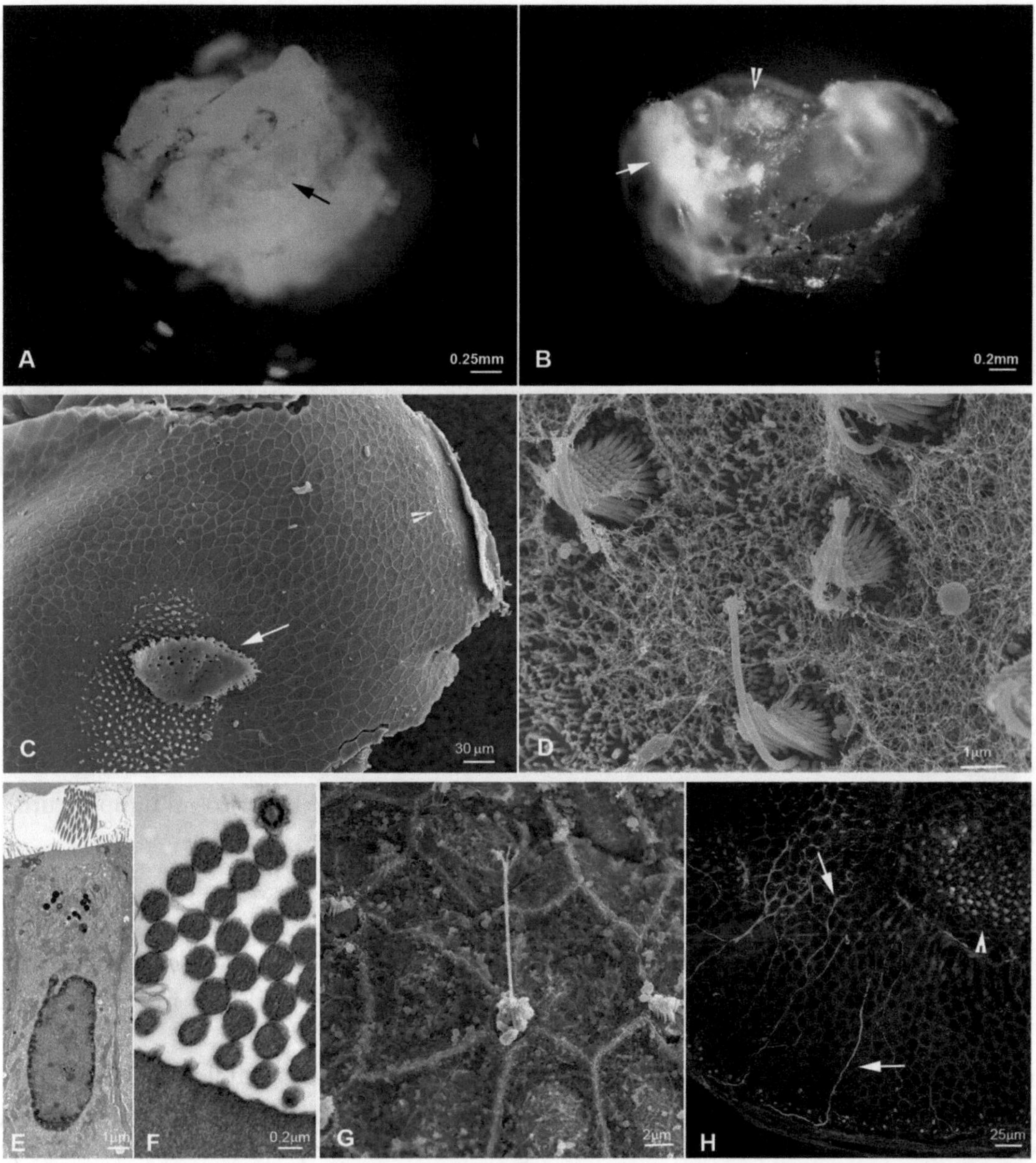

Fig. 1 (**a**) The bony otic capsule with oval window opened (*arrow*). (**b**) Labyrinth removed from otic capsule with underside of the utricle shown to the *left-hand side* (*arrow*) and saccule *uppermost* (*arrowhead*). One ampulla with crista, *right-hand side*. (**c** and **d**) Scanning electron micrographs of saccule. Hair bundles of the saccular macula are partially covered with otoconial membrane (*arrow*) and extramacular hair bundles at the very outer edge of the saccule are present (*arrowhead*). (**d**) Higher magnification of hair cell bundle showing the graduated length of stereocilia and long kinocilium. (**e** and **f**) Transmission electron micrograph of hair cell and cross section of bundle. (**g**) Extrastriolar region of utricular macula after demembranation: small hair bundles are present. One bundle retains its long kinocilium. Its microtubular filaments have begun to splay. Adjacent hair bundle has recess only at position of kinocilium. (**h**) Immunofluorescence of saccule. Neurofilaments are labelled with antibody to neurofilament (*green*) and extend from the central macular hair cell region out to extramacular hair cells at the edge of the saccule (*arrows*). Phalloidin labels the f-actin at cell junctions and the actin-rich hair bundles (*arrowhead*, *red*) (color figure online)

7. The membranous labyrinth will be exposed and can be freed from the bony capsule by inserting a fine spatula beneath the saccule. Care should be taken not to damage the sensory epithelia.
8. The inner ear epithelia (Fig. 1b) can be further dissected to separate the individual end organs, i.e., the saccule can be trimmed with fine iris scissors from the utricle and the semicircular canals removed so that the ampullae can be opened to access the cristae.
9. Gently remove any remaining otoconia and otoconial membrane by lifting with a fine forceps from the saccule and utricle so that the hair bundles and surface of the maculae are visible under the scanning electron microscope (Fig. 1c).
10. Place the dissected epithelium in specimen holders within small glass bottles (*see* **Note 6**).
11. The dissected inner ear tissues are processed through the repeated TCH-OsO_4 procedure to enhance preservation of the morphology and provide a conductive coat [6].
12. Wash the samples in distilled water.
13. Incubate the tissues in a fresh, filtered, saturated solution of TCH for 20 min at room temperature.
14. Wash 6×, in distilled water 5 min each.
15. Incubate in 1 % aqueous OsO_4 for 1 h at room temperature.
16. Wash 6× in distilled water.
17. Repeat incubation in TCH and OsO_4 as in **steps 13–15**.
18. Dehydrate the samples through an ethanol series from 30 to 90 % incubating for 10 min each on a rotator.
19. Dehydrate the samples 2× in 100 % ethanol for 10 min each (*see* **Note 7**).
20. Samples are then critical point dried from liquid CO_2.
21. Mount the dried samples with the epithelium side up onto SEM stubs using silver conductive paint (*see* **Notes 8** and **9**).
22. Leave the samples to fully dry before sputter coating.
23. Sputter-coat with platinum as per device manufacturer's protocol.
24. The samples are ready for observation under a scanning electron microscope.

3.3 SEM of Demembranated Inner Ear Epithelia Tissue

Prior to fixation, the surface membranes can be removed to reveal the underlying cellular ultrastructural organization. Thus, details of the cytoskeletal arrangement that contribute to the overall architecture can be examined using a high resolution SEM. In addition to visualizing structural components, immunogold labelling can be applied to localize proteins and actin filaments can be labelled with gold particles by exposing the sample to gold-conjugated phalloidin. Secondary gold markers are imaged using backscattered electron detection (*see* **Note 10**).

1. Immediately after isolation, incubate the opened otic capsule for 7 min at room temperature in 0.25 % Triton X-100 with 0.25 % saponin in CPP buffer. Agitate gently on a rotator or rocker.
2. Rinse the sample three times in cytoskeletal buffer.
3. Fix in 2.5 % glutaraldehyde in CPP buffer (without protease inhibitor) for 2.5 h for morphological examination.
4. Briefly rinse samples in 0.1 M cacodylate buffer.
5. Post-fix in OsO_4 for 90 min.
6. Rinse 3× with cacodylate buffer.
7. Dissect the saccule and utricle from the labyrinth.
8. Process tissue through the thiocarbohydrazide-OsO_4 repeated procedure before dehydration and critical point drying (*see* Subheading 3.2, **steps 11–21**).
9. Do not sputter-coat. The tissue is examined in a field emission SEM operating at 3 or 5 kV.

3.4 Demembranation and Gold Labelling of Tissue

1. Remove plasma membrane with extraction technique (*see* Subheading 3.3, **step 1**).
2. Rinse three times in CPP buffer.
3. Fix tissue in 4 % paraformaldehyde with 0.05 % glutaraldehyde.
4. Block nonspecific antigens with blocking solution.
5. Incubate tissue in primary antibody for 3 h at room temperature.
6. Rinse with PBS wash solution 4×, 10 min each.
7. Incubate in secondary antibody (goat anti-rabbit IgG conjugated to 10 or 15 nm gold particles) overnight at 4 °C.
8. Thoroughly rinse in PBS wash solution.
9. Postfix in 2.5 % glutaraldehyde for 1 h at room temperature.
10. Processing the tissues for SEM (*see* Subheading 3.3, **steps 4–9**).
11. Carbon-coat tissue (*see* **Note 11**).
12. Use backscatter detection in the SEM to visualize the gold labelling of the sample.

3.5 Transmission Electron Microscopy

For complete sections of the inner ear labyrinth, intact otic capsules can be sectioned following glutaraldehyde fixation and decalcification (*see* **Note 12**). However, end organs can be dissected from the otic capsule following fixation and post-fixation with osmium and processed for TEM as for intact otic capsules.

1. Fix and process samples (*see* Subheading 3.2, **steps 1–10**).
2. Dehydrate samples in an ethanol series to 70 % ethanol.
3. *En bloc* stain in saturated uranyl acetate in 70 % ethanol overnight (*see* **Note 13).**
4. Remove uranyl acetate.
5. Completely dehydrate the samples through to 100 % ethanol.
6. Incubate tissue in propylene oxide for 15 min.
7. Repeat **step 6**.
8. Incubate in propylene oxide and plastic resin 3:1 for 2 h.
9. Leave the tissue overnight in 1:1 propylene oxide; resin (*see* **Note 14**).
10. Infiltrate in pure resin.
11. For embedding the tissue, place the sample infiltrated by pure resin in a mold and orient the tissue as desired under a stereomicroscope.
12. Polymerize by incubating for 24 h at 60 °C.
13. Ultrathin sections are cut from the polymerized block using a diamond knife. Sections of 80 nm thick are collected on to uncoated copper grids (*see* **Note 15**).
14. Stain the sections by floating the grids (section-side down) on droplets of 1 % aqueous uranyl acetate for 15 min.
15. Wash 3× on droplets of deionised water.
16. Stain with lead citrate solution as in **step 14** (*see* **Note 16**).
17. Wash 3× deionized water.
18. Examine the sections in a transmission electron microscope.

3.6 Immunofluorescence for Hair Cell Detection

After fixation and opening of the otic capsule from culture, tissues can be processed for immunofluorescence either as an intact labyrinth or the maculae and cristae can be dissected out at the onset. To prevent loss of the samples processing is carried out using 72 or 96 well plates. During change of solutions damage to the epithelium should be avoided by gently aspirating fluid from around the tissue or if tissue is manipulated from one well to the next a fine spatula is used to lift the sample from underneath.

1. Fix inner ear epithelia in situ within the otic capsule using 4 % paraformaldehyde for 90 min. Initially, a small amount of fixative is gently perfused through the oval window using a syringe to ensure that the fixative reaches the entire labyrinth rapidly.
2. Rinse the otic capsule 3× for 5 min each with PBS.
3. Transfer the otic capsule to a 35 mm petri dish with PBS and place under a dissecting microscope.

4. Open otic capsule using a stainless steel probe or fine dental instrument, such as a dental excavator, and remove the labyrinth with a fine forceps.
5. Dissect the maculae and cristae and remove any remaining otoconial membrane.
6. Incubate the samples in 0.5 % Triton X-100 for 20 min to permeabilize the tissue.
7. Transfer the tissues to blocking solution for 40 min.
8. Replace the blocking solution with a primary antibody for hair cell or neurofilament detection (antibodies to otoferlin, 1:200; calretinin, neurofilament, 1:200 diluted in antibody dilution buffer).
9. Incubate the tissues for 3 h at room temperature on a rocker or overnight at 4 °C.
10. Remove the primary antibody and add PBS.
11. Wash in PBS 4× for 10 min each in PBS.
12. Incubate the tissue is in a 1:100 dilution of the secondary antibody conjugated to either FITC or TRITC in blocking solution (*see* **Note 17**). At this step phalloidin conjugated to an alternative fluorophore can be added to the secondary antibody at 1:1,000 dilution to detect filamentous actin.
13. Remove the secondary antibody and replace with PBS.
14. Wash 3× for 5 min each in PBS.
15. Mount the tissue using antifade mountant with or without DAPI to label nuclei on a multispot glass slide. Carefully manipulate the tissue to ensure that the epithelium is facing upwards.
16. Coverslip and seal with nail varnish.
17. View the samples under a fluorescence microscope to locate the sensory epithelium. If necessary, use confocal microscope to acquire z-stack images.

4 Notes

1. Osmium tetroxide is volatile and a hazard to human health. Handle OsO_4 inside a fume hood (waste OsO_4 must be collected in a marked bottle for suitable disposal).
2. Demembranation with immunogold label: following demembranation in buffer at pH 6.0 and fixation the tissue needs to be processed in cytoskeletal buffer at pH 7.0.
3. Care should be taken to wear gloves when handling the newts to prevent Salmonella transmission or fungal contamination.

Both ears can be used for all experiments but it makes good practice to use one ear as a control without ablating the hair cells so that any effects of culturing can be considered. When both ears are used for regeneration study, it is useful to fix one for immunohistochemical analysis and the other for electron microscopy. Incubators for amphibian cell culture should be set at 25 °C with 2 % CO_2.

4. All further procedures should be carried out in a laminar flow hood with sterile tools and medium to prevent infection.
5. Extensive hair cell loss is seen at 2 days post-gentamicin treatment and hair cell regeneration begins around 12 days post-treatment.
6. As an alternative to specimen holders specifically produced for electron microscopy, we have found that plastic capsules with an open lattice base (such as those used to hold desiccant) suitable as they allow fluids to be drained from the samples without damage or loss of tissue. To prevent loss during the critical point drying stage cover the capsules with copper mesh or lids if available.
7. If required, the process can be halted at 70 % ethanol stage and left overnight at 4 °C.
8. Care should be taken that sufficient wet paint is applied to attach the sample but that paint does not coat any of the upper surface of the sample.
9. It is not always easy to determine orientation for mounting specimens. Both saccule and utricle tend to be concave and on careful inspection of the saccule the opening to the amphibian papilla can be seen towards one edge of the epithelium. Nerve tracts are often visible on the underneath of both end organs.
10. Biotinylated phalloidin and gold-conjugated streptavidin can be used in conjunction with the demembranation protocol as per immunogold labelling.
11. Carbon coat tissue by either using appropriately equipped sputter coater or by evaporation using appropriate device.
12. We have found that otic capsules need to be incubated for several days (up to 5) in 4.13 % EDTA to decalcify sufficiently for sectioning. They should become more translucent and when pressed with forceps be soft and pliable.
13. Uranyl acetate is mildly radioactive and should be handled with care. Waste uranyl acetate should be retained in a dedicated bottle for specialist disposal.
14. This infiltration technique allows for the plastic to be initially introduced in a less viscous form and eventually all of the sample will be infiltrated with viscous plastic support material.

15. Training and experience are required to cut ultrathin sections and technical staff usually perform this task.
16. Care needs to be taken to avoid carbon dioxide in the atmosphere interacting with lead citrate, which would result in precipitates of lead carbonate over the sections. This can be done with pellets of sodium hydroxide placed locally within the staining dish.
17. Choice of secondary antibody depends upon the host in which the primary antibody has been raised.

Acknowledgement

Many thanks to Professor Andrew Forge for sharing his extensive knowledge of electron microscopy techniques. This work was supported by Action on Hearing Loss (formerly RNID and Deafness Research, UK).

References

1. Forge A, Wright T (2002) The molecular architecture of the inner ear. Brit Med Bull 63:5–24
2. Li L, Forge A (1997) Morphological evidence for supporting cell to hair cell conversion in the mammalian utricular macula. Int J Dev Neurosci 15:433–446
3. Forge A, Li L, Nevill G (1998) Hair cell recovery in the vestibular sensory epithelia of mature guinea pigs. J Comp Neurol 397:69–88
4. Taylor RR, Forge A (2005) Hair cell regeneration in sensory epithelia from the inner ear of a urodele amphibian. J Comp Neurol 484: 105–120
5. Forge A, Schacht J (2000) Aminoglycoside antibiotics. Audiol Neuro-otol 5:3–22
6. Davies S, Forge A (1987) Preparation of the mammalian organ of Corti for scanning electron microscopy. J Microsc 147:89–101

Part IV

Transgenesis in Salamanders

Chapter 21

Transgenesis in Axolotl (*Ambystoma mexicanum*)

Shahryar Khattak and Elly M. Tanaka

Abstract

Transgenic animals have been indispensable in elucidating and deciphering mechanisms underlying various biological phenomena. In regeneration, transgenic animals expressing fluorescent protein genes have been crucial for identifying the source cells for regeneration and the mechanism of blastema formation. Animals are usually generated by manipulating their genome using various techniques at/in one cell embryo/fertilized egg stage. Here, we describe the generation of germline transgenic axolotls (*Ambystoma mexicanum*) using the I-SceI meganuclease and Tol2 transposase.

Key words Salamander, I-SceI meganuclease, Tol2 transposase, Transgenics

1 Introduction

The ability to generate transgenic vertebrate animals has transformed the molecular analysis of development, adult physiology, and disease. Recently, generation of transgenic animals via the random integration of plasmid or BAC constructs into the genome has been applied to zebrafish and axolotl systems for the purposes of studying regeneration. The implementation of these animals has had a fundamental role in elucidating the cell sources of regenerating structures, the mechanisms by which regenerating tissue forms, as well as analysis of molecular pathways operating during regeneration [1–8].

Such insights have been gained in salamander limb and spinal cord regeneration via the generation of transgenic axolotls. Axolotls are the most easily bred salamander species in laboratory setting throughout the year with an average spawn size of 500 eggs per mating. Adult axolotls are kept in water with an average temperature of 18–20 °C (±0.5 °C) with a day–night cycle of 12 h. The conductivity of water is in the range of 500–750 microSiemens (μS)/cm with a pH between 7.0 and 8.0. The level of NO_2 in water should be zero and that of NO_3 should be kept below 100 mg/l. The water should be dechlorinated by overnight evaporation,

Anoop Kumar and András Simon (eds.), *Salamanders in Regeneration Research: Methods and Protocols*, Methods in Molecular Biology, vol. 1290, DOI 10.1007/978-1-4939-2495-0_21, © Springer Science+Business Media New York 2015

chemical conditioning or carbon filtering and changed three times per week. Freshly hatched axolotl larvae are fed daily with brine shrimps (artemia) until reaching 5 cm in size (snout-to-tail). Animals from 6 to 20 cm are fed small (3 mm) fish pellets and artemia everyday. The amount of artemia is gradually decreased so the animals get only fish pellets when they reach around 10 cm in size. Adults (>20 cm in size) are fed big (5 mm) fish pellets three times per week.

Male axolotls reach sexual maturity between 9 and 12 months while females are ready to be mated between 12 and 15 months of age. Males can be mated every 4 weeks, whereas females are ready to mate again in 8 weeks from the previous spawn. For mating, the male and female are put together in a covered mating tank (temperature 15–16 °C). The animals should not be disturbed until the female starts to lay eggs. Plastic leaves in the tank help the females to lay eggs on and facilitate the harvesting of eggs.

Transgenic axolotls have been generated using I-SceI meganuclease and Tol2 transposase [9–11]. Both these methods involve collecting one cell stage fertilized eggs followed by injecting the collected eggs with a mixture of DNA and I-SceI meganuclease or DNA and Tol2 transposase enzymes. Germline establishment of axolotl transgenic lines is dependent on identifying and growing up the founder animals. Founder animals ideally express the transgene in all cells of its body and show no signs of morphological abnormality. The founder animals are grown to sexual maturity (12–15 months) and mated with a non-transgenic mate to determine the germline transmission. Both Tol2 and I-SceI mediate integration of the transgene into the genome but Tol2 seems to be more efficient than I-SceI enzyme. Because of being highly efficient, founders generated via Tol2 method sometimes have non-uniform expression of transgene between cells of the same animal. This is presumably due to multiple integrations of the DNA between different cells of the same animal. On the other hand, the I-SceI meganuclease method generates fewer but uniformly expressing founders. Although the best way is to verify the expression of transgene in each cell is by cutting, sectioning and/or immunostaining the limb or tail of the founder animal, this is time-consuming and not a visual representation of the transgene expression. An easy and efficient way is to insert a fluorescent reporter in the DNA being injected. This allows easy visualization and selection of the putative founder animal.

2 Materials

2.1 Equipment

1. Axolotls (can be purchased from the Ambystoma Genetic Stock Center).
2. Artemia and fish pellets.

3. I-SceI enzyme and I-SceI enzyme buffer.
4. Plasmids: pBSII-SK-mTol2-MCS and T7-TPase (available from Addgene).
5. DNA Plasmid Maxi Kit.
6. Scalpels to scrape the eggs from the leaves.
7. Containers for collecting eggs and to keep animals.
8. Cups to keep small animals.
9. Sieve to rinse the eggs.
10. Fish nets (8, 10, and 20 cm size).
11. Three pairs of forceps: two to dejelly the eggs and one to break the needle for injection.
12. Petri dishes to collect eggs (145, 94, 60, and 30 mm).
13. 24-well plates to keep the embryos for the first week.
14. Sterile plastic transfer pipettes (2 mL) to transfer the embryos.
15. Slide with micrometric scale, 0.1 mm to determine the volume of the injection-droplet.
16. Mold for making injection plates (Fig. 1).
17. Stereomicroscope with a light source.
18. PV830 Pneumatic Pico Pump.
19. Micromanipulator and microelectrode holder for injection.
20. Glass capillaries with 1.00 mm O.D. × 0.58 mm.
21. Micropipette Puller.
22. mMESSAGE mMACHINE T7 ULTRA Kit.

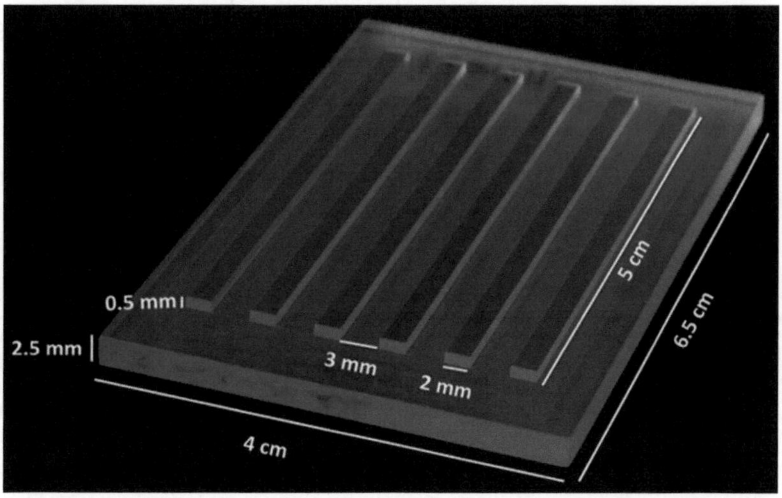

Fig. 1 Shows the plastic mold for making the injection plate. The mold shown here is for making a 10 cm injection plate. Reproduced from ref. [11]

2.2 Solutions

1. 10× MMR: Mix 400 mL of 5 M NaCl, 40 mL of 1 M KCl, 20 mL of 1 M $MgCl_2$, 40 mL of 1 M $CaCl_2$, 4 mL of 0.5 M EDTA, 100 mL of 1 M HEPES. Fill up to 1.5 L with deionized water. Adjust the pH to 7.8 with 10 M KOH. Fill up to 2 L with deionized water. Autoclave and store at room temperature for up to 6 months.
2. 1× MMR/pen-strep: mix 100 mL of 10× MMR with 13 mL of Penicillin–Streptomycin and fill up to 1 L with deionized water. Filter-sterile and store at room temperature for up to 2 months.
3. 1× MMR/20 % (wt/vol) Ficoll/pen-strep: mix 25 mL of 10× MMR, 50 g of Ficoll and 7 mL of Penicillin–Streptomycin. Fill up to 250 mL with deionized water. Filter-sterile and store at 4 °C for up to 2 months.
4. 0.1× MMR/5 % (wt/vol) Ficoll/pen-strep: mix 10 mL of 10× MMR, 50 g of Ficoll and 13 mL of Penicillin–Streptomycin. Fill up to 1 L with deionized water. Filter-sterile and store at 4 °C for up to 2 months.
5. 0.1× MMR/pen-strep: mix 20 mL of 10× MMR and 25 mL of Penicillin–Streptomycin. Fill up to 2 L with deionized water. Filter-sterile and store at Room Temperature for up to 2 months.
6. Holtfreter's Solution (400 % concentration): mix 2.875 g of KCl, 5.36 g of $CaCl_2{\cdot}2H_2O$, 11.125 g of $MgSO4{\cdot}7H_2O$ and 158.4 g of NaCl. Fill up to 10 L with distilled water. Store at room temperature for up to 6 months.
7. 10× TBS: For 1 L, add 24.2 g of TRIS and 90 g of NaCl in 990 mL of distilled water. Mix well with stir bar and adjust the pH to 8.0 by adding 10 mL of HCl (37 %). Store at room temperature for up to 6 months.
8. Benzocaine 10 % (wt/vol): mix 50 g of Benzocaine in 500 mL of 100 % Ethanol. Store at room temperature for up to 12 months.
9. Anesthetic for Axolotls: 0.03 % (vol/vol) Benzocaine: To make 10 L mix 500 mL of 10× TBS, 500 mL of 400 % Holtfreter's and 30 mL of 10 % Benzocaine. Fill up to 10 L with distilled water and store at room temperature for up to 6 months.

3 Methods

3.1 Preparing DNA and Tol2 mRNA for Injection

1. Clone the transgene cassette in the multiple cloning sites of the "pBSII-SK-mTol2-MCS" plasmid. Cloning the cassette in the multiple cloning site allows the use of I-SceI meganuclease OR the Tol2 transposase method to generate transgenic axolotls (Fig. 2).

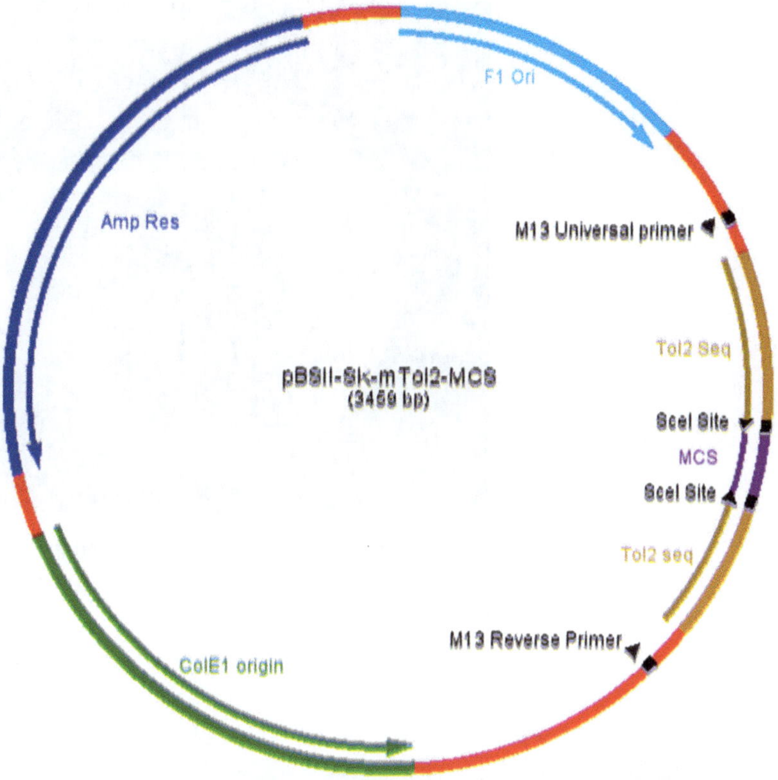

Fig. 2 Graphic map of "pBSII-SK-mTol2-MCS" plasmid. MCS: Multiple cloning site to clone the cassette of interest for making transgenic axolotls. Note that the MCS is surrounded by both I-SceI meganuclease as well as Tol2 transposase sites/sequences. This gives the user flexibility to use one or the other method or do a direct comparison between the two techniques using the same plasmid. Reproduced from ref. [11]

2. Maxiprep the resulting plasmid using standard kits and sequence verify that the miniTol2, I-SceI sites and the transgenic expression cassettes are intact and harbor the correct expected sequence.
3. Prepare Tol2 transposase mRNA by linearizing the "T7-TPase" plasmid with BspEI enzyme and utilizing it as template for in vitro transcription by following manufacturer's instruction of mMESSAGE mMACHINE T7 Ultra Kit.

Make aliquots of DNA (**step 2**) and in vitro transcribed RNA (**step 3**) and store them at −20 °C and −80 °C respectively (*see* **Note 1**).

3.2 Plate and Needles

1. Pour 10 mL of 2 % (wt/vol) agar solution (made in distilled water) in a 10 cm plastic petri dish.
2. Place the injection mold (upside down) (Fig. 1) into the agar solution and let it solidify at room temperature.

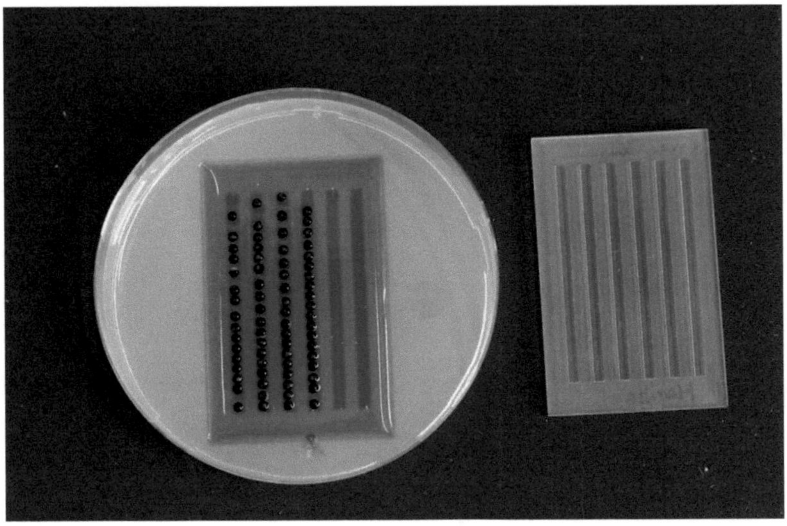

Fig. 3 Depicts a 10 cm injection plate and the mold used to make this injection plate. Axolotl eggs are aligned in the grooves/channels of the injection plate to be injected with DNA injection mix

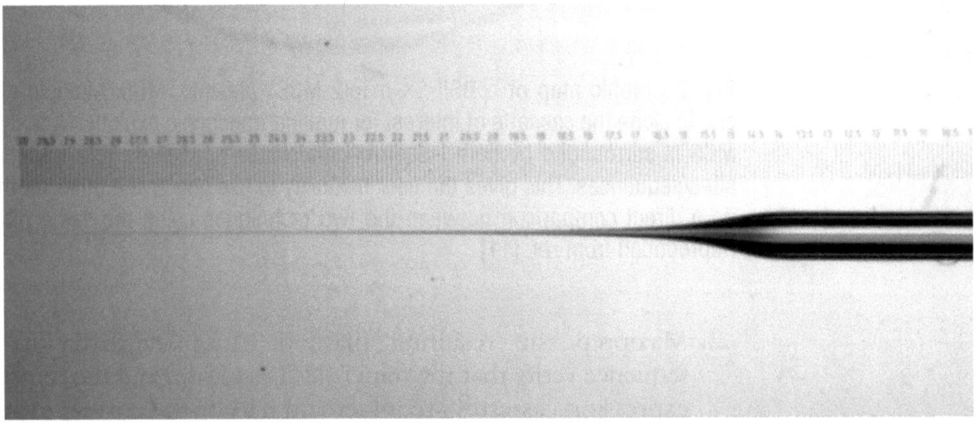

Fig. 4 The injection needle is made by pulling a glass capillary with a micropipette puller

3. Carefully remove the mold. This will result in channels for holding the eggs in a soft agar bed (Fig. 3).
4. Prepare multiple plates and seal them with Parafilm and store at 4 °C until use.
5. For injection needles, pull the glass capillaries (1.00 mm O.D. × 0.58 mm) using a micropipette puller and store them for future use (Fig. 4).

3.3 Egg Collection for Injection

1. Collect the eggs from the plastic leaves by gently scraping them off with a scalpel and place them in a metal sieve.
2. Rinse the eggs with 70 % Ethanol for approximately 15–20 s followed by rinsing thoroughly (but gently) with sterile water.
3. Place the eggs in a petri dish containing 1× MMR/pen-strep solution using a transfer pipette.
4. Under stereomicroscope, manually dejelly the eggs using sharp forceps/tweezers.

Transfer the dejellyed eggs into a clean petri dish containing 1× MMR/pen-strep solution (*see* **Note 2**).

3.4 Injection of DNA into Eggs

1. Pour 1× MMR/20 % Ficoll/pen-strep solution in the injection plate (**step 4**, Subheading 3.2).
2. Gently align the dejellyed eggs (**step 5**, Subheading 3.3) in the grooves of the injection plate (Fig. 3).
3. Prepare the injection mix in a 1.5 mL tube and keep it on ice until use.

For Tol2 mediated transgenesis, inject a volume of 5 nL per egg with a total plasmid DNA concentration of 0.05 ng and 0.25 ng of Tol2 RNA (*see* **Note 3**).

For I-SceI transgenesis, inject a total of 0.5 ng DNA in 5 nL volume per egg. To prepare 10 μL injection mix, take 1 μg DNA, 1 μL of 10× I-SceI buffer, 2 μL I-SceI (5,000 U/mL) I-SceI enzyme and fill it up to 10 μL with DNAse–RNAse-free water.

4. Fill up the pulled glass needle (**step 5**, Subheading 3.2) with the injection mix and attach the needle to micromanipulator and pneumatic injection pump.
5. Break the tip of the needle using the slide with micrometric scale under the stereomicroscope such that 5 nL of injection mix is released when pressure is applied through the pneumatic injection pump (Fig. 5).
6. Gently insert the needle in the animal pole of the eggs and release the 5 nL of injection mix half way into the animal pole.
7. By moving the injection plate, inject all the aligned eggs one by one. Leave some the eggs as uninjected control.
8. Transfer the eggs in a petri dish containing 1× MMR/20 % Ficoll/pen-strep solution and leave at room temperature for 2 h.
9. After 2 h, transfer the injected eggs into a fresh petri dish containing 0.1× MMR/5 % Ficoll/pen-strep solution and leave them undisturbed for 24 h at room temperature.

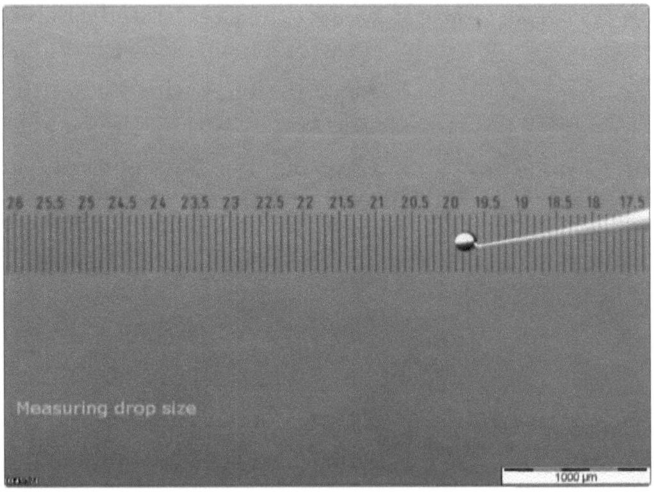

Fig. 5 A 5 nL drop size is measured using the DNA injection mixed filled injection needle using the micrometric scale under the stereomicroscope. If the drop size is not optimal, break the needle more or adjust the pressure of the pneumatic injection pump to obtain the 5 nL volume drop size

10. Transfer the healthy eggs into a sterile 24-well plate (one egg per well) containing 0.1× MMR/ pen-strep solution. Let the eggs develop for 7–8 days in the 24-well plate at room temperature.
11. Using a transfer pipette, gently transfer (one by one) the normally developed larvae from the wells of the 24-well plate to a 145 mm petri dish containing 0.1× MMR/pen-strep solution.
12. The larvae will start swimming after 2–3 days.

3.5 Transgenic Larvae Screening

1. Dilute the benzocaine to 0.01 % with distilled water. This is used to anesthetize the larvae.
2. Cut the tip of the transfer pipette to make it wide bore for picking up the swimming larvae.
3. Place the wide bore transfer pipette in front of the larva such that larvae will swim into it.
4. Gently place the larva from the transfer pipette into the 0.01 % benzocaine and wait for it to anesthetize.
5. Using the same pipette, gently pick up the anesthetized larva and place it in a drop of water in a petri dish and observe under the fluorescence microscope for transgene expression (*see* **Note 4**).

4 Notes

1. The in vitro transcription of Tol2 transposase should be performed in a RNAse free environment. Wipe the bench with RNAse away or similar products. Use only DNAse-, RNAse-free

filter tips and tubes. We usually make aliquots of 4.1 μL of 500 ng/μL concentrated in vitro transcribed RNA in 0.2 mL PCR tubes for storing at −80 °C.

2. Pre-screen the eggs for signs of damage or abnormal developing eggs. A healthy egg would contain two regions/poles. A dark brown animal pole and a beige vegetal pole. Care should be taken not to injure the egg and touch it with the sharp forceps/tweezers.
3. The amount mentioned for injection is for an 8,000 bp size plasmid. We recommend doing a pilot experiment for every new plasmid that needs to be injected. The pilot experiment should be to determine the optimal/best combination of plasmid DNA vs. Tol2 RNA to obtain a founder that expresses the transgene in majority of its cells in the body.
4. While screening, only anesthetize 1–2 larvae, as the freshly hatched larvae are very sensitive to benzocaine. As soon as the screening is over, place it back into 0.1× MMR/pen-strep solution. Then anesthetize the next 1–2 larvae for screening and so on.

References

1. Kragl M, Knapp D, Nacu E, Khattak S, Maden M, Epperlein HH, Tanaka EM (2009) Cells keep a memory of their tissue origin during axolotl limb regeneration. Nature 460:60–65
2. McHedlishvili L, Mazurov V, Grassme KS, Goehler K, Robl B, Tazaki A, Roensch K, Duemmler A, Tanaka EM (2012) Reconstitution of the central and peripheral nervous system during salamander tail regeneration. Proc Natl Acad Sci U S A 109: E2258–E2266
3. Sandoval-Guzman T, Wang H, Khattak S, Schuez M, Roensch K, Nacu E, Tazaki A, Joven A, Tanaka EM, Simon A (2014) Fundamental differences in dedifferentiation and stem cell recruitment during skeletal muscle regeneration in two salamander species. Cell Stem Cell 14:174–187
4. Khattak S, Schuez M, Richter T, Knapp D, Haigo SL, Sandoval-Guzman T, Hradlikova K, Duemmler A, Kerney R, Tanaka EM (2013) Germline transgenic methods for tracking cells and testing gene function during regeneration in the Axolotl. Stem Cell Reports 1:90–103
5. McHedlishvili L, Epperlein HH, Telzerow A, Tanaka EM (2007) A clonal analysis of neural progenitors during axolotl spinal cord regeneration reveals evidence for both spatially restricted and multipotent progenitors. Development 134:2083–2093
6. Chen CH, Durand E, Wang J, Zon LI, Poss KD (2013) zebraflash transgenic lines for in vivo bioluminescence imaging of stem cells and regeneration in adult zebrafish. Development 140(24):4988–4997
7. Gupta V, Poss KD (2012) Clonally dominant cardiomyocytes direct heart morphogenesis. Nature 484:479–484
8. Choi WY, Gemberling M, Wang J, Holdway JE, Shen MC, Karlstrom RO, Poss KD (2013) In vivo monitoring of cardiomyocyte proliferation to identify chemical modifiers of heart regeneration. Development 140:660–666
9. Sobkow L, Epperlein HH, Herklotz S, Straube WL, Tanaka EM (2006) A germline GFP transgenic axolotl and its use to track cell fate: dual origin of the fin mesenchyme during development and the fate of blood cells during regeneration. Dev Biol 290:386–397
10. Khattak S, Richter T, Tanaka EM (2009) Generation of transgenic axolotls (*Ambystoma mexicanum*). Cold Spring Harb Protoc 2009(8):pdb prot5264. doi:10.1101/pdb.prot5264
11. Khattak S, Murawala P, Andreas H, Kappert V, Schuez M, Sandoval-Guzman T, Crawford K, Tanaka EM (2014) Optimized axolotl (Ambystoma mexicanum) husbandry, breeding, metamorphosis, transgenesis and tamoxifen-mediated recombination. Nat Protoc 9:529–540

Chapter 22

Generating and Identifying Axolotls with Targeted Mutations Using Cas9 RNA-Guided Nuclease

G. Parker Flowers and Craig M. Crews

Abstract

The CRISPR/Cas9 RNA-guided nuclease now enables a reverse genetics approach to investigate the function of genes of interest during regeneration in the axolotl. The process of generating the constructs necessary for targeting a gene of interest is considerably less labor intensive than for other methods of targeted mutagenesis such as Zinc finger nucleases or Transcription activator-like effector nucleases. Here, we describe the identification of targetable sequences in the gene of interest, the construction of unique guide RNAs, the microinjection of these RNAs with Cas9-encoding mRNA, the selection of well-injected animals, and an inexpensive, PCR-based method for identifying highly mutagenized animals.

Key words Transgenesis, Salamander, CRISPR, Limb regeneration, Guide RNA

1 Introduction

The axolotl has a long history of use as a genetic model organism, but in recent decades it has been surpassed in use by other amphibian and aquatic model systems such as Xenopus species and zebrafish [1]. While a number of mutant axolotl lines have been identified, the comparatively long generation time and exceptionally large, currently unsequenced, genome hinder genetic studies in the axolotl. Nonetheless, better-characterized genetic model organisms lack the remarkable adult regenerative ability of the axolotl, and because of this fascinating ability the axolotl continues to be a premiere vertebrate model organism for regeneration research [2]. Numerous recent studies identifying transcripts and proteins specifically upregulated in various tissues during regeneration are providing greater insight into the molecular mechanisms governing regeneration in the axolotl [3–9], though the tools for investigating the function of these genes in vivo have been relatively blunt, with ectopic overexpression and transient knockdown via

Anoop Kumar and András Simon (eds.), *Salamanders in Regeneration Research: Methods and Protocols*, Methods in Molecular Biology, vol. 1290, DOI 10.1007/978-1-4939-2495-0_22, © Springer Science+Business Media New York 2015

morpholino the primary tools for analysis. Techniques in which selected genes can be completely functionally ablated can provide more definitive evidence of whether genes implicated in regeneration are essential for this process.

In recent years, a number of advances in targeted mutagenesis now permit the generation of cell lines and whole organisms with mutations at targeted sites. Zinc-finger nucleases (ZFN) [10], Transcription activator-like effector nucleases (TALENS) [11], and RNA-guided nucleases (RGNs) derived from bacterial Clustered Regularly Interspersed Short Palindromic Repeats (CRISPR) [12] all act to induce double strand breaks at targeted sequences which are inefficiently repaired by nonhomologous end-joining mechanisms. The most recently developed of these systems, CRISPR-derived RGNs, are particularly attracting the attention of researchers as, rather than requiring the generation of a unique protein to target a sequence of interest, RGNs use short guide RNAs (sgRNA) to direct the Cas9 nuclease to the targeted sequence. We find that the comparative ease of assembly of RGN constructs as well as their efficiency at generating high mutation rates at targeted sites make them very well suited for work in the axolotl [13]. This innovation, combined with the growing body of data regarding the axolotl transcriptome, allows the use of a reverse genetic approach to probe the function of genes during tissue regeneration.

While the CRISPR RGN system has been demonstrated to work very well in a number of organisms, including the axolotl, there are a number of challenges to consider when applying this technique to the axolotl. The axolotl genome is currently unsequenced, though multiple accessible sources of cDNA sequencing are available [7, 14]. These data are sufficient for targeting a large number of genes implicated in regeneration; however, several steps should be taken to confirm the identity of targetable sequences and that successful targeted mutagenesis can subsequently be verified. Using currently available sequence, one is only typically able to amplify short sequences of 100–200 bases around targeted sites, a product size insufficiently large to be used in several of the endonuclease-based assays that are frequently used to screen for targeted mutagenesis in other organisms. Thus, we have adapted a protocol for rapidly screening for mutagenesis frequency using very small, fluorescently labeled PCR products and capillary electrophoresis. Despite several challenges, the rapid assembly of constructs for this system, combined with our novel screening protocol, allows one to proceed from selection of a gene to target to identifying highly mutagenized animals in less than a week.

2 Materials

2.1 sgDNA Construction Components

1. *pDR274* expression Vector (Addgene Plasmid 42250).
2. BsaI restriction enzyme (NEB).
3. QIAquick PCR purification kit (Qiagen).
4. Two oligonucleotides containing target sequence and 5′ ends compatible with *pDR274*.
5. 10× Annealing buffer (0.4 M Tris–HCl, 0.2 M $MgCl_2$, 0.5 M NaCl, 10 mM EDTA, pH 8.0).
6. Thermocycler.
7. Quick Ligase (NEB).
8. Chemically competent Dh5α cells.
9. QIAprep Spin Miniprep kit (Qiagen).
10. 0.8 % agarose TAE gel.

2.2 sgRNA and Cas9 mRNA Synthesis Components

1. *pMLM3613* Cas9 expression vector (Addgene Plasmid 42251).
2. DraI restriction enzyme (NEB).
3. PmeI restriction enzyme (NEB).
4. Phenol–chloroform–isoamyl alcohol (25:24:1).
5. Chloroform.
6. RNase-free 100 % ethanol.
7. 3 M sodium acetate pH 5.2.
8. RNase-free water.
9. 100 % isopropanol.
10. T7 Maxi script kit (Life Technologies).
11. T7 mMessage mMachine kit (Life Technologies).
12. 20× SB buffer (720 mM boric acid, 200 mM NaOH).
13. RNA loading dye.
14. ssRNA ladder (NEB).
15. GlycoBlue Coprecipitant (Life Technologies, optional).

2.3 Mating Components

1. 10 L plastic tub.
2. Novaqua (Kordon).
3. Amquel (Kordon).
4. Modified 40 % Holtfreter's solution (20 mM NaCl, 0.2 mM KCl, 0.8 mM $NaHCO_3$, 0.2 mM $CaCl_2$, 4 mM $MgSO_4$, 0.1 mL/L Amquel, 0.1 mL/L Novaquel, pH to 7.5).
5. Freezer bags.
6. Sealable plastic bags.

7. Rounded stones.
8. Plastic Foliage.
9. Male and female axolotl.
10. Fish nets.

2.4 RNA Microinjection Components

1. 25 × 75 × 1 mm microscope slides.
2. 100 mm × 15 mm petri dishes.
3. Agarose.
4. 10× Marc's Modified Ringers (1 M NaCl, 20 mM KCl, 10 mM $MgSO_4$, 20 mM $CaCl_2$, 50 mM HEPES, pH 7.4, autoclaved)—use to make 1× MMR with 50 U/L Penicillin–Streptomycin, 0.1× MMR with 50 U/L Penicillin–Streptomycin, 1× MMR with 20 % Ficoll and 50 U/L Penicillin–Streptomycin, and 0.1× MMR with 50 U/L Penicillin–Streptomycin.
5. Ficoll PM 400 (Sigma-Aldrich).
6. Penicillin–Streptomycin.
7. Fine forceps (e.g., Dumont #55 Fine Forceps, Fine Science Tools).
8. Micropipette Puller (e.g., Flaming/Brown P-87, Sutter).
9. 1 mm glass capillary with filament (e.g., TW100F-4, World Precision Instruments).
10. Microloader tips.
11. Microinjector (e.g., Picospritzer III, Parker-Hannifin).
12. Dissecting microscope with fluorescence.
13. Hemocytometer or scale slide.

2.5 Genotyping Components

1. DNeasy Blood and Tissue Kit (Qiagen).
2. Tricaine methanesulfonate.
3. 35 mm petri dish.
4. Scalpel blades.
5. 100 mM NaOH.
6. 1 M Tris–HCl pH 7.4.
7. PCR-confirmed gene specific primers.
8. One gene specific primer with TCCCAGTCACGACGT-sequence at 5′ end.
9. Primer with m13 (TCCCAGTCACGACGT) sequence and 5′ 6-FAM modification (available through IDT, Invitrogen, and Operon).
10. Taq Polymerase.
11. dNTPs.
12. Hi-Di Formamide (Life Technologies).

3 Methods

3.1 sgDNA Construction

1. Linearize 4 μg of *pDR274* with 10 μL 10× CutSmart buffer and 5 μL of BsaI enzyme and bring to 100 μL with water. Digest at 37 °C for 1 h.
2. Purify using QIAquick PCR purification kit or other method, determine concentration by spectrophotometry, and adjust concentration to 25 ng/μL. This stock of linear *pDR274* should be sufficient for generating hundreds of unique constructs.
3. Identify targetable sites in your gene of interest (Fig. 1a) (*see* **Note 1**).
4. Obtain two oligonucleotides corresponding to your site. The sequence of oligonucleotide 1 should be TA+20 base target sequence (up to, but not including the NGG protoadaptor motif). Oligonucleotide 2 should also be 22 bases, beginning with 5′-AAAC and followed by the reverse complement of the last 18 bases of the target sequence (Fig. 1b). When annealed, these primers will have two overhangs, one 5′-TAGG and the other 5′AAAC that are compatible with the overhangs of BsaI-digested *pDR274* (Fig. 1c).
5. Resuspend your oligonucleotides in water to bring to a final concentration of 100 μM.
6. Add 1 μL of each oligonucleotide to 5 μL 10× annealing buffer and bring to 50 μL with H_2O in a PCR tube.
7. Anneal primers by placing in a thermocycler and running the following program: 98 °C for 0: 30 s, –5 °C per cycle for 17 cycles.

a 5′-GGGT**GGTGCCCATCCTGGTCGAGC**TGGACGGCACGTAAA-3′
3′-CCCA**CCACGGGTAGGACCAGCTCG**ACCTGCCGTGCATTT-5′

b Oligo 1- 5′-TA**GGTGCCCATCCTGGTCGAGC**-3′

Oligo 2- 5′-AAAC**GCTCGACCAGGATGGGCA**-3′

c Annealed Oligos- 5′-TA**GGTGCCCATCCTGGTCGAGC-3′**
5′-AAAC**GCTCGACCAGGATGGGCA**-3′

Fig. 1 Schematic of genomic RGN target and oligonucleotides used to construct targeting vector. (**a**) Putative genomic sequence with target sequence in *blue* and the protoadaptor motif in *red*. This sequence conforms to the GG(N_1...N_{18})NGG constraints. (**b**) The sequence of the two oligonucleotides used to insert the targeting sequence into the pDR274 vector. (**c**) After annealing, the oligos will have two 5′ overhangs compatible with the BsaI-cut ends of pDR274. One end will have a TAGG overhang, where the GG corresponds to the first two bases of the target site. The other end will have an AAAC overhang

8. Mix 1 μL of annealed primers, 1 μL of BsaI-linearized *pDR274* (25 ng), 7 μL H_2O, 10 μL 2× NEB Quick Ligase Buffer, and 1 μL T4 DNA ligase. Incubate at room temperature for 10 min.
9. Use 1 μL of ligation to transform 50 μL of Dh5α cells, and plate on 50 μg/mL kanamycin plates.
10. The following day pick four to eight colonies from plate and grow up in 1 mL LB overnight with 50 μg/mL kanamycin.
11. The next day, carry out minipreps of selected colonies using QIAprep Spin Miniprep kit or preferred method.
12. Screen for correct ligations by digesting 2 μL of miniprepped DNA, with 0.5 μL BsaI in 1 μL of 10× CutSmart buffer, and 6.5 μL H_2O for 1 h at 37 °C. The proper ligation of the target sequence into *pDR274* eliminates BsaI cut sites.
13. Run the digests on 0.8 % agarose gel in TAE. Failed ligations produce a single linearized band, while correct products will produce multiple undigested bands migrating at different speeds.
14. Submit one or two plasmids for sequencing with the forward m13 primer (5′-GTAAAACGACGGCCAGT-3′, *see* **Note 2**). Two minipreps of these constructs should provide template to synthesize enough RNA to inject thousands of embryos.
15. Upon confirmation that the construct contains the correct targeting sequence, linearize the DNA. Digest 20 μL of the miniprepped plasmid DNA with 5 μL of DraI in 10 μL of CutSmart buffer and 65 μL H_2O for 1 h at 37 °C.
16. Add 100 μL of RNase-free H_2O and 200 μL of phenol–chloroform–isoamyl alcohol. Vortex.
17. Spin at top speed (<15,000×*g*) on a benchtop centrifuge for 5 min.
18. Collect just under 200 μL from the upper aqueous phase, being very careful to avoid collecting any of the lower phase. Add 200 μL of chloroform and vortex.
19. Spin at top speed (<15,000×*g*) on a benchtop centrifuge for 3 min.
20. Collect just under 200 μL from the upper aqueous phase, being very careful to avoid collecting any of the lower phase (**steps 18** through **20** may be repeated for optimal purification).
21. Add 75 μL of 3 M Sodium Acetate pH 5.2 and 500 μL 100 % ethanol. Adding 1 μL of GlycoBlue can facilitate the precipitation and visualization of the pellet, but is not necessary. Store at −20 °C for at least 1 h.
22. Spin at top speed (<15,000×*g*) in centrifuge for 15 min.
23. Remove ethanol and wash pellet with 500 μL of chilled 70 % ethanol. Spin at top speed (<15,000×*g*) in centrifuge for 5 min.

24. Remove as much ethanol as possible without dislodging pellet and air-dry to remove residual ethanol.
25. Resuspend in 10 μL of RNase-free H_2O.
26. Quantify DNA concentration by spectrophotometry.

3.2 sgRNA Synthesis

1. Using linearized DNA and T7 MaxiScript kit, mix 1 μg DNA, 2 μL 10× Transcription Buffer, 1 μL each of 10 mM ATP, CTP, GTP, and UTP, 2 μL T7 Enzyme mix, and bring to 20 μL with RNase-free H_2O. Incubate for 1–2 h at 37 °C.
2. Add 1 μL of Turbo DNase and incubate for 15 min 37 °C.
3. Mix 20 μL sgRNA transcription reaction with 15 μL ammonium acetate, 115 μL water, and 150 μL of phenol–chloroform–isoamyl alcohol. Vortex.
4. Spin at top speed (<15,000 ×*g*) on a benchtop centrifuge for 5 min.
5. Collect just under 150 μL from the upper aqueous phase, being very careful to avoid collecting any of the lower phase.
6. Add 150 μL of chloroform and vortex.
7. Spin at top speed (<15,000 ×*g*) on a benchtop centrifuge for 3 min.
8. Collect just under 150 μL from the upper aqueous phase, being very careful to avoid collecting any of the lower phase.
9. Add 150 μL of 100 % isopropanol and 1 μL of GlycoBlue. Freeze at −20 °C for 1 h.
10. Spin at top speed (<15,000 ×*g*) on a benchtop centrifuge for 15 min.
11. Remove all liquid, being careful not to dislodge the pellet. Let air-dry at room temperature.
12. Resuspend in 20 μL of RNase-free H_2O. Mix well and make 4 μL aliquots to store at −80 °C.
13. Retain 1 μL for analysis using agarose gel and spectrophotometry.
14. Wipe electrophoresis equipment and gel combs with RNase Away. Prepare 1 % agarose gel in 1× SB buffer.
15. Mix 0.5 μL of sgRNA with 4.5 μL H_2O and 5 μL of RNA loading dye. Prepare 0.5 μL of ssRNA ladder with 4.5 μL H_2O and 5 μL of RNA loading dye. Denature both samples at 70 °C for 10 min.
16. Run ladder and sample for 20–30 min at 150 V. A single distinct band of approximately 100 bp indicates correct sgRNA synthesis.
17. *Assess sgRNA concentration by spectrophotometry.*

3.3 Cas9 mRNA Synthesis (See Note 3)

1. Linearize 20 μL of *pMLM3613* miniprepped DNA with 5 μL of PmeI in 10 μL of Cutsmart Buffer and 65 μL of H_2O for 1 h at 37 °C.
2. Purify linearized DNA using phenol–chloroform extraction followed by ethanol precipitation as above (Subheading 3.2, **steps 16–24**) and resuspend air-dried pellet after 70 % ethanol wash in 20 μL H_2O.
3. Assess DNA concentration by spectrophotometry.
4. Using linearized *pMLM3613* DNA and T7 mMessage mMachine kit, mix 1 μg DNA, 10 μL of 2× NTP mix , 2 μL of 10× Transcription buffer, 2 μL of T7 polymerase 1 μL each and bring to 20 μL with RNase-free H_2O. Incubate for 1–2 h at 37 °C.
5. Add 1 μL of TurboDNase and incubate for 15 min 37 °C.
6. Mix 20 μL *pMLM3613* transcription reaction with 15 μL ammonium acetate, 115 μL water, and 150 μL of phenol–chloroform–isoamyl alcohol. Vortex to mix.
7. Spin at top speed (<15,000 × *g*) on a benchtop centrifuge for 5 min.
8. Collect just under 150 μL from the upper aqueous phase, being very careful to avoid collecting any of the lower phase. Add 150 μL of chloroform and vortex.
9. Spin at top speed (<15,000 × *g*) on a benchtop centrifuge for 3 min.
10. Collect just under 150 μL from the upper aqueous phase, being very careful to avoid collecting any of the lower phase. Add 150 μL of 100 % isopropanol and 1 μL of GlycoBlue. Freeze at −20 °C for 1 h.
11. Spin at top speed (<15,000×*g*) on a benchtop centrifuge for 15 min.
12. Remove all liquid, being careful not to dislodge the pellet. Let air-dry at room temperature.
13. Resuspend in 20 μL of RNase-free H_2O. Mix well and make 4 μL aliquots to store at −80 °C. Retain 1 μL for analysis using agarose gel and spectrophotometry.
14. Wipe electrophoresis equipment and gel combs with RNase away. Prepare 1 % agarose gel in 1× SB buffer.
15. Run ladder and sample for 20–30 min at 150 V. Correct sgRNA synthesis will be indicated by a single distinct band of approximately 5 kb.

3.4 Obtaining and Microinjecting Embryos (See Note 4)

1. Before injection, prepare ramps by melting 1 g of agarose in 100 mL of 1× MMR. Pour this hot agarose mixture into three petri dishes. Place a 25 × 75 × 1 mm microscope slide diagonally in the petri dish, propping the edge of the slide at a shallow

angle along the edge of the dish, to make a ramp in the agarose. Let the agarose solidify, remove the slide, and cut away any overhanging agarose with a razor blade.

2. Prepare injection needles. Using a P-87 Flaming/Brown Needle Puller we use Pressure = 200, Pull = 90, Velocity = 70; however, these parameters must be optimized for the filament being used. This produces a needle with a very long, thin tip.
3. 48–36 h before start of injection, fill lidded plastic tank with 10 L of 40 % Modified Holtfreter's solution.
4. Line bottom of tank with rounded stones to prevent spermatophores from sticking to surface of tank.
5. Place plastic foliage such as leaves and plants in tank (optional).
6. Place two small freezer bags in sealed plastic bags in tank to reduce water temperature.
7. Place sexually mature female and male in tank (*see* **Note 5**).
8. The following morning, check for successful spermatophore deposition (*see* **Note 6**).
9. Remove the male.
10. Place the female in a fresh tub in 40 % Modified Holtfreter's solution. It will typically begin to lay eggs that evening or the following morning.
11. Remove the tip of a 3 mL transfer pipette with scissors to create an opening larger than an axolotl egg. Collect eggs by placing the female in a fresh tub of 40 % Modified Holtfreter's solution. Scrape eggs from surface and transfer to new container with transfer pipette (*see* **Note** 7).
12. Remove excess liquid from collected eggs and place in container of 1× MMR with pen-strep on ice.
13. Transfer a few eggs to a petri dish at a time, leaving the remaining eggs to soak in 1× MMR.
14. Under a dissecting microscope, use two very sharp forceps to remove jelly and membrane. Take care to not damage the eggs with your forceps.
15. Discard damaged eggs, load the dejellied eggs onto the ramp, and remove all fluid using a transfer pipette. Cover the embryos with several milliliters of 20 % Ficoll in 1× MMR with Pen-Strep (*see* **Note 8**).
16. Prepare RNA injection mix. As a starting concentration, we mix together the RNAs to a final concentration of approximately 150 ng/μL of *cas9* mRNA, 30 ng/μL sgRNA, and 10–50 ng/μL of *nls-GFP* mRNA just prior to injecting. Store on ice. Using microloader tips, load an injection needle with 2–3 μL of this mixture (*see* **Note 9**).

17. Using fine forceps, break the tip of your injection needle to create an opening. Insert the needle into the microinjector capillary holder and tightly seal.
18. Using a hemocytometer or other scale bar, inject one droplet of your injection mix over the scale. Adjust the pulse duration so that a single pump produces a droplet of a diameter of 180–200 μm (*see* **Note 10**).
19. Orient the injection ramp so that you insert the needle into the animal pole of the embryos along the sloped side of the ramp. A micromanipulator is not necessary for these injections. To confirm that the needle has not become clogged during the procedure, occasionally inject a droplet outside of the injection media. If necessary, adjust the pulse duration and/or clip the needle tip to produce the proper sized droplet.
20. Very carefully place the injected embryos in an incubator at 16–18 °C.
21. Two to three hours post-injection gently fill the injection plate with 0.1× MMR with pen-strep. While some leakage is normal, remove any damaged or excessively leaking embryos (*see* **Note 11**).
22. The following morning, gently transfer the embryos into fresh 0.1× MMR with pen-strep.
23. Continue to monitor the embryos daily, removing abnormal embryos, and replacing media (*see* **Note 12**).
24. Depending upon the concentration of injected tracer mRNA, the embryos will be translucent enough to assess the uniformity of injection after about 1 week. Using a fluorescence microscope, separate embryos displaying ubiquitous fluorescent tracer expression from those displaying little or no expression (*see* **Note 13**).

3.5 DNA Extraction

1. If numbers of well-injected embryos are sufficiently large, sacrifice at least two whole embryos to assess background mutation frequency. Extract DNA from individual embryos using QIAgen DNeasy Blood and Tissue kit.
2. After several weeks, a small quantity of tissue can be removed from the tail of an axolotl larva without deleterious effects. Anesthetize the larva by placing it in 0.1 g/L Tricaine in 40 % Modified Holtfreter's solution in a small petri dish.
3. When the larva stops responding to touch, using a scalpel, slice off transparent tissue in the tail fin beyond the posterior end of the notochord while the embryo is still submerged in Tricaine.
4. Immediately transfer the larva to a fresh container of 40 % Modified Holtfreter's solution without Tricaine.

5. Under the microscope, retrieve the tail clip and using forceps, place in a 50 μL PCR tube.
6. Repeat for as many larvae as necessary.
7. Using a fine pipette tip, remove excess liquid from fin clip.
8. Add 50 μL of 100 mM NaOH to fin clip. Vortex well, and centrifuge briefly.
9. In a thermocycler, heat at 95 °C for 20 min.
10. Vortex well.
11. Add 15 μL of 1 M Tris–HCl pH 7.4 to each tube and mix well by pipetting.

3.6 Fluorescent PCR to Detect Somatic Mutations (See Note 14)

1. Identify primers around target site that produce a band (excluding primer dimers) of the predicted size by PCR (*see* **Note 15**).
2. If primary PCR produces a band of the proper size and additional bands, obtain internal nested primers and confirm that secondary amplification produces single product of the intended size. The primer pair producing the single, final nested product must have melting temperatures greater than 55 °C (*see* **Note 16**).
3. For one of the primers producing the single nested PCR product, obtain a new copy with forward m13 sequence (5′-TCCCAGTCACGACGT-3′) attached to the 5′ end.
4. Set up each primary PCR as follows:
   (a) 2.5 μL 10× *Taq* buffer.
   (b) 0.5 μL 10 mM dNTPs.
   (c) 0.5 μL 10 μM forward primer.
   (d) 0.5 μL 10 μM reverse primer.
   (e) 0.125 μL *Taq* DNA polymerase.
   (f) 1 μL extracted axolotl genomic DNA.
   (g) To 25 μL with nuclease-free water.

In each genotyping run, include at least one PCR using genomic DNA extracted from an uninjected larva or embryo and one PCR without DNA template to test for contamination.

5. Run the PCR on a thermocycler as follows:
   (a) 95 °C for 30 s.
   (b) 30 cycles of 95 °C for 30 s, 55 °C for 20 s, 68 °C for 15 s.
   (c) 68 °C for 1 min (*see* **Note 17**).
6. Confirm that the PCRs are successful and that there is no contamination by running 5 μL of product on 1 % SB gel (*see* **Note 18**).
7. If primary PCR is successful, proceed to secondary PCR. First, prepare 1–1,000 dilutions of your primary PCR products (including no DNA control) in nuclease-free water.

8. Set up each secondary PCR as follows:
   (a) 2.5 μL 10× *Taq* buffer.
   (b) 0.5 μL 10 mM dNTPs.
   (c) 0.5 μL 100 nM nested m13-tailed forward primer.
   (d) 0.5 μL 10 μM nested reverse primer.
   (e) 0.5 μL 10 μM 6-FAM m13 primer.
   (f) 0.125 μL *Taq* DNA polymerase.
   (g) 1 μL 1:1,000 dilution of primary PCR (*see* **Note 19**).
9. Run the PCR on a thermocycler as follows:
   (a) 95 °C for 30 s.
   (b) 10 cycles of 95 °C for 30 s, 55 °C for 20 s, 68 °C for 15 s.
   (c) 20 cycles of 95 °C for 30 s, 53 °C for 20 s, 68 °C for 15 s.
   (d) 68 °C for 1 min.
10. Confirm that the PCRs are successful and that there is no contamination by running 5 μL of product on 1 % SB gel.
11. Prepare PCR products for fragment analysis by mixing with HiDI formamide and, if necessary, an appropriate size standard. Liz-500 is compatible with 6-FAM-labeled products and leads to accurate size prediction of products under 300 bp (*see* **Note 20**).
12. Using appropriate analysis software, such as Gene Marker, identify peaks. Proper amplification from control DNA should result in single peak of appropriate size. Mutagenized animals usually display multiple peaks of differing sizes (Fig. 2) (*see* **Note 21**).
13. To develop homozygous mutant lines, raise candidate mutagenized animals to sexual maturity (*see* **Note 22**).

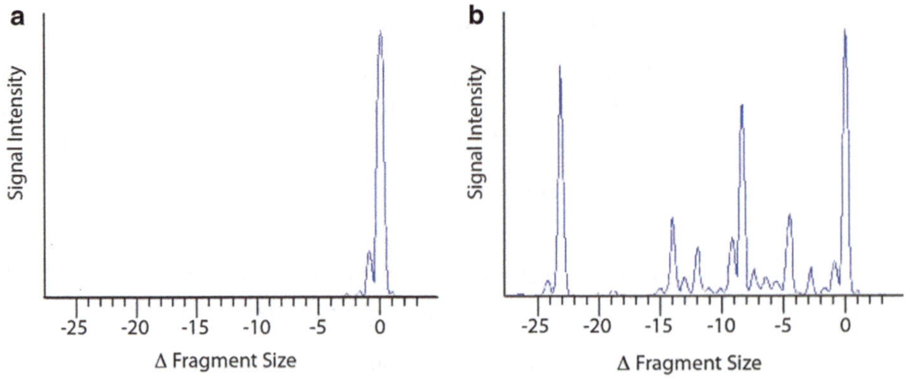

Fig. 2 Analysis of fluorescent fragment sizes after PCR of targeted gene from uninjected (**a**) and RGN-injected (**b**) animals. While a PCR of the region of interest from DNA from a control animal produces a single peak of the expected size (**a**), PCR of DNA extracted from an RGN-injected animal reveals multiple products indicating a high deletion frequency in the target site

4 Notes

1. Several online sources provide useful EST sequencing and cDNA sequences that can be used to identify potential target sites. These include GenBank, Sal-site (www.ambystoma.org), and Axolomics (www.axolomics.org). There are several constraints on the selection of genomic targets in the axolotl. First, the CRISPR-derived RGN system can be used to target any sequence preceding an NGG protoadaptor motif (PAM); however, the in vitro synthesis of sgRNAs using the T7 polymerase will result in the attachment of GG to the 5′ end of the sgRNA. Thus, the method described here will allow one to target any sequence that conforms to the structure of 5′GG(N_1...N_{18})NGG (Fig. 1a). Such targets occur once in every 128 bases of random sequence, and they can be identified in a sequence using the online resource http://zifit.partners.org/ZiFiT/ . If such constraints do not permit the identification of targets, alternative strategies may be used, including, but not limited to, synthesizing sgRNAs using Sp6 polymerase which attaches GA to the 5′ end of sgRNAs, synthesizing sgRNAs with shorter targeting sequences [15], or permitting mismatches between the 5′ end of the sgRNA and the target sequence [16]. Second, RGN targets should be capable of being amplified for genotyping with available genomic information. Intron-exon boundaries are not clearly delineated in cDNA-derived sequences and the average intron size in the axolotl genome is larger than 10 kb [17]; thus, one should first confirm that a target can be amplified with available sequence data before proceeding to subsequent steps. Furthermore, targets should be spaced sufficiently far from primers used for genotyping to detect deletions that may extend beyond the targeted site.
2. Samples that are not digested by BsaI almost always contain the correct product.
3. We strongly recommend assessing the quality of injection by coinjecting an mRNA encoding a fluorescent protein with Cas9 mRNA and the sgRNA to confirm that injected RNAs do not undergo degradation and are uniformly distributed throughout the embryo. A number of plasmids are publically available from which such mRNAs encoding fluorescent proteins can be easily synthesized. Assuming such plasmids contain a polyadenylation signal sequence, synthesis of such mRNAs should follow the same protocol as below. Linearize the plasmid with an enzyme that cuts after the signal sequence and use a mMessage mMachine kit with the proper polymerase for synthesis.

4. Alternative, more detailed methods describing microinjection of axolotl embryos can be found elsewhere [18].
5. We find that the probability of successful mating is increased if the lights of the room are turned off 1 or 2 h in advance of the normal dark cycle. If females can be easily discerned from one another (e.g., different pigmentation pattern), we find that one male can simultaneously be used for matings with two females without reduction of success.
6. Spermatophores appear as white, jellylike clouds in the tank, either stuck to rocks or floating in the water. While deposition of multiple larger spermatophores is most favorable, we have seen successful matings with very trace evidence of spermatophore deposition. We find that the sperm loses motility within a couple of hours after deposition and there is no need to leave the female in the water for an extended period of time. The female typically will not begin to lay eggs until the evening or following morning. If a female lays at an inconvenient time, the rate of laying can be slowed by keeping the female at 12 °C overnight.
7. While storing freshly laid eggs at 4–8 °C delays the first cleavage, we do not recommend storing eggs at this temperature for more than 2 h as this can lead to chromosomal abnormalities and reduced viability [19].
8. Ficoll 400 is an expensive reagent but, in our experience, is the best substance to prevent leakage of yolk from injected embryos. Ficoll use can be minimized by preparing injection ramps deep enough so that the eggs can be completely submerged in Ficoll without entirely covering the surface of the petri dish.
9. These are recommended starting concentrations of sgRNA and *cas9*. If subsequent genotyping suggests low mutagenesis frequency, increase the concentration or volume of injected sgRNA while maintaining a fixed amount of *cas9* [20].
10. A droplet of this diameter has a volume of 3–4 nL. The diameter of such a droplet is about 1/5th the radius of an axolotl egg. If a pulse duration less than 100 ms produces a droplet of this size, consider using a new needle with a finer tip. The ideal needle must be rigid enough to enter the eggs without breaking, and small enough to not cause excessive yolk leakage.
11. Prolonged storage (more than several hours) in Ficoll and 1× MMR is harmful to the embryos.
12. The viability of different clutches of embryos and larvae varies greatly after embryonic manipulation, with high rates of abnormalities in some clutches, and almost no abnormalities in others. Furthermore, the viability of embryos produced at the beginning of a spawning can differ from those produced at the end.

Thus, low viability after a single injection is not necessarily an indicator of inherent toxicity of the *cas9-sgRNA* mixture.

13. Our observations of the localization of fluorescent signal after coinjection of *EGFP* mRNA with *cas9* and sgRNAs suggest that the uniformity of the distribution of injected RNAs is highly variable. While some sets of injections produce a high percentage of animals with uniform distributions of EGFP, other injections performed under very similar conditions produce a high percentage of animals with unilateral or punctate EGFP expression. This variability in mRNA distribution likely explains much of the variability in mutagenesis frequency after injection, and selecting for uniformly injected animals in the manner suggested can greatly reduce the labor required for subsequent genotyping.
14. While there are numerous methods to detect mutation frequency, the axolotl genome possesses particular challenges that make many methods used in other organisms less practical than the fragment analysis method described here. Such methods require amplification of large (>300 bp products) which most often requires obtaining additional sequencing information than currently publically available. Cloning and sequencing PCR products can provide a rough assessment of mutation frequency, but this is cost-prohibitive and too labor intensive to do on a large scale. The fragment analysis of fluorescent PCR products described here provides an affordable means to screen large numbers of embryos using smaller amplicon sizes.
15. Due to the axolotl's large genome size, genomic PCRs from uninjected embryos often produce unintended products. Even products of proper size should be confirmed by gel extraction, TA cloning, and sequencing of several clones. If such PCRs are found to produce unintended products, the most expedient way to eliminate such amplifications is to carry out a second round of amplification of the primary PCR product using nested, internal primers.
16. If the primary PCR produces a single discrete band of the appropriate size from genomic DNA, one may obtain a 6-FAM modified form of one of the primers used for this PCR and skip genotyping **steps 7** through **10**. The process described here utilizes a universal 6-FAM labeled primer for the purpose of cost reduction. This method is explained in greater detail elsewhere [21].
17. These conditions only provide a loose framework, and PCR conditions should be individually optimized. A fast, high fidelity polymerase may be substituted for *Taq* polymerase.
18. Do not proceed with genotyping if there is any evidence of PCR contamination, as any contamination makes subsequent mutation frequency determination impossible.

19. Note that the ratio of the concentration of the nested, m13-tailed primer to the reverse gene specific primer is 1:1,000. Higher concentrations of the nested, m13-tailed primer can result in failure of production of sufficient quantities of 6-FAM-labeled product for analysis.
20. Numerous academic research facilities provide fragment analysis for a low cost compared to standard sequencing methods. These include the DNA Analysis Facility at Yale University (http://dna-analysis.research.yale.edu).
21. While this method provides a rapid and inexpensive assessment of successful mutagenesis, definitive confirmation of mutagenesis can only be achieved through cloning and sequencing of individual PCR products.
22. Several considerations should be made in the development of mutant lines. As RGNs may have sequence-specific off-target effects, crossing animals that have been mutagenized using the same sgRNA may result in illusory, off-target genotype–phenotype associations. When possible, several distinct sgRNAs should be used to produce mutants for each gene of interest.

References

1. Malacinski GM (1989) Developmental Genetics. In: Malacinski GM, Armstrong JB (eds) Dev. Biol. Axolotl. Oxford University Press, New York, NY, pp 103–109
2. Voss SR, Epperlein HH, Tanaka EM (2009) Ambystoma mexicanum, the axolotl: a versatile amphibian model for regeneration, development, and evolution studies. Cold Spring Harb Protoc 2009:pdb.emo128. doi:10.1101/pdb.emo128
3. Monaghan JR, Athippozhy A, Seifert AW, Putta S, Stromberg AJ, Maden M, Gardiner DM, Voss SR (2012) Gene expression patterns specific to the regenerating limb of the Mexican axolotl. Biol Open 1:937–948. doi:10.1242/bio.20121594
4. Campbell LJ, Suárez-Castillo EC, Ortiz-Zuazaga H, Knapp D, Tanaka EM, Crews CM (2011) Gene expression profile of the regeneration epithelium during axolotl limb regeneration. Dev Dyn 240:1826–1840
5. Knapp D, Schulz H, Rascon CA, Volkmer M, Scholz J, Nacu E, Le M, Novozhilov S, Tazaki A, Protze S, Jacob T, Hubner N, Habermann B, Tanaka EM (2013) Comparative transcriptional profiling of the axolotl limb identifies a tripartite regeneration-specific gene program. PLoS One 8:e61352. doi:10.1371/journal.pone.0061352
6. Holman EC, Campbell LJ, Hines J, Crews CM (2012) Microarray analysis of microRNA expression during axolotl limb regeneration. PLoS One 7:e41804. doi:10.1371/journal.pone.0041804
7. Stewart R, Rascón CA, Tian S, Nie J, Barry C, Chu L-F, Ardalani H, Wagner RJ, Probasco MD, Bolin JM, Leng N, Sengupta S, Volkmer M, Habermann B, Tanaka EM, Thomson JA, Dewey CN (2013) Comparative RNA-seq analysis in the unsequenced axolotl: the oncogene burst highlights early gene expression in the blastema. PLoS Comput Biol 9:e1002936. doi:10.1371/journal.pcbi.1002936
8. Wu C-H, Tsai M-H, Ho C-C, Chen C-Y, Lee H-S (2013) De novo transcriptome sequencing of axolotl blastema for identification of differentially expressed genes during limb regeneration. BMC Genomics 14:434. doi:10.1186/1471-2164-14-434
9. Monaghan JR, Epp LG, Putta S, Page RB, Walker JA, Beachy CK, Zhu W, Pao GM, Verma IM, Hunter T, Bryant SV, Gardiner DM, Harkins TT, Voss SR (2009) Microarray and cDNA sequence analysis of transcription during nerve-dependent limb regeneration. BMC Biol 7:1. doi:10.1186/1741-7007-7-1
10. Porteus MH, Carroll D (2005) Gene targeting using zinc finger nucleases. Nat Biotechnol 23:967–973
11. Wood AJ, Lo T-W, Zeitler B, Pickle CS, Ralston EJ, Lee AH, Amora R, Miller JC, Leung E,

Meng X, Zhang L, Rebar EJ, Gregory PD, Urnov FD, Meyer BJ (2011) Targeted genome editing across species using ZFNs and TALENs. Science 333:307

12. Hwang WY, Fu Y, Reyon D, Maeder ML, Tsai SQ, Sander JD, Peterson RT, Yeh J-RJ, Joung JK (2013) Efficient genome editing in zebrafish using a CRISPR-Cas system. Nat Biotechnol 31:227–229
13. Flowers GP, Timberlake AT, McLean KC, Monaghan JR, Crews CM (2014) Highly efficient targeted mutagenesis in axolotl using Cas9 RNA-guided nuclease. Development 141:2165–2171
14. Smith JJ, Putta S, Walker JA, Kump DK, Samuels AK, Monaghan JR, Weisrock DW, Staben C, Voss SR (2005) Sal-Site: Integrating new and existing ambystomatid salamander research and informational resources. BMC Genomics 6:181. doi:10.1186/1471-2164-6-181
15. Fu Y, Sander JD, Reyon D, Cascio VM, Joung JK (2014) Improving CRISPR-Cas nuclease specificity using truncated guide RNAs. Nat Biotechnol 32:279–284
16. Hwang WY, Fu Y, Reyon D, Maeder ML, Kaini P, Sander JD, Joung JK, Peterson RT, Yeh J-RJ (2013) Heritable and precise zebrafish genome editing using a CRISPR-Cas system. PLoS One 8:e68708. doi:10.1371/journal.pone.0068708
17. Smith JJ, Putta S, Zhu W, Pao GM, Verma IM, Hunter T, Bryant SV, Gardiner DM, Harkins TT, Voss SR (2009) Genic regions of a large salamander genome contain long introns and novel genes. BMC Genomics 10:19. doi:10.1186/1471-2164-10-19
18. Khattak S, Murawala P, Andreas H, Kappert V, Schuez M, Sandoval-Guzmán T, Crawford K, Tanaka EM (2014) Optimized axolotl (Ambystoma mexicanum) husbandry, breeding, metamorphosis, transgenesis and tamoxifen-mediated recombination. Nat Protoc 9: 529–540
19. Humphrey RR, Fankhauser G (1957) The origin of spontaneous and experimental haploids in the Mexican axolotl (Siredon—or Ambystoma—Mexicanum). J Exp Zool 134: 427–447
20. Guo X, Zhang T, Hu Z, Zhang Y, Shi Z, Wang Q, Cui Y, Wang F, Zhao H, Chen Y (2014) Efficient RNA/Cas9-mediated genome editing in Xenopus tropicalis. Development 141: 707–714
21. Schuelke M (2000) An economic method for the fluorescent labeling of PCR fragments. Nat Biotechnol 18:233–234

Chapter 23

Gene Manipulation for Regenerative Studies Using the Iberian Ribbed Newt, *Pleurodeles waltl*

Toshinori Hayashi and Takashi Takeuchi

Abstract

Newts provide a unique model animal for regenerative studies. An experimental model system for molecular genetics in newts is needed in order to clarify the mechanisms of regeneration from the perspective of gene functions. We have identified that Iberian ribbed newt (*Pleurodeles waltl*) is a suitable model animal for such studies. Here we describe protocols for gene manipulation using *Pleurodeles waltl*.

Key words Newt, Salamander, Molecular genetics, Regeneration, Transgenesis, TALENs, Genome editing

1 Introduction

Regeneration of a lost tissue in an animal is an important issue. Urodele amphibians, such as newts have the remarkable capability to regenerate organs, and various body parts, such as limbs, ocular tissues, brain, spinal cord, intestine, and these animals have been used for regeneration studies for centuries [1, 2]. However, the identities of the genes that regulate regeneration are still unclear.

To elucidate gene function during newt regeneration, it is particularly important to establish transgenic and knockout lines, and systematically cross these lines to study the functions of genes. For this purpose, we chose the Iberian ribbed newt (*Pleurodeles waltl*) as a model animal. Previously we have shown that *P. waltl* can spawn fertilized eggs all year around in the laboratory [3]. Each female can lay more than 150 eggs per spawning and they spawn every 2–4 weeks (Table 1). In this chapter we describe methods for artificial insemination, I-SceI mediated transgenesis, and genome editing using *P. waltl*.

Anoop Kumar and András Simon (eds.), *Salamanders in Regeneration Research: Methods and Protocols*, Methods in Molecular Biology, vol. 1290, DOI 10.1007/978-1-4939-2495-0_23, © Springer Science+Business Media New York 2015

Table 1
Comparison of the breeding properties of *P. waltl* is comparable to *X. tropicalis*

		P. waltl	*X. tropicalis*
Maturation	♂	6 months ~	5 months ~
	♀	9 months ~	6 months ~
Adult size	♂	12 cm ~	4–4.5 cm
	♀	14 cm ~	5–6 cm
Egg size		1.2–1.5 mm	0.7–0.8 mm
Eggs per spawning		150–600 (5,000~/year)	1,000–9,000

Modified from Hayashi et al. 2013

2 Materials

2.1 Reagents

1. 10× Holtfreter's solution (HF): Dissolve 35 g NaCl (600 mM), 0.5 g KCl (7 mM), and 2 g $NaHCO_3$ (24 mM) in 1 L of distilled water (*see* **Note 1**).
2. 1× HF solution: Add 100 mL 10× HF solution to 900 mL distilled water. Sterilize by autoclaving and allow it to cool at room temperature. Add 5 mL of 1 M HEPES (5 mM), 1.7 mL of 1 M $CaCl_2$ (1.7 mM), and 1.7 mL 1 M $MgSO_4$ solution (1.7 mM) to this solution. Store at room temperature (*see* **Note 2**).
3. 0.25× HF solution: Add 25 mL 10× HF solution to 900 mL distilled water and sterilize by autoclaving. Allow the solution to cool at room temperature and add 5 mL of 1 M HEPES, 1.7 mL of 1 M $CaCl_2$, 1.7 mL 1 M $MgSO_4$, and 5 mL of (50 mg/mL) gentamicin sulfate. Make up the solution to 1 L and store at 4 °C (*see* **Note 2**).
4. 0.1 % MS222 (for anesthesia): Add 0.4 g to 400 mL 0.25× HF solution (*see* **Note 3**).
5. Human chorionic gonadotropin (hCG): Dissolve 1,000 units in 1 mL 1× HF solution (*see* **Note 4**).
6. Injection medium: Dissolve 4 g of Ficoll in 100 mL 0.25× HF solution containing gentamicin (*see* **Note 5**).
7. De-jellying solution (2 % Cysteine solution or 2 % Sodium thioglycolate): Dissolve 1 g of cysteine or sodium thioglycolate in 50 m L 25× HF solution, adjust to pH 7.6 (pH 7.9 for sodium thioglycolate) with 5 N NaOH (*see* **Note 3**).
8. PCR reagents and primers for genotyping (optional).

2.2 Vectors/Plasmids

1. Expression vectors for transgenesis: use I-SceI site-containing vector (*see* **Note 6**).
2. Plasmids cording TALENs, which recognize the target gene (*see* **Note 7**).

2.3 Equipment

1. Microinjector (*see* **Note 8**).
2. Micromanipulator (*see* **Note 9**).
3. Stereomicroscope with flexible lighting (in common with frog and newt).
4. Low-temperature incubator (temperature range; 8–16 °C).
5. Puller and glass capillaries (for making micro needles) (*see* **Note 10**).
6. Injection dish (Nunc mini tray, 60 wells) (*see* **Note 11**).
7. Transfer pipettes/3 mL soft plastic pipettes (*see* **Note 12**).
8. 1 mL disposable syringes. 27-gauge needles.

3 Methods

3.1 Maintenance of a Newt Colony

1. Housing: Iberian ribbed newts (*Pleurodeles waltl*) should be kept in a tap-water aquarium at 25–26 °C under natural light cycles (*see* **Note 13**).
2. Feeding: Animals should be fed at least five times per week with hatched brine shrimp for larvae and compound feeds for juvenile/adult newts (*see* **Note 14**).

3.2 Artificial Insemination

1. For egg collection, inject 14–18 h prior to egg collection, inject 100–200 units hCG subcutaneously into the lower jaw of mature females using 1 mL syringe and 27-gauge needle.
2. The next day, when the female starts laying eggs, transfer her into 0.1 % MS-222.
3. When the animal is anesthetized, wipe the animal lightly to remove water drops.
4. Push the abdomens gently and collect the eggs in a dry petri dish (Fig. 1a) (*see* **Note 15**).
5. For sperm collection, anesthetize sexually mature males (older than 6 months post fertilization) with 0.1 % MS222. When the animal is anesthetized, wipe the animal lightly to remove water drops.
6. Squeeze out the sperm by pressing the abdomen of mature males (Fig. 1b) and place it into 1× HF solution to 2,000–5,000 sperm/μL (*see* **Notes 16** and **17**). The sperm suspensions can be stored for up to 8 h at room temperature.

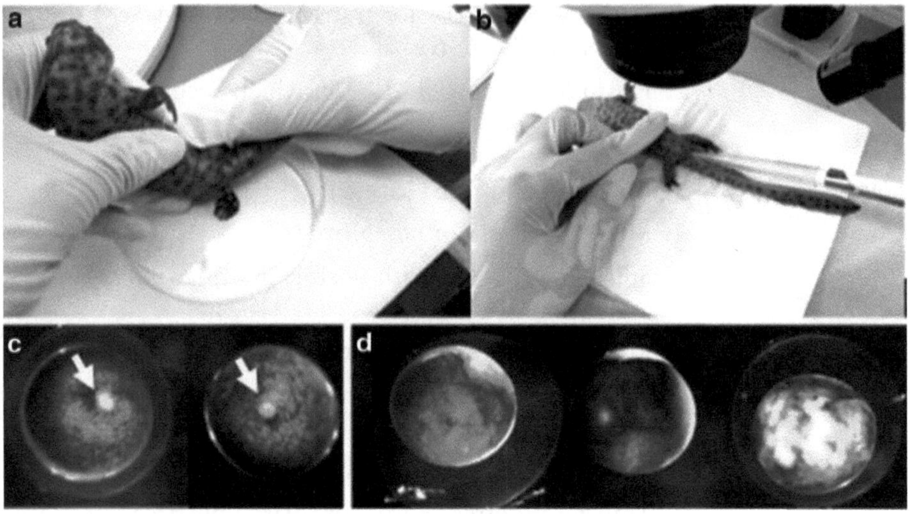

Fig. 1 Egg (**a**) and sperm (**b**) collection for the artificial insemination. (**c** and **d**) De-jellied fertilized eggs. (**c**) Good quality eggs with (*left*)/ without (*right*) an eggshell. *Arrows* indicate white spots on the animal pole. (**d**) Examples of low quality eggs. *Arrows* indicate leak of cytoplasm

7. Mix eggs with diluted sperm (approximately 5–20 μL per 20 eggs) gently using pipette tips.
8. Incubate inseminated eggs at room temperature for 15 min (*see* **Note 18**).
9. Pour 0.25× HF solution into the dish and incubate eggs for a further 15 min to allow water absorption by the jelly.
10. Soak the fertilized eggs (single-cell stage embryos) in de-jelly solution for 5 min with gentle shaking (*see* **Note 19**).
11. Rinse de-jellied eggs 3–5 times in 25 % HF solution.
12. Sort good eggs (Fig. 1c, d) under the stereomicroscope and discard low quality eggs (*see* **Note 20**).
13. Transfer the eggs into an injection tray filled with injection medium (*see* **Note 21**).
14. Remove eggshells using fine forceps and store at 8 °C until the microinjection (*see* **Note 22**).

3.3 Preparation of Plasmid Vectors

1. Prepare the plasmid vectors using an ion exchange column from *E. coli* culture (*see* **Note 23**).
2. Dissolve the plasmid in water (100–200 ng/μL) (*see* **Note 4**).
3. Linearize 200 ng of plasmid vectors with 25 units of I-SceI in 20 μL total volume of 1× I-SceI buffer at 37 °C for 40 min (*see* **Notes 24** and **25**).

3.4 Preparation of TALEN mRNAs (See Note 26)

1. Linearize 2–10 μg of plasmid vectors with appropriate restriction enzymes.
2. Extract the plasmids by phenol/chloroform. Precipitate the plasmids by ethanol precipitation, then dissolve in RNAse-free water (500 ng/μL).
3. Synthesize TALEN mRNAs by in vitro transcription using RNA polymerase (*see* **Note 27**).
4. Purify the mRNAs by ethanol precipitation, and dissolved in nuclease-free water (100–500 ng/μL). Stored at -80 °C.
5. Just before use, dilute RNAs with RNAse-free water to 10–400 ng/μL (*see* **Note 28**).

3.5 Microinjection (See Note 29)

1. Prepare micro needles (diameter of outer tip: 10–20 μm) (*see* **Note 30**).
2. Set up the microinjector and micromanipulator (Fig. 2). Refer to the operating instructions.
3. Load the I-SceI digested plasmid solution or TALEN mRNAs into a micro needle.
4. Set the injection volume between 9 and 20 nL (*see* **Note 31**).
5. Support the egg with the forceps and prick the animal pole side with the micro needle.
6. Push the "Inject" button and then pull out the needle.
7. Leave the injected eggs and move to the next one.
8. After injection of plasmid DNA, incubate the eggs (embryos) at 8–10 °C for 4–8 h to delay the first cleavage, and then keep at 16 °C overnight (*see* **Note 32**).

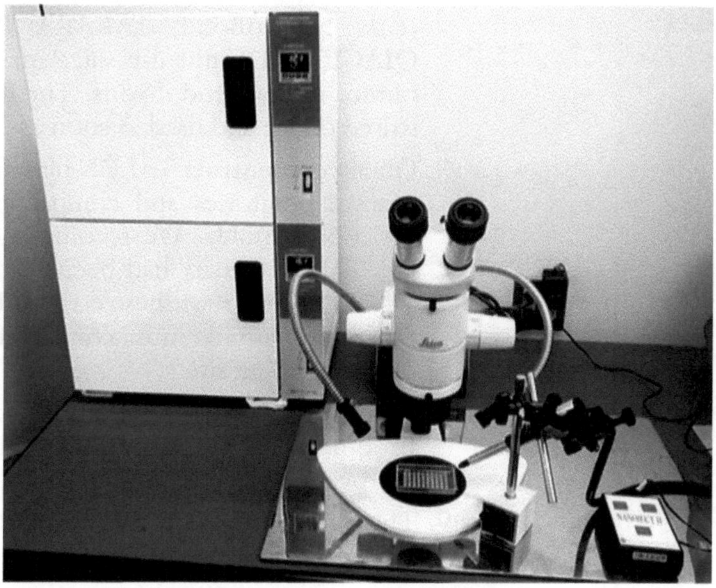

Fig. 2 Setting up the equipment for the microinjection

3.6 Sorting the Embryos and Rearing

1. Return the temperature to 25–26 °C (*see* **Note 33**).
2. Transfer the embryos into petri dishes or plastic cups (*see* **Note 21**).
3. Rinse the embryos three times in 0.25× HF solution to remove Ficoll (*see* **Note 34**).
4. Incubate the embryos in 0.25× HF solution for 5–7 days (hatching-stage) and then transfer them into tap water. Start feeding with brine shrimps after 10–12 days of post fertilization.
5. Sort intended transgenic or knockout animals by expression of the reporter gene or genotyping (*see* **Note 35**).
6. Raise these animals to sexual maturation. Inter cross the animals to establish the transgenic/knockout lines.

4 Notes

1. Sterilize by autoclaving and store at room temperature.
2. After autoclaving, add HEPES, $CaCl_2$, $MgSO_4$, and gentamicin sulfate.
3. Prepare this solution just before use.
4. Store at −20 °C. It should be stable after several times of freeze-thawing.
5. Sterilize by filtration using membrane filter and store in a clean bottle at 4 °C.
6. For efficient integration into the newt genome, sub-clone the construct between 2 I-SceI recognition sites [4, 5]. The plasmid DNA should be prepared using an ion exchange column (e.g., NucleoBond, TAKARA Bio or Plasmid Midi kit, QIAGEN). Do not use silica membrane columns since they cannot remove endotoxins. The digested plasmids should be stored on ice and used as soon as possible.
7. Obtain or construct TALEN plasmids for your targeting genes. Many laboratories and companies provide various types of TALEN scaffolds. We recommend Platinum Gate TALENs [6], as high activity had been confirmed in the new embryos [7]. In order to synthesize the mRNAs by in vitro transcription, the plasmids must contain T7, T3, or SP6 RNA polymerase binding site.
8. We prefer to use a NANOJECT II microinjector (Drummond, Broomall, PA). As this injector is driven not by compressed air but by an electric motor, it is compact and easy to operate.
9. The injector head should be mounted in a suitable micromanipulator. Although MM33 would be suitable for NANOJECT

II, products from other companies can also be used (such as M-152, NARISHIGE, Japan) with an attachment supplied with the injector.

10. A puller is a device for pulling a glass capillary under heating to make glass micro needles. Select appropriate type of capillary for your microinjector according to manufacturer's instruction.
11. As the de-chorionized eggs settle in the well at the bottom, it helps to hold the egg during the injection.
12. For handling the eggs and embryos, cut the tip to adjust the opening to egg size (3–5 mm diameter).
13. To increase the growth rate of the animals, the optimal temperature range is 25–26 °C. However, the animals can be maintained in a healthy state at 16–24 °C. Brief filtering helps to reduce the frequency of changing the water. Mature males and females should be housed separately in order to avoid natural mating.
14. *P. waltl* newts eat compound feeds available for *Xenopus*, catfish, and trout. We prefer to use feeds for aquarium fishes rather than for hatchery fishes. Aquarium feeds are less likely to soil the water in the tank.
15. Pushing lightly. After squeezing the eggs, start the insemination immediately. You may collect the eggs 2–3 times from the same animal at 2–3 h intervals.
16. By pushing lightly, urine is pressed out first. If you push too strongly, the bladder will be crushed and the newt will die. Sperm can be collected from the same animals every 2–4 weeks.
17. Mix the sperm suspension by gentle pipetting.
18. The lid of the petri dish should be shut to maintain humidity.
19. Keep 5 min or less, as de-jerry solution is toxic. And then rinse well at the next step. The eggshells (Fig. 1c arrowheads) still remained on the most eggs.
20. White spot on the animal pole is a sign for good eggs (arrows in Fig. 1c). The animal pole side are clearly pigmented. There is no leak of cytoplasm. On the other hand, it is difficult to see the white spot on the low quality eggs. They also show leak of cytoplasm (Fig. 1d, arrows).
21. Use the transfer pipet. De-jerried eggs and embryos are fragile.
22. To obtain full transgenic or full knockout animals, it is important to perform the microinjection at single-cell stage. The fertilized eggs of *P. waltl* require 5–6 h for the first cleavage at 25 °C. You can delay the timing of the first cleavage at 8 °C; however, a good result will be obtained if injection is performed as quickly as possible after fertilization.

23. The ion exchange columns are available as a kit (e.g., NucleoBond, TAKARA Bio or Plasmid Midi kit, QIAGEN). Do not use silica membrane columns since they cannot remove endotoxins.
24. Keep 40 min, and then transfer the tube on ice until use.
25. In most cases, we obtained a good result by injection of 100 pg/egg plasmid DNA. When you inject two vectors at the same time, mix them (100 ng each, total 200 ng in 20 μL reaction).
26. Keep the working area and reagents RNAse-free.
27. As the 5′ ends of RNAs must be caped, it is easy to use a kit for caped-mRNA synthesize (e.g., mMESSAGE mMACHINE T7 Ultra Kit, Life Technologies).
28. Optimize the appropriate amounts of TALENs, because it depends completely on their levels of activity. The mRNA should be tested between total 12–400 pg/egg. Use equal amounts of the Right- and Left-TALENs.
29. This injection protocol is modified from I-SceI methods for *Xenopus* [4]. You may also refer to [8, 9] that were developed for urodeles.
30. Pull the glass capillaries using a Puller according to manufactures instruction. The micro needle tips should not be ground. Prepared spear needles and stocked for future use.
31. We obtained a good result with injection volume in this range. Refer to the operating instructions of your microinjector for calibration of the injection volume.
32. To increase the ratio of the full transgenic animals, integration of the transgene into the newt genome should be occur at single-cell stage. Therefore, it is important to delay cell cleavage by keeping them at low temperature. After injection of TALEN mRNAs, this stem is not required. 25 °C is the optimal temperature for TALENs.
33. The embryos should be at morula stage. The criteria for the developing stages were defined by Shi and Boucaut [10].
34. Ficoll prevents gastrulation. Rinse out completely.
35. For genotyping, collect the tail tips or limb tips under anesthesia. 0.01 % MS222 is suitable for the larvae.

Acknowledgments

We would like to thank Prof. Takashi Yamamoto and Dr. Tetsushi Sakuma (Hiroshima University) for providing the TALENs. We also thank Profs. Andras Simon and Anoop Kumar for their valuable help and advice. Kyorin Corporation (Hyogo, Japan) kindly provided the feed for the newts. This research was supported by MEXT/JSPS KAKENHI (grant numbers, 26650082, 25124706, and 25840086).

References

1. Brockes JP, Kumar A (2002) Plasticity and reprogramming of differentiated cells in amphibian regeneration. Nat Rev Mol Cell Biol 3:566–574
2. Agata K, Inoue T (2012) Survey of the differences between regenerative and non-regenerative animals. Dev Growth Differ 54: 143–152
3. Hayashi T, Yokotani N, Tane S, Matsumoto A, Myouga A, Okamoto M, Takeuchi T (2013) Molecular genetic system for regenerative studies using newts. Dev Growth Differ 55: 229–236
4. Ogino H, McConnell WB, Grainger RM (2006) High-throughput transgenesis in Xenopus using I-SceI meganuclease. Nat Protoc 1:1703–1710
5. Pan FC, Chen Y, Loeber J, Henningfeld K, Pieler T (2006) I-SceI meganuclease-mediated transgenesis in Xenopus. Dev Dyn 235: 247–252
6. Sakuma T, Ochiai H, Kaneko T, Mashimo T, Tokumasu D, Sakane Y, Suzuki K, Miyamoto T, Sakamoto N, Matsuura S, Yamamoto T (2013) Repeating pattern of non-RVD variations in DNA-binding modules enhances TALEN activity. Sci Rep 3:3379
7. Hayashi T, Sakamoto K, Sakuma T, Inoue T, Kawaguchi E, Agata K, Yamamoto T, Takeuchi T (2014) Molecular genetic system for regenerative studies using newts. Dev Growth Differ 56:115–121
8. Casco-Robles MM, Yamada S, Miura T, Nakamura K, Haynes T, Maki N, Del Rio-Tsonis K, Tsonis PA, Chiba C (2011) Expressing exogenous genes in newts by transgenesis. Nat Protoc 6:600–608
9. Khattak S, Murawala P, Andreas H, Kappert V, Schuez M, Sandoval-Guzmán T, Crawford K, Tanaka EM (2014) Optimized axolotl (Ambystoma mexicanum) husbandry, breeding, metamorphosis, transgenesis and tamoxifen-mediated recombination. Nat Protoc 9:529–540
10. Shi DL, Boucaut JC (1995) The chronological development of the urodele amphibian Pleurodeles waltl (Michah). Int J Dev Biol 39:427–441

Part V

Gene Expression

Chapter 24

Transcriptomics Using Axolotls

S. Randal Voss, Antony Athippozhy, and M. Ryan Woodcock

Abstract

Microarray and RNA-sequencing technology now exists for the characterization of the *Ambystoma mexicanum* transcriptome. With sufficient replication, these tools give the opportunity to truly investigate gene expression in a variety of experimental paradigms. Analysis of data from the Amby002 array and RNA-sequencing technology can identify genes that change expression levels in concert with each other, which in turn may reveal mechanisms associated with biological processes and molecular functions.

Key words Microarray, RNA sequencing, Transcriptomics, Statistics, RNA

1 Introduction

Transcription is a fundamental molecular process that is common to all organisms. During transcription, an RNA molecule is synthesized from a template DNA sequence (gene). The synthesis of RNA molecules differs in time, type, and amount across cells. The RNAs that derive from transcription directly or indirectly orchestrate development and maintain cellular structures and functions. At higher levels of biological organization, RNAs regulate physiology and behavior. Thus, the analysis of transcription is relevant to all life scientists, and it can be studied in any organism using common molecular methods, analytical techniques, and instrumentation.

In model genetic organisms, transcription methodologies are routinely used to explore the functions of genes and phenotypes, to monitor experiments, and to classify and define biological processes. Such applications will expand as costs decline, and new approaches are developed to allow greater flexibility of experimental design and greater accuracy of RNA measurement. In reality, transcription studies are in their infancy, and relatively few temporal and spatial analyses of gene expression have been accomplished in model organisms. At the same time, it is now possible with next-generation nucleotide sequencing technology to obtain the complete genome or transcriptome of any organism.

Anoop Kumar and András Simon (eds.), *Salamanders in Regeneration Research: Methods and Protocols*, Methods in Molecular Biology, vol. 1290, DOI 10.1007/978-1-4939-2495-0_24, © Springer Science+Business Media New York 2015

With such information, the same gene expression techniques that are used in genetic models can be applied to any organism. It therefore seems likely that the next 10 years will see a renaissance of the comparative experimental approach that dominated biological research in the twentieth century. A focus of this approach will be gene expression.

Therefore, it is critical that researchers have access to equipment and methods that can enable gene expression studies. Here, we present methods for applying Affymetrix microarray and RNA-sequencing approaches for the Mexican axolotl (*Ambystoma mexicanum*). The axolotl has been the subject of genome resource development for over a decade. In 2005, a first generation Affymetrix GeneChip with relatively few probesets (~5,000) was developed and used to profile gene expression during thyroid hormone-induced metamorphosis and spinal cord regeneration [1]. In 2012, a higher content (~20,000 probesets) second-generation Affymetrix GeneChip (Amby_002) was made available to the community [2]. The Amby_002 GeneChip has been used to quantify mRNAs sampled from several regenerating tissues, including brain, iris, dorsal root ganglia, and limb. Notably, a recent study yielded a very detailed readout of the first 28 days of limb regeneration, and these data can be accessed from a gene expression database at Sal-Site (www.ambystoma.org) (Fig. 1).

Several groups have also begun to use RNA sequencing to estimate gene expression in the axolotl [3, 4]. Unfortunately, all studies to date have not replicated samples, and thus the reliability of the expression estimates is unclear. We recently completed a study that sequenced biological replicates that were technically replicated using Amby_002. We describe here methods for obtaining RNA-sequencing estimates from replicate samples and how also to compare these estimates to annotated probesets from Amby_002.

2 Materials

2.1 Animals

Genetically homogeneous axolotls are available from the *Ambystoma* Genetic Stock Center, University of Kentucky.

2.2 RNA Extraction

1. Homogenization equipment: Pestles and 1.5 mL nuclease-free microtubes or 1 mL syringes and 18–26-G needles.
2. Triazol (Life Technologies, #15596018).
3. RNeasy Mini Kit (QIAGEN, Cat no. 74104) with RNase-free DNase Set (QIAGEN, Cat no. 79254).
4. Nanodrop ND-1000 or 2100 Bioanalyzer (Agilent Technologies).

2.3 Affymetrix Microarray Analysis

1. GeneChips are available from Affymetrix (Part No. 520748).
2. GeneChips are processed and normalized estimates of gene expression are provided by the University of Kentucky Microarray Core Facility.

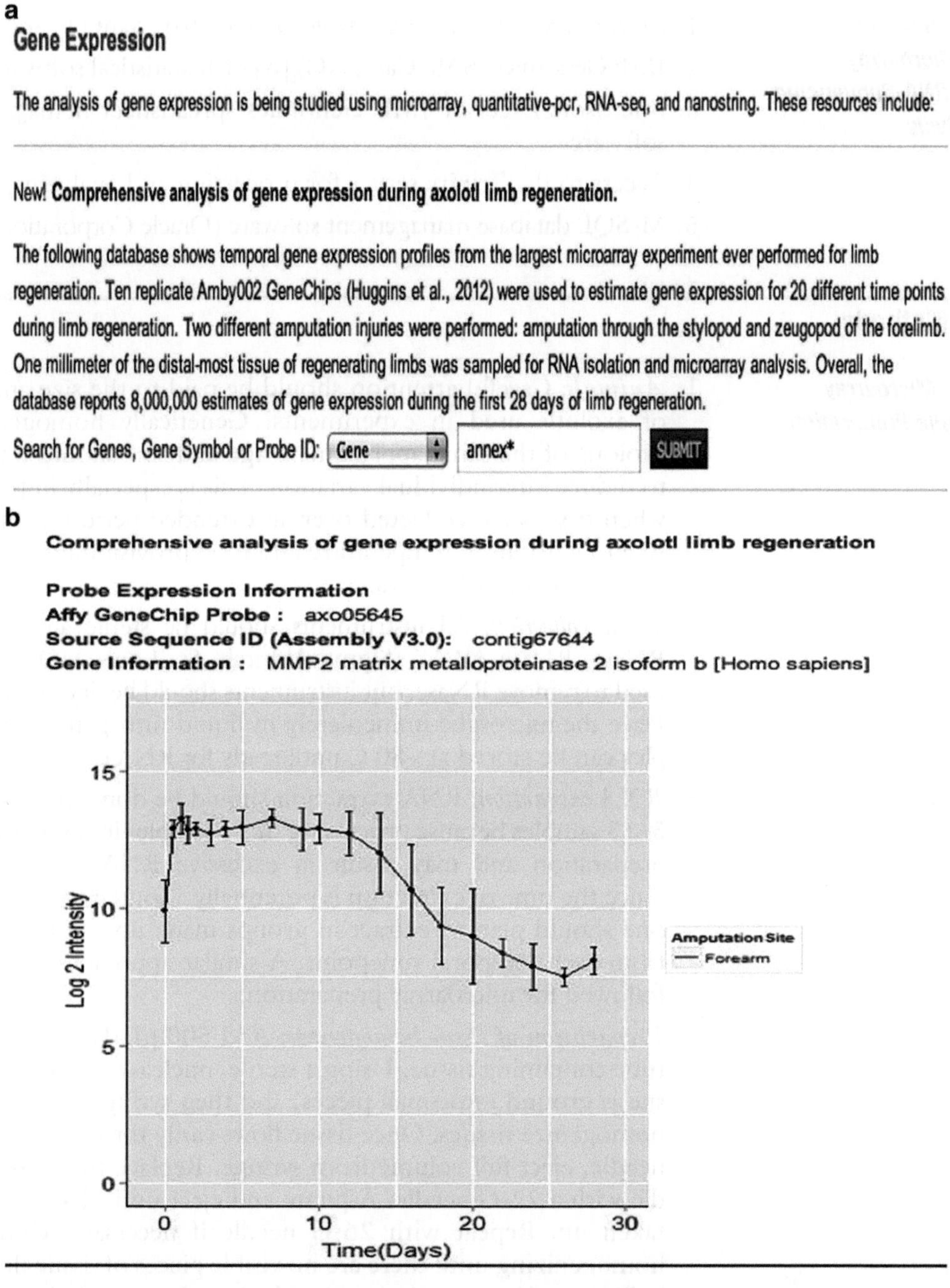

Fig. 1 (**a**) Query tool within Sal-Site used to search for microarray data for an individual probeset. Probesets may be queried by Affymetrix probeset ID or gene name. (**b**) Display of expression profile of a sample probeset. Data are transformed so that measurements approximate a normal distribution and thus can be analyzed by parametric analysis. *Log2* intensity values from the microarray are plotted as a function of time for probeset axo05645, which has been annotated as matrix metalloproteinase 2

2.4 RNA-Sequencing Analysis

Total RNA samples are submitted to a genome sequencing center for Illumina paired-end sequencing.

2.5 Software for Microarray and RNA-Sequencing Analysis

1. Affymetrix Expression Console (Affymetrix, Santa Clara, CA).
2. JMP Genomics (SAS, Cary, NC) [5] or R statistical software [6].
3. Microsoft Excel or JMP Genomics spreadsheet management software.
4. Access to the Trinity server for processing and analyzing data.
5. MySQL database management software (Oracle Corporation) [7].

3 Methods

3.1 Microarray Sample Preparation

1. *Animals*: Careful attention should be paid to the size and age of axolotls used in experiments. Genetically homogeneous axolotls of the same approximate age and size should be used to reduce interindividual variation. This is especially important when tissues are collected over an extended period of time as small variations in temperature and food provisioning can alter developmental schedules.
2. *Tissue collection*: All instruments should be sterile and free of RNase. RNase AWAY (Sigma Aldritch, St. Louis MO) can be used to remove RNase, but instruments should be dry before use. Place the microtube immediately in liquid nitrogen. The samples can be stored at −80 C until ready for RNA extraction.
3. *RNA extraction*: RNA extraction should be done in groups of 5–15 samples because processing more samples lengthens tissue preparation and may result in excessive RNA degradation. Since the time of extraction is potentially a source of variability, one should plan to extract in groups made up of one sample from each temporal timepoint. A similar approach should be followed for microarray preparation.
4. *Preparation of tissue homogenate*: Add 500 μL Triazol to each tube containing tissue. Using a sterile, nuclease-free pestle, tissue is ground into small pieces, and then syringes are used to homogenize tissues. Once tissue flows easily through the 18-G needle, eject full volume from syringe. Replace the 18-G needle with a 22-G needle. Aspirate and eject until tissue is easily taken up. Repeat with 26-G needle if necessary. Continue homogenizing until there are no visible pieces of tissue floating in Triazol. Homogenization can be lengthy, so to help prevent RNA degradation, tubes should be kept on ice.
5. *Pellet nonhomogenized tissue*: Spin tubes at 9,280 × g for 10 min at 4 °C. Check tubes for tissue pellets. Thoroughly homogenized tissue should not produce pellets as a result of this centrifugation. Move supernatant to a new labeled tube if a pellet is present.
6. Add 100 μL chloroform per 500 μL Triazol to each tube. Some Triazol may have been lost during tissue homogeniza-

tion. If the volume is less than 500 μL, fresh Triazol can be added to reach 500 μL. Invert and flick tube 20 times or until solution is a consistent light-pink color. Incubate at room temperature for 10 min.

7. Centrifuge at 15,680 × *g* for 10 min at 4 °C.
8. Transfer the aqueous phase (top layer) to a new labeled 1.5 mL tube. Do not disturb the white precipitate at the immiscible layer.

3.2 Purification of RNA Using the QIAGEN RNeasy Mini Kit

1. Add 700 μL of RLT buffer and 500 μL of 200 proof ethanol to the tubes containing supernatant. Mix by pipetting.
2. Label RNeasy spin columns. Transfer 700 μL of sample to RNeasy spin column placed in a 2 mL collection tube.
3. Centrifuge at 9,280 × *g* for 15 s. Discard the flow-through in the collection tube, and then add remaining sample to the spin column.
4. Centrifuge again at 9,280 × *g* for 15 s and discard the flow-through.
5. Wash each column with 350 μL of RW1 wash buffer. The spin column can be turned on its side and rotated to wash sides of column.
6. Centrifuge at 9,280 × *g* for 15 s. Discard the flow-through.
7. Aliquot 80 μL of DNase and RDD buffer mix to each column. Incubate at room temperature for 15 min.
8. Prepare the DNase and RDD buffer as a master mix containing 11 μL DNase and 77 μL RDD buffer per sample.
9. Centrifuge at 9,280 × *g* for 15 s. Discard the flow-through.
10. Add 500 μL of RPE to each spin column. Discard flow-through.
11. Place the column in a new collection tube. Add 500 μL of RPE to each spin column. Centrifuge for 2 min to remove excess RPE. Discard flow-through. Label new, nuclease-free 1.5 mL tubes.
12. Transfer column to 1.5 mL tube. Add 30 μL of RNase-free water to the center of the column. Incubate for 1 min at room temperature, and then spin for 1.5 min at 9,280 × *g*. Discard the column.
13. Use 1.5 μL aliquot of the eluted RNA sample for spectrophotometric analysis by a Nanodrop spectrophotometer. The ratios A260/A280 and A260/A230 should be higher than 1.8. Lower A260/A230 ratios may indicate incomplete RLT removal but should not affect RNA integrity or microarray analysis.
14. To evaluate RNA integrity, an Agilent 2100 Bioanalyzer may be used. The RNA integrity number (RIN) should ideally be greater than 8.

15. The standard procedure for the Affymetrix 3′ IVT Express kit is used for microarray analysis. A minimum of 50 ng total RNA is needed per GeneChip.

3.3 Basic Statistical Analysis

1. Upon finishing the microarray assays, data are made available in the form of .CEL files. These files contain intensity values for the analyzed probes sorted by an *X*/*Y* coordinate system. Intensity values are on a scale that ranges from 0 to several thousands. A CLF file and a CDF file are then used to map the location of the intensity measures to specific probes and probesets. More information about Affymetrix file formats can be found at (http://media.affymetrix.com/support/developer/powertools/changelog/FILE-FORMATS.html). Microarrays can be analyzed and processed according to standard Affymetrix procedures. Affymetrix software is utilized for generating and extracting data from Affymetrix CEL files, such as Affymetrix GeneChip Command Console and Affymetrix Expression Console. The necessary Affymetrix library files are available through www.ambystoma.org. We have found that using the RMA algorithm [8] for data preparation works well. The RMA algorithm consists of three steps. First, microarrays are quantile normalized, which fixes the distribution of measurements so that measurements are comparable across all arrays. Second, the samples are summarized by means of Tukey's median polish [9] so that all probes associated with the same probeset are combined into a single expression measure. Finally, data are transformed so that measurements can approximate a normal distribution, and the data can be analyzed by means of parametric analysis.
2. Once the raw hybridization data has been extracted into a text file, any software capable of handling spreadsheets, such as Microsoft Excel or JMP Genomics can be used for further analyses. Alternatively, the researcher can program tools in languages such as R if finer control over methodology is desired.
3. To increase statistical power, researchers may wish to remove gene expression estimates that are low across all samples, as such detected changes in gene expression are likely to be biologically insignificant. Although the *Ambystoma* GeneChip does not have mismatch probes present, researchers can determine presence/absence calls based on the magnitude of expression estimates. A histogram can be generated to view the distribution of expression profiles to identify a value where estimates below that value are unlikely to be expressed (Fig. 2). One commonly used method is to remove all probesets that express values below a given quartile [10–12].
4. Statistical treatment of the data remains at the discretion of the researcher and is fully dependent on experimental design.

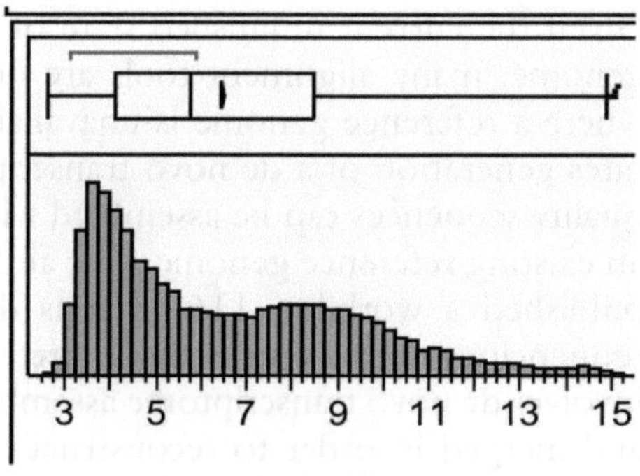

Fig. 2 Representative frequency distribution of *Log2* transformed intensity values for a single Affymetrix GeneChip generated in JMP Genomics version 5.0. The bottom quartile is identified at 4.12. The distribution can be used to define unexpressed and poorly expressed probesets

Typically, in studies with high replication, parametric analyses such as ANOVA work well and provide statistical power. These tests can be done in a JMP Genomics workflow or by means of manually defining the formulas in Excel or in R. To determine the magnitude of an effect, fold changes should be calculated remembering that values extracted from the RMA algorithm are log 2 transformed. Thus, a difference of 1 equates to a twofold change in signal intensity.

5. A number of multiple testing procedures are available for analyzing microarray data sets, although the false discovery rate (FDR) [13] is the most commonly used. These calculations can be automated in JMP Genomics or R or calculated by formulas in Microsoft Excel. An FDR of 0.05 is often used to determine statistical significance, but in practice this value can often be overly stringent.

6. Upon retrieving a list of differentially expressed genes based on defined statistical criteria (such as *p*-values from ANOVAs and fold changes), a variety of tools exists to explore the relationships between gene expression profiles. Gene lists can be uploaded into gene ontology analysis software such as DAVID [14] to identify ontologies and pathways that are overrepresented. In turn, the groupings of genes identified by overrepresentation analysis can then be used to generate new hypotheses regarding the previous experiment. In addition, Pearson's correlation can be used to group genes with similar expression profiles. Average linkage clustering using Pearson's correlation as a distance metric has been a commonly applied tool in microarray data analysis [15], and branches on the resulting dendrogram can be used to identify groups of genes with highly similar expression profiles.

3.4 RNA Sequencing of the Mexican Axolotl

3.4.1 Use of the Trinity Platform for Processing and Analyzing Data Assembly

Given the current unfinished state of the *Ambystoma mexicanum* genome, many alignment tools are not yet usable. In such cases where a reference genome is unavailable, transcriptomics necessitates generation of a de novo transcriptome, where a set of high-quality sequences can be assembled without the aid of mapping to an existing reference genome. The authors of Trinity have recently published a workflow [16] that is directly applicable to RNA-sequencing experiments without a reference genome. The method involves de novo transcriptome assembly, wherein reads are aligned and merged in order to reconstruct original sequences. Notably, this de novo transcriptome production accounts for the presence of structural variations among transcripts, such as those that originate from alternative splicing (isoforms), and also resolves recently duplicated (paralogous) genes. In its initial step, Trinity assembles reads into contigs recovering only a single (best) representative for a set of alternative variants that share a sequence of a fixed length of k nucleotides ("k-mer," with k considerably shorter than the read length) due to alternative splicing, gene duplication, or allelic variation. Often the full-length transcripts recovered represent the dominant isoform. Thereafter, Trinity groups related contigs into clusters and constructs a de Bruijn graph for each cluster with each graph reflecting the complexity of overlap between variant contigs. Lastly, Trinity processes each cluster's individual graph in parallel and reports full-length transcripts for all plausible transcript sequences. Ultimately, the researcher recovers a sizeable set of components (putative genes) and their associated transcript forms.

The Trinity platform will accommodate RNA-sequencing data in the form of either single or paired-end reads. In our most recent studies, we have utilized Illumina paired-end sequencing data for gene expression analyses given that paired-end sequencing increases the depth of sequencing, improves de novo assembly efficiency, and offers the advantages for gene discovery and characterization of novel splice isoforms [17]. Additionally, Trinity de novo sequence assembly may be supplemented with a long read set such as an *A. mexicanum* transcriptome of deep sequenced contigs (V4.0 currently available through Sal-Site). Separate sequencing runs may not be directly comparable prior to normalization. For example, one run may have more total reads than another, meaning that a higher number of counts for a given transcript could be associated with the increased total number of reads instead of a relative increase in the presence of that particular transcript [18, 19]. Additionally, correction may be necessary to accommodate for analytical and methodological biases due to transcript length variation [20, 21]. Therefore, prior to sequence assembly, paired reads may be processed through the Trinity in silico normalization algorithm according to user-defined settings for k-mer size (25 default).

3.4.2 Transcript Expression Analysis Using RSEM/EBSeq

Transcript and isoform abundance estimation (expression values) for a de novo transcriptome may be computed using RSEM software (version 1.2.9) [22]. Typically, a run of RSEM is a two-step process. Initially, a set of reference transcript sequences are generated and preprocessed for secondary use in the RSEM workflow. Thereafter, a set of RNA-sequencing reads are aligned to the reference transcripts, and the resulting alignments are used to estimate abundances and their credibility intervals. Notably, RSEM software is most suitable for de novo transcriptome expression analysis as RSEM does not require a reference genome for transcript mapping; rather, a de novo transcriptome may be used as its own set of reference transcripts. However, RSEM requires that all alignments of the same read group be grouped together, and in the case of paired-end reads, the two mates of any alignment must be input adjacently. Each RSEM result file recovers an expression value for both transcripts per million (TPM) and fragments per kilobase of exon per million fragments mapped (FPKM). Both of these are normalized values that attempt to correct for total number of reads, but TPM is preferred [22].

Differential expression analyses across two or more biological conditions in an RNA-sequencing experiment may be conducted using EBSeq, which is packaged within RSEM [23]. In practice, EBSeq identifies statistically significant expression changes of genes and isoforms by assigning each identified transcript a "pattern" and a posterior probability. In addition, as part of the EBSeq analysis, expected counts from the RSEM output are median normalized [23].

3.4.3 Sequence Annotation Aided by Nucleotide BLAST

It is not straightforward to annotate transcripts from RNA-sequencing data because transcripts/isoforms and genes may not show a 1:1 relationship. To principally address this complexity in the absence of a sequenced genome, we suggest selecting a representative reference transcript for each component (gene). Reference transcripts may be identified from either the isoform of greatest length or from the isoform with highest average expression values (TPM or FPKM) across all experimental conditions. In practice, RSEM expression values from the transcriptome assembly may be imported into a database (such as MySQL) and then queried to determine the average isoform values across all experimental conditions.

Reference transcriptome annotation may be facilitated by means of nucleotide BLAST [24], where the reference transcriptome is queried against a BLAST database of known annotation (such as sequences corresponding to genes on the *Ambystoma* microarray probeset). Annotation can be easily assigned for BLAST results with perfect 1:1 query-target relationships. However, deviations from this later pattern with redundant query hits may require manual curation and sorting to determine the best BLAST match.

The most likely annotation can be determined where target matches have relatively long alignments with high percent identity. RNA-sequence annotation enables interface and cross-comparison to data from different experimental platforms (e.g., microarrays).

3.5 Downstream Analyses of Gene Lists

3.5.1 Overrepresentation Analysis of Ontologies

Once gene lists have been generated from statistical analysis of normalized data (e.g., using EBSeq), overrepresentation analysis of gene lists can be performed in the same manner as in microarray data analysis. These tools typically perform Fisher's exact tests to identify groups of genes (by assigning them based on involvement with a specific biological term known as an ontology) that appear in a list more often than expected by chance if a random selection of genes of the same size as the gene list was tested. Detection of changes in large numbers of genes all associated with the same process can suggest a change in the process itself, which can be further tested with downstream experiments. Recommended sites for performing this type of analysis include DAVID [14] and Panther [25–27].

3.5.2 Clustering Analysis of RNA-Sequencing Data

Clustering analysis can be performed on RNA-sequencing data once care has been taken to normalize the data. A number of packages in R exist for the purpose of investigating RNA-sequencing data, for example, DESeq [28] and edgeR [29]. The authors of Trinity have utilized the edgeR package to analyze RNA-sequencing data and prepare the data for clustering analysis. Note that the default parameters within Trinity utilize Euclidean distance and complete linkage clustering [30]. We recommend changing the parameters to "Pearson" and "average" to detect similarities between expression patterns while overlooking differences in magnitude of expression.

References

1. Page RB et al (2007) Microarray analysis identifies keratin loci as sensitive biomarkers for thyroid hormone disruption in the salamander Ambystoma mexicanum. Comp Biochem Physiol C Toxicol Pharmacol 145:15–27
2. Huggins P et al (2012) Identification of differentially expressed thyroid hormone responsive genes from the brain of the Mexican Axolotl (Ambystoma mexicanum). Comp Biochem Physiol C Toxicol Pharmacol 155: 128–135
3. Stewart R et al (2013) Comparative RNA-seq analysis in the unsequenced axolotl: the oncogene burst highlights early gene expression in the blastema. PLoS Comput Biol 9:e1002936. doi:10.1371/journal.pcbi.1002936
4. Wu CH, Tsai MH, Ho CC, Chen CY, Lee HS (2013) De novo transcriptome sequencing of axolotl blastema for identification of differentially expressed genes during limb regeneration. BMC Genomics 14:434. doi:10.1186/1471-2164-14-434
5. JMP Genomics Version 5.0. (1989–2007) SAS Institute, Cary, NC
6. R: a language and environment for statistical computing. (2012) R Foundation for Statistical Computing, Vienna, Austria
7. MySQL AB (2008–2009) Sun Microsystems Inc
8. Irizarry RA et al (2003) Exploration, normalization, and summaries of high density oligonucleotide array probe level data. Biostatistics 4:249–264
9. Holder D, Raubertas RF, Pikounis VB, Svetnik B, Soper K (2001). Statistical analysis of high density oligonucleotide arrays: a SAFER

approach. In GeneLogic Workshop on Low Level Analysis of Affymetrix GeneChip Data. Proceedings of the ASA annual meeting August 5-9, 2001.

10. Page RB et al (2008) Effect of thyroid hormone concentration on the transcriptional response underlying induced metamorphosis in the Mexican axolotl (Ambystoma). BMC Genomics 9:78. doi:10.1186/1471-2164-9-78
11. Fleet JC, Wang L, Vitek O, Craig BA, Edenberg HJ (2003) Gene expression profiling of Caco-2 BBe cells suggests a role for specific signaling pathways during intestinal differentiation. Physiol Genomics 13:57–68
12. Shah SH, Pallas JA (2009) Identifying differential exon splicing using linear models and correlation coefficients. BMC Bioinformatics 10:26. doi:10.1186/1471-2105-10-26
13. Benjamini Y, Hochberg Y (1995) Controlling the false discovery rate: a practical and powerful approach to multiple testing. J R Stat Soc Series B 57:289–300
14. Dennis G Jr et al (2003) DAVID: database for annotation, visualization, and integrated discovery. Genome Biol 4:P3
15. Eisen MB, Spellman PT, Brown PO, Botstein D (1998) Cluster analysis and display of genome-wide expression patterns. Proc Natl Acad Sci U S A 95:14863–14868
16. Haas BJ et al (2013) De novo transcript sequence reconstruction from RNA-seq using the Trinity platform for reference generation and analysis. Nat Protoc 8:1494–1512
17. Maher CA et al (2009) Chimeric transcript discovery by paired-end transcriptome sequencing. Proc Natl Acad Sci U S A 106: 12353–12358
18. Robinson MD, Oshlack A (2010) A scaling normalization method for differential expression analysis of RNA-seq data. Genome Biol 11:R25. doi:10.1186/gb-2010-11-3-r25
19. Mortazavi A, Williams BA, McCue K, Schaeffer L, Wold B (2008) Mapping and quantifying mammalian transcriptomes by RNA-Seq. Nat Methods 5:621–628
20. Oshlack A, Wakefield MJ (2009) Transcript length bias in RNA-seq data confounds systems biology. Biol Direct 4:14. doi:10.1186/1745-6150-4-14
21. Bullard JH, Purdom E, Hansen KD, Dudoit S (2010) Evaluation of statistical methods for normalization and differential expression in mRNA-Seq experiments. BMC Bioinformatics 11:94. doi:10.1186/1471-2105-11-94
22. Li B, Dewey CN (2011) RSEM: accurate transcript quantification from RNA-Seq data with or without a reference genome. BMC Bioinformatics 12:323. doi:10.1186/1471-2105-12-323
23. Leng N et al (2013) EBSeq: an empirical Bayes hierarchical model for inference in RNA-seq experiments. Bioinformatics 29:1035–1043
24. Altschul SF, Gish W, Miller W, Myers EW, Lipman DJ (1990) Basic local alignment search tool. J Mol Biol 215:403–410
25. Thomas PD et al (2006) Applications for protein sequence-function evolution data: mRNA/protein expression analysis and coding SNP scoring tools. Nucleic Acids Res 34:W645–W650
26. Mi H, Muruganujan A, Casagrande JT, Thomas PD (2013) Large-scale gene function analysis with the PANTHER classification system. Nat Protoc 8:1551–1566
27. Mi H, Muruganujan A, Thomas PD (2013) PANTHER in 2013: modeling the evolution of gene function, and other gene attributes, in the context of phylogenetic trees. Nucleic Acids Res 41:D377–D386
28. Anders S, Huber W (2010) Differential expression analysis for sequence count data. Genome Biol 11:R106. doi:10.1186/gb-2010-11-10-r106
29. Robinson MD, McCarthy DJ, Smyth GK (2010) edgeR: a Bioconductor package for differential expression analysis of digital gene expression data. Bioinformatics 26:139–140
30. Grabherr MG et al (2011) Full-length transcriptome assembly from RNA-Seq data without a reference genome. Nat Biotechnol 29: 644–652

Chapter 25

Sal-Site: Research Resources for the Mexican Axolotl

Nour W. Al Haj Baddar, M. Ryan Woodcock, Shivam Khatri, D. Kevin Kump, and S. Randal Voss

Abstract

Sal-Site serves axolotl research efforts by providing Web access to genomic data and information, and living stocks that are reared and made available by the Ambystoma Genetic Stock Center (AGSC). In this chapter, we detail how investigators can search for genes of interest among Sal-Site resources to identify orthologous nucleotide and protein-coding sequences, determine genome positions within the *Ambystoma* meiotic map, and obtain estimates of gene expression. In the near future, additional genomic resources will be made available for the axolotl, including a listing of genes that are partially or wholly contained within Bacterial Artificial Chromosome (BAC) vectors, a prioritized collection of deeply sequenced BAC clones, chromosome-specific assemblies of genomic DNA, and transgenic axolotls that are engineered using TALENs and CRISPRs. Also, services provided by the AGSC will be expanded to include microinjection of user constructs into single cell embryos and distribution of axolotl tissues, DNA, and RNA. In conclusion, Sal-Site is a useful resource that generates, shares, and evolves *Ambystoma* associated information and databases to serve research and education.

Key words Salamander, Limb regeneration, Transcriptome, Analysis, Genomic resources

1 Introduction

Salamanders are important models in basic and biomedical research. Over the last 100 years, salamanders have figured prominently in studies of development, neurophysiology, ecology, evolution, and tissue regeneration. To enhance efforts in these and other research areas, genome resource development is underway for the primary salamander model—the Mexican axolotl (*Ambystoma mexicanum*). A bioinformatics Web-base—Sal-Site—was created to make axolotl genomic data and information available to the community. Here we describe Sal-Site content and best practices for extracting useful information.

Anoop Kumar and András Simon (eds.), *Salamanders in Regeneration Research: Methods and Protocols*, Methods in Molecular Biology, vol. 1290, DOI 10.1007/978-1-4939-2495-0_25, © Springer Science+Business Media New York 2015

2 Sal-Site Website

2.1 Homepage

The Sal-Site homepage provides hyperlinks to access axolotl genome resources and information (Fig. 1). The header tab (Fig. 1A) indexes major sections of the website: axolotl research, genome resources, *Ambystoma* Genetic Stock Center (AGSC), education, about, and photo gallery. In addition, an orientation video (Fig. 1B) introduces the Mexican axolotl (*Ambystoma mexicanum*) and highlights its remarkable ability to regenerate complex tissues and organs. The video also details general functions of the AGSC. This unique facility houses, breeds, and maintains a large inbred axolotl population, providing axolotls nationally and internationally. The homepage also displays additional summary information regarding the axolotl's type location (Lake Xochimilco, Mexico), general phenotypic characteristics, and Sal-Site support mechanisms (Fig. 1C). The Fun Facts section (Fig. 1D) provides quick informational statements regarding axolotls that periodically change. Finally, the home page provides quick links for the most highly accessed Sal-Site links (Fig. 1E).

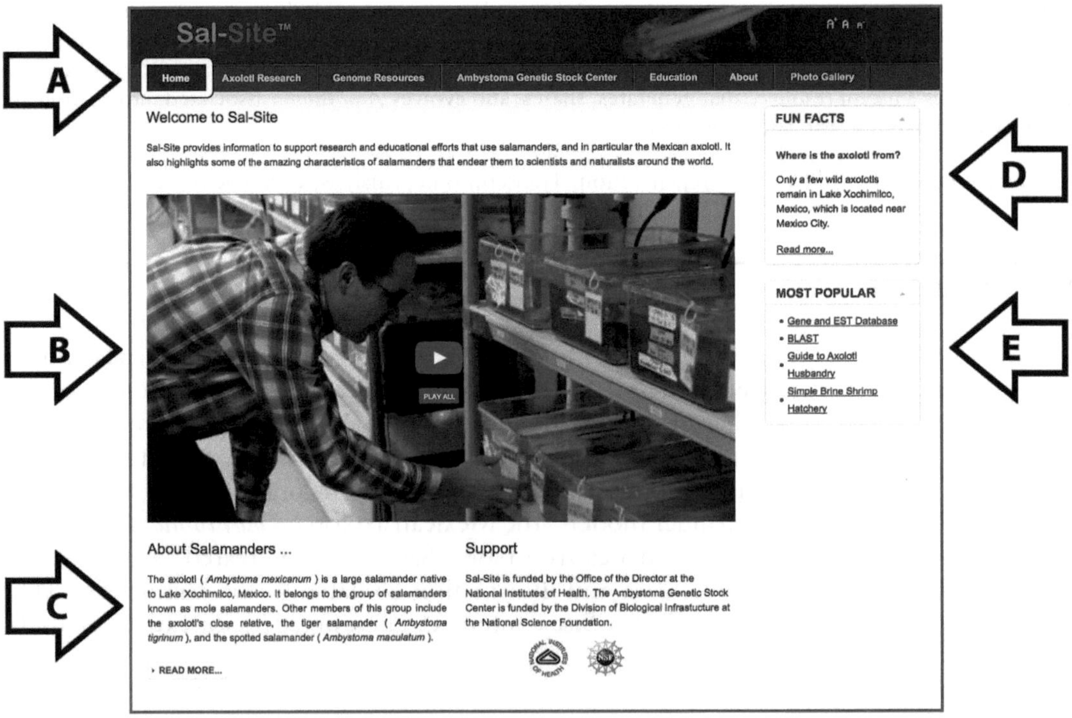

Fig. 1 The Sal-Site homepage. This hompage (***A***) has hyperlinks to orientation video (***B***) that introduces *Ambystoma mexicanum*, general information about Sal-Site funding mechanisms (***C***), salamander fun facts (***D***), and the most highly accessed links of Sal-Site (***E***)

2.2 Axolotl Research

The significance of the axolotl model will grow with genome resource development and further advances in the field. The axolotl research section of Sal-Site (Fig. 2) details projects and publications associated with axolotl resource development. Over the last 10 years, the axolotl transcriptome has been deeply sequenced, microarrays have been developed and used to study regeneration and metamorphosis, and a meiotic map was constructed. Current projects include transgenic axolotl development, sequencing of whole axolotl chromosomes, and creation of large insert DNA libraries using Bacterial Artificial Chromosome (BAC) vectors. With respect to the BAC project, a workflow is in place to screen BAC super pools by next generation sequencing to create an inventory of potential candidate genes for isolation. This initiative is developing regulatory sequence information for key developmental genes, information that is useful for developing reporter constructs for transgenic development (Fig. 3). This and the other initiatives will enable sequence-based methods of inquiry that are

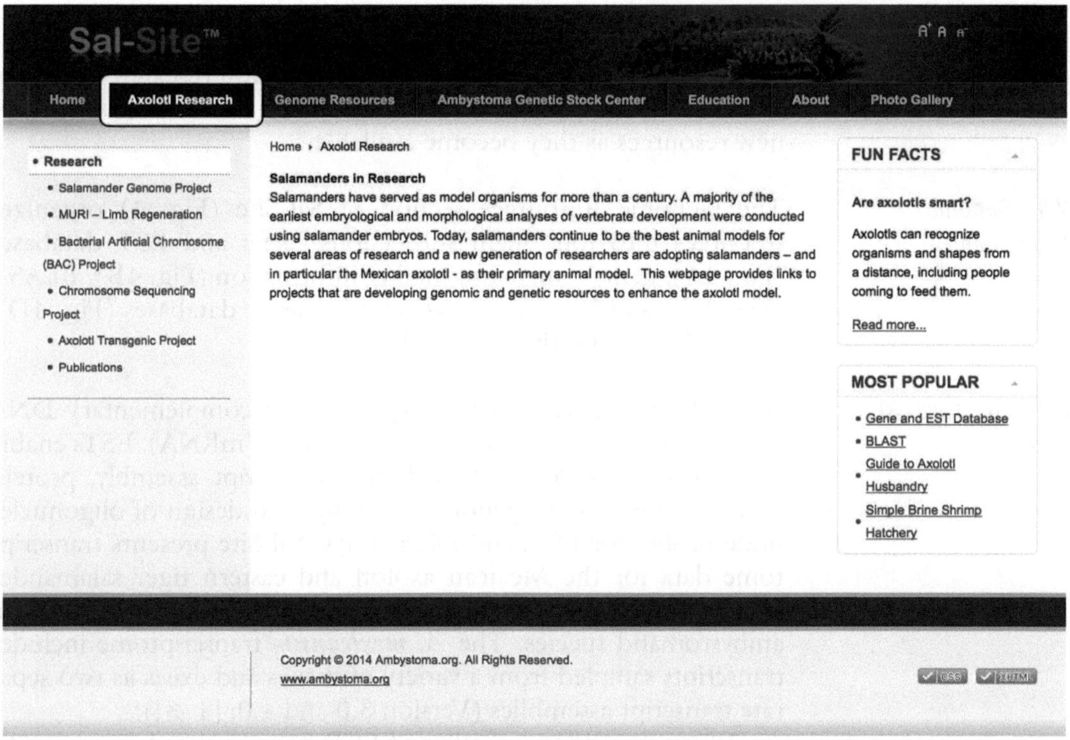

Fig. 2 Axolotl research webpage provides hyperlinks to axolotl research projects and selected publications. The axolotl research projects include the salamander genome project, axolotl Bacterial Artificial Chromosome (BAC) project, axolotl chromosome sequencing project, and axolotl transgenic project

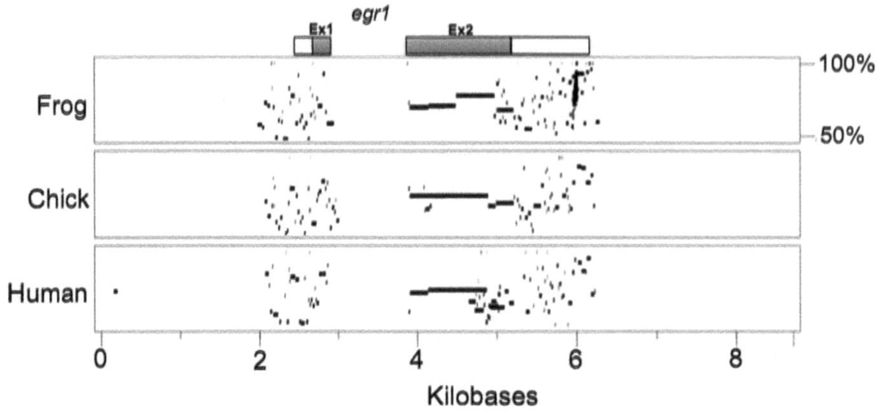

Fig. 3 Axolotl BAC project. Alignment of genomic sequences from frog (*Xenopus tropicalus*), chick (*Gallus gallus*), and human (*Homo sapiens*) to an axolotl BAC that contains early growth response 1 (*egr1*). The *horizontal black bars* show regions of sequence similarity where nucleotide alignments are greater than 50 % identical. The *white* and *gray bars* at the top show untranslated regions and exons (Ex), respectively

typical of genetic model organisms. The axolotl research section of Sal-Site will continually evolve to keep the community abreast of new resources as they become available.

2.3 Genome Resources

The Genome Resources section of Sal-Site (Fig. 4) organizes resources into four main subsections: gene and EST databases (Fig. 4A), genetic map and marker information (Fig. 4B), BLAST (Fig. 4C), and limb regeneration microarray databases (Fig. 4D). Each subsection is discussed below.

2.3.1 Gene and EST Database

An EST represents a partial sequence of a complementary DNA (cDNA) synthesized from a messenger RNA (mRNA). ESTs enable multiple lines of research, including transcript assembly, protein sequence annotation, genome mapping, and design of oligonucleotide probes for PCR and microarrays. Sal-Site presents transcriptome data for the Mexican axolotl and eastern tiger salamander (*A. tigrinum tigrinum*) (Fig. 5), and in the future, for additional ambystomatid species. The *A. mexicanum* transcriptome includes transcripts sampled from a variety of tissues and exists as two separate transcript assemblies (Version 3.0 and 4.0; [1–3]).

When a sufficient number of ESTs are generated for a specific mRNA, it is possible to assemble these into a contiguous nucleotide sequence that provides a model of the protein-coding sequence. However, if EST depth is not sufficient to build a contiguous sequence (contig), which is problematic in working with expanded genic regions of the axolotl genome [4], a contig may only partially cover an mRNA, or independent contigs may map to

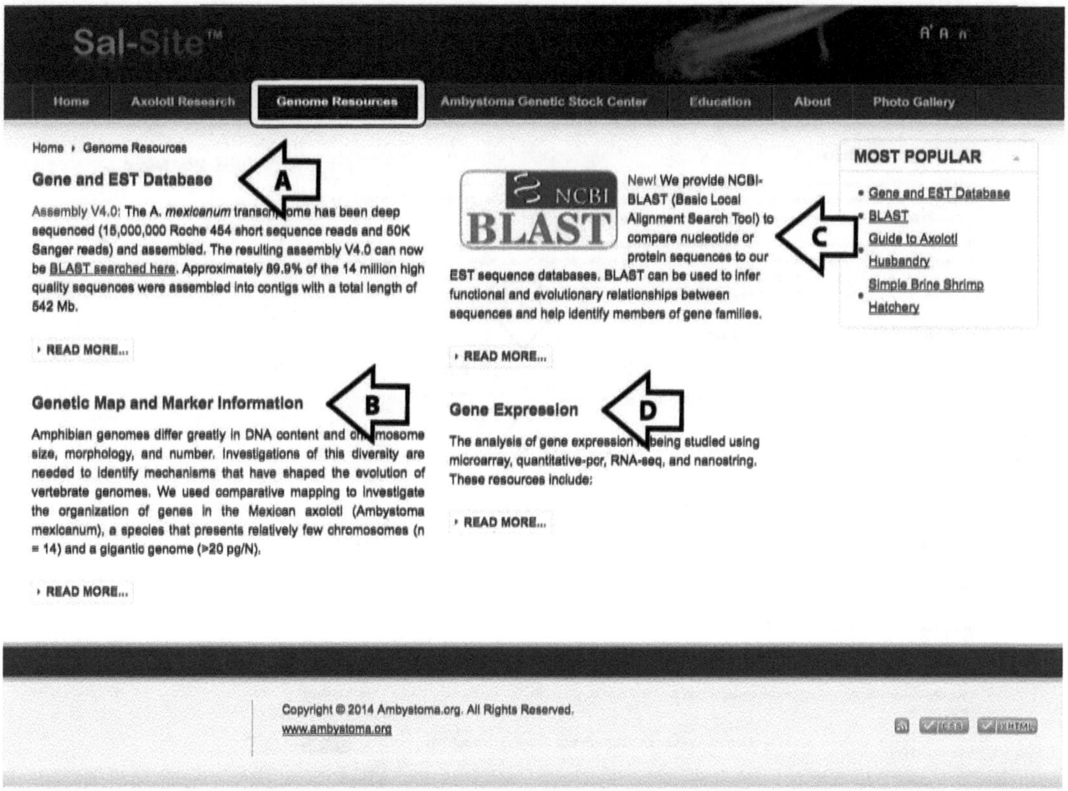

Fig. 4 Genome resources main webpage. This webpage provides hyperlinks to resources and tools. (***A***) Gene and EST Database, (***B***) Genetic Map and Marker Information, (***C***) NCBI-BLAST tool, and (***D***) Gene Expression

nonoverlapping portions of the same mRNA. ESTs may also assemble into multiple overlapping reads that are suggestive of splice variants. When there is sequence contiguity among a group of contigs, they are classified into a common isogroup. For a given isogroup, any inferred splice variants (which would putatively represent individual transcripts) are reported as isotigs. The Sal-Site genome resources webpage enables visitors to search for genes of interest in *A. mexicanum* or *A. t. tigrinum* salamander transcriptome databases. The search for genes in assembly V3.0 and V4.0 (Fig. 5) can be conducted by typing a specific contig (Fig. 5A), isotig (Fig. 5B), or gene symbol identifier (Fig. 5C) in the appropriate field, and by selecting a reference database from the drop-down menu (Fig. 5D).

Figure 6 shows steps that an investigator would follow to identify a gene of interest in the gene and EST database. For example, if a user is interested in searching the *A. mexicanum* V4.0 database for *early growth response 1*, the official gene symbol (*egr1*) is typed in the appropriate field (Fig. 6A) using one of the reference databases (Fig. 6B). After clicking the submit button, a results page appears to show axolotl contigs (hits) that are annotated for *egr1*

Fig. 5 Gene and EST database. Genes of interest can be searched using the axolotl V4.0 and V3.0 transcriptome assemblies as reference databases. Searches can be conducted by typing a specific contig (***A***), isotig (***B***), or gene symbol identifier (***C***) in the appropriate field, followed by selecting a reference database from the drop-down menu (***D***)

(Fig. 6C). It is important to note that more than one contig hit may be reported for a query; these may represent split or nonoverlapping contigs for the same gene model, or incomplete transcript models with low quality sequence. Moreover, when more than one contig shows sequence identity to the same region of a query, it is possible that the contigs are paralogous genes. In this example, two contigs have been assembled for *egr1*. By clicking on the hyperlink provided for the first contig (contig316872), a gene card appears (Fig. 7A). The gene card offers an additional hyperlink to open a NCBI page for the corresponding orthologous human protein (Fig. 7B). The gene card page also includes a hyperlink to PubMed that reports some of the scientific articles that are related to the gene of interest (Fig. 7C). Clicking the contig ID hyperlink opens a sequence annotation page that provides detailed information about the contig, including the number of ESTs in the contig (Fig. 7D), contig nucleotide sequence (Fig. 7E), and the longest computer predicted open reading frame (ORF) with protein

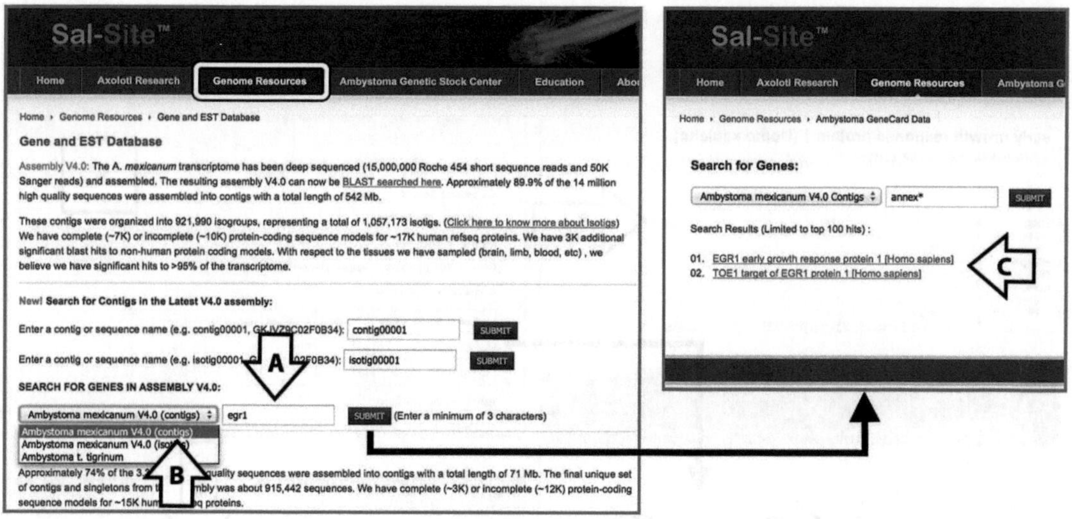

Fig. 6 An example search of the Gene and EST database. The gene symbol for *early growth response—egr1—* is typed in the appropriate field (***A***) and a reference database is selected; *Ambystoma mexicanum* V4.0 contigs in this example (***B***). After clicking the submit button, a result webpage appears showing axolotl hits (contigs) annotated for *egr1* (***C***)

translation (Fig. 7F). Additionally, this sequence annotation page contains data for BLAST hits linked to human sequences curated by NCBI, including sequence similarity and alignment statistics (Fig. 7G). In this example, EST depth, sequence similarity, and sequence coverage provide very good evidence that contig316872 corresponds to *egr1*.

2.3.2 Genetic Map and Marker Information

Although the Mexican axolotl genome is large, it has been possible to order genes into linkage groups and identify conserved gene orders in comparison to other vertebrate models [3–5]. Sal-Site provides information about the ordering of 917 genetic markers among 17 linkage groups (LG)—as an example, part of LG3 is shown—(Fig. 8A). The number of linkage groups approaches the *Ambystoma* haploid chromosome number ($N=14$) and is comprised mostly of EST-based markers. A conserved synteny table provides the positions of orthologous loci from human, axolotl, chicken, and *Xenopus tropicalis* genomes (Fig. 8B). Continuing with the example above, the genome position of *egr1* has not been determined, however it is possible to determine the most likely position by comparing conserved synteny blocks between the axolotl and chicken gene maps. In the chicken genome, *egr1* is found on chromosome 13. Because all but two of the chicken chromosome 13 genes that have been mapped in the axolotl locate to axolotl LG3, the location of these genes within LG3 provide the most likely position for *egr1*. The *Ambystoma* linkage map has been used to establish orthology of key developmental genes [6, 7] and identify quantitative trait loci [8–11].

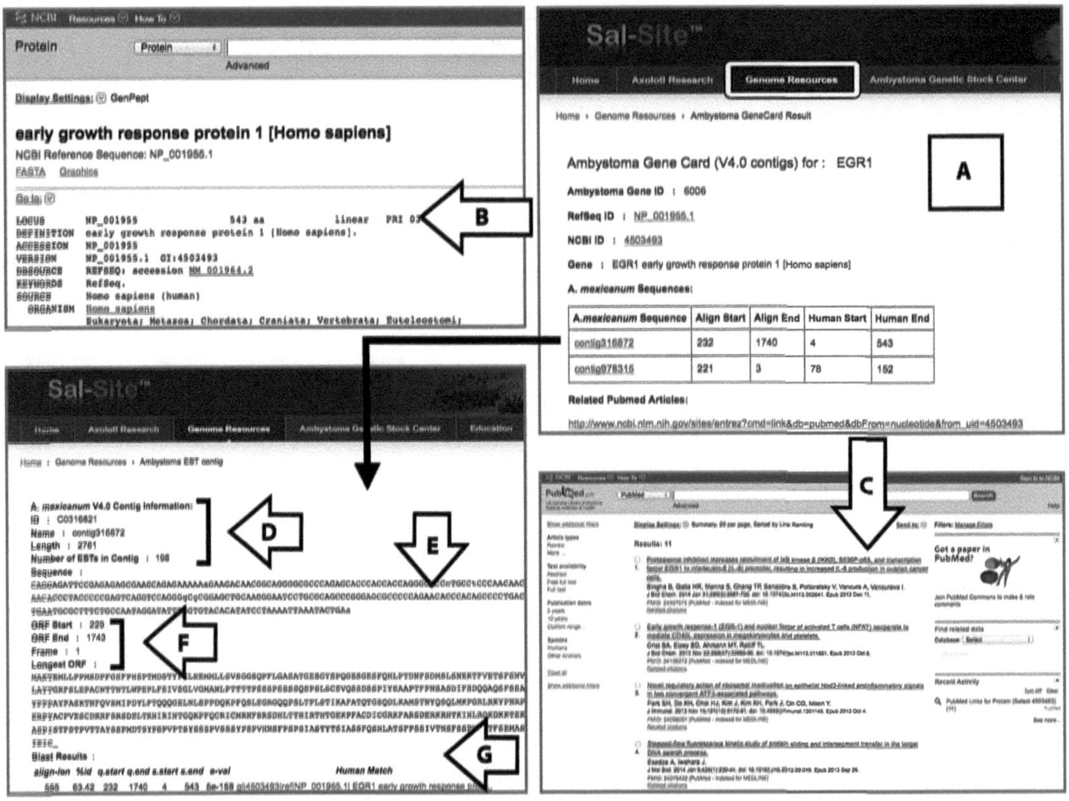

Fig. 7 *An example gene card for Ambystoma mexicanum erg1*. (***A***) The Sal-Site gene card for *egr1* provides a hyperlink to human EGR1 at NCBI (***B***). The gene card page also includes a hyperlink to access related PubMed scientific articles (***C***). The gene card also provides links of the axolotl contigs annotated for *erg1*. Clicking the first link (contig316872) accesses an annotation webpage that details the number of ESTs associated with the contig (***D***), contig nucleotide sequence (***E***), and the longest computer predicted open reading frame (ORF) with protein translation (***F***). Additionally, this sequence annotation webpage contains data for BLAST hits linked to human sequences curated by NCBI, including sequence similarity and alignment statistics (***G***). In this example, EST depth and sequence coverage provide very good evidence that contig316872 corresponds to *egr1*

2.3.3 BLAST

The Basic Local Alignment Tool (BLAST) is a powerful tool to identify sequence similarities among nucleotide sequences [12]. Sal-Site provides NCBI-BLAST tools to allow sequence similarity searches of *A. mexicanum* and *A. t. tigrinum* nucleotide databases (Fig. 9). To illustrate, we will use human *egr1* (gi|31317226) as a query sequence to identify the axolotl sequences that share nucleotide similarity with it. The input sequence (query) can be pasted as a FASTA formatted sequence, (Fig. 9A) or uploaded from a file (Fig. 9B). The BLAST search can be conducted using a user-selected algorithm (available options include: blastn, blastp, blastx, tblastx, tblastn; Fig. 9C) against a chosen *Ambystoma* database from the drop-down menu (Fig. 9D). Clicking the search button submits the sequence data and opens a BLAST results report. Example portions of BLAST reports are displayed in Figs. 10 and 11.

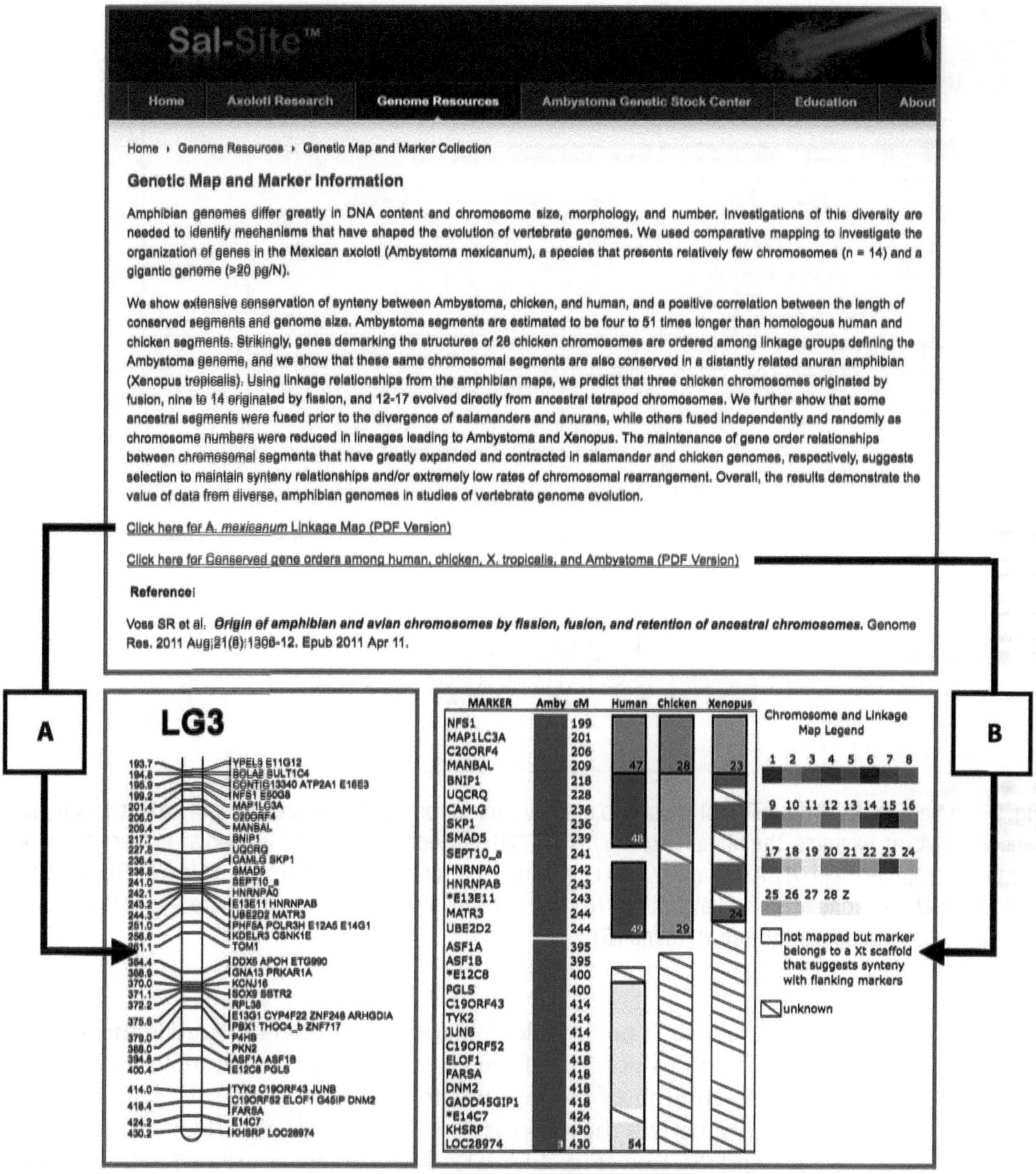

Fig. 8 Axolotl genetic map and marker information. This webpage details information about the ordering of 917 genetic markers among 17 linkage groups (LG). A part of LG3 is shown as an example (***A***). A conserved synteny table provides the positions of orthologous loci from human, axolotl, chicken, and *Xenopus tropicalis* genomes (***B***)

The first portion of BLAST output contains header information (Fig. 10 I) that reports the algorithm used for the BLAST search task (e.g., blastn) and the database that has been searched. The second portion of BLAST output (Fig. 10 II) displays the length of the query input sequence and a graphical representation of the BLAST results summarizing the top hits (i.e., a sequence

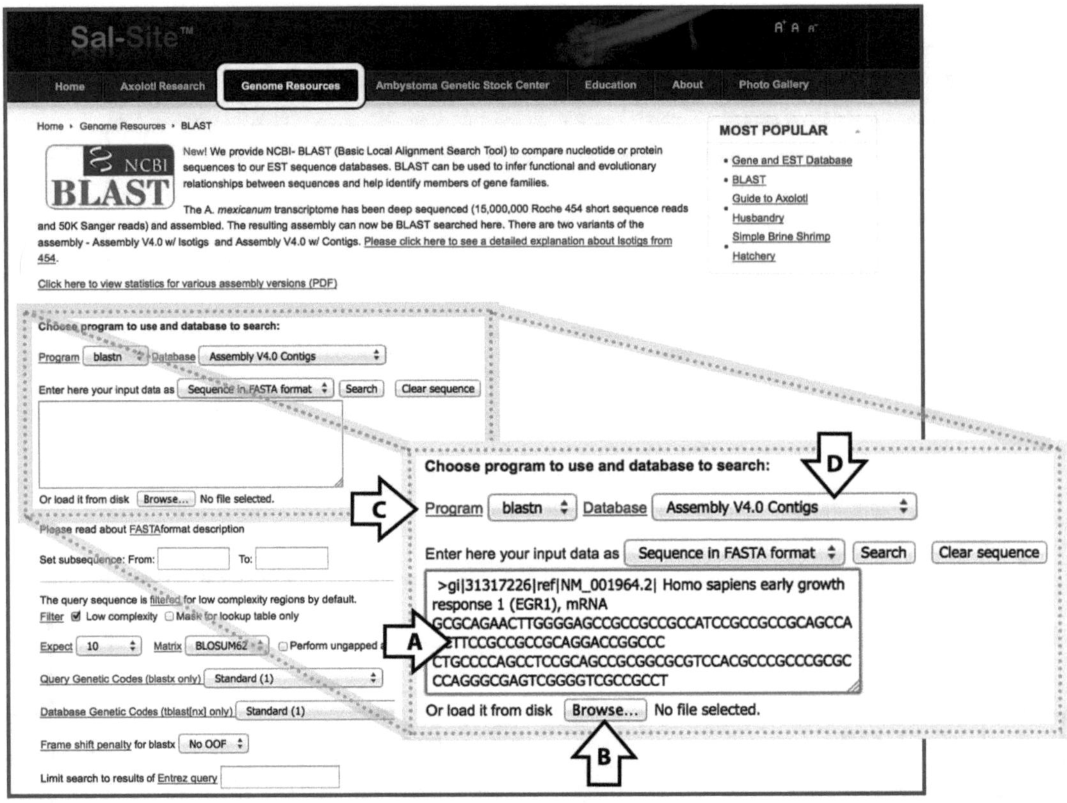

Fig. 9 The Sal-Site NCBI-BLAST tool is used to identify axolotl contigs that share sequence identity with query sequences. As an example, the human *egr1* (gi|31317226) nucleotide sequence is used as a query and pasted into the appropriate field (***A***). Alternatively, the query sequence can be uploaded from a file (***B***). The BLAST can be performed using one of several algorithms (***C***); the BLASTN algorithm is selected here. The reference database is chosen through a drop-down menu (***D***)

that shows nucleotide similarity to the query) found in the database. The third portion of the BLAST output report includes a one-line description of the best hits found (Fig. 11 III). Herein, every line summarizes the information about the contig ID, its length, score, and *E*-value (Fig. 11A–D). The sequences are ordered in this section from greatest to least similar, relative to the query sequence. The score (Fig. 11C) measures similarity between the query and the hit sequences: the greater the value, the greater the similarity between hit sequence and query. The *E*-value (Fig. 11D) is an index of the statistical significance of the hit reported: the lower the number, the greater the similarity between sequence and query. Note that the best axolotl hit recovered for human *egr1* is contig316872 (Fig. 11A), as it has the highest score and the lowest *E*-value. Clicking on the score hyperlink (Fig. 11C) of a selected contig takes the user to the final portion of the BLAST output (Fig. 11 IV). This part gives a detailed pairwise alignment of the query to each of the recovered hits. The header (Fig. 11E) is a repetition of the header information in the previous output section

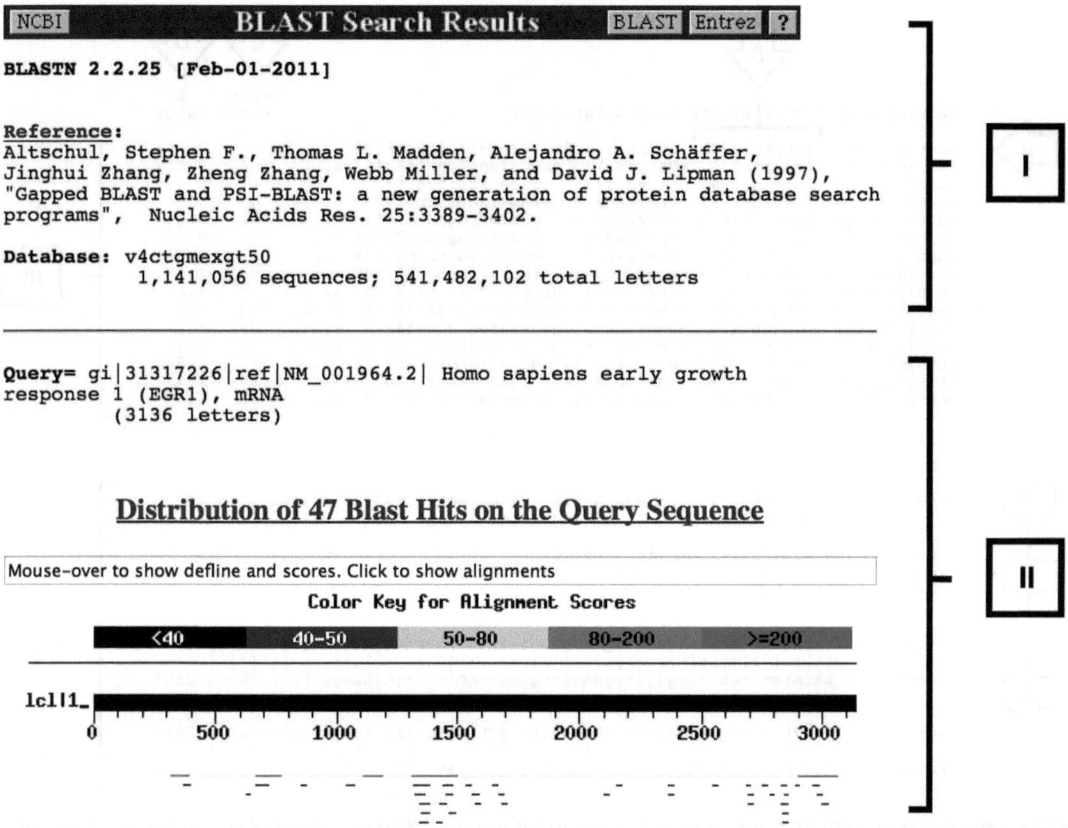

Fig. 10 BLAST output. The human *egr1* nucleotide sequence (gi|31317226) was used as a query to identify the most likely axolotl orthologous sequence in the V4 transcript assembly. **(I)** The report shows the algorithm that was used for the BLAST search and the database that was searched. **(II)** A graphical representation of results that summarizes the top hits to the query sequence. The color bar shows the alignment score. Each of the bars below the query corresponds to a hit. By moving the cursor over these hit bars, it is possible to display a short summary of the sequence ID and alignment

(Fig. 11 III). In this example "identities (89 %)" represent the percentage of nucleotide positions that is fully identical between contig316872 and human *egr1* sequence in the region of alignment. The pairwise alignment (Fig. 11G) provides the coordinates of the query and subject (or hit) sequences that are aligned (i.e., nucleotides positions) in addition to a vertical line series that represent identical nucleotides shared between both sequences.

2.3.4 Gene Expression

Gene expression technologies have advanced nearly all areas of biological research, providing, for example, greater insight into molecular pathways underlying biological processes. A number of advanced molecular tools for gene expression analysis (including microarray, quantitative-pcr, RNA-seq, and Nanostring) are being used in axolotl studies. Sal-Site provides information that is useful for conducting microarray analyses using a custom Affymetrix GeneChip that is available for purchase from Affymetrix (part

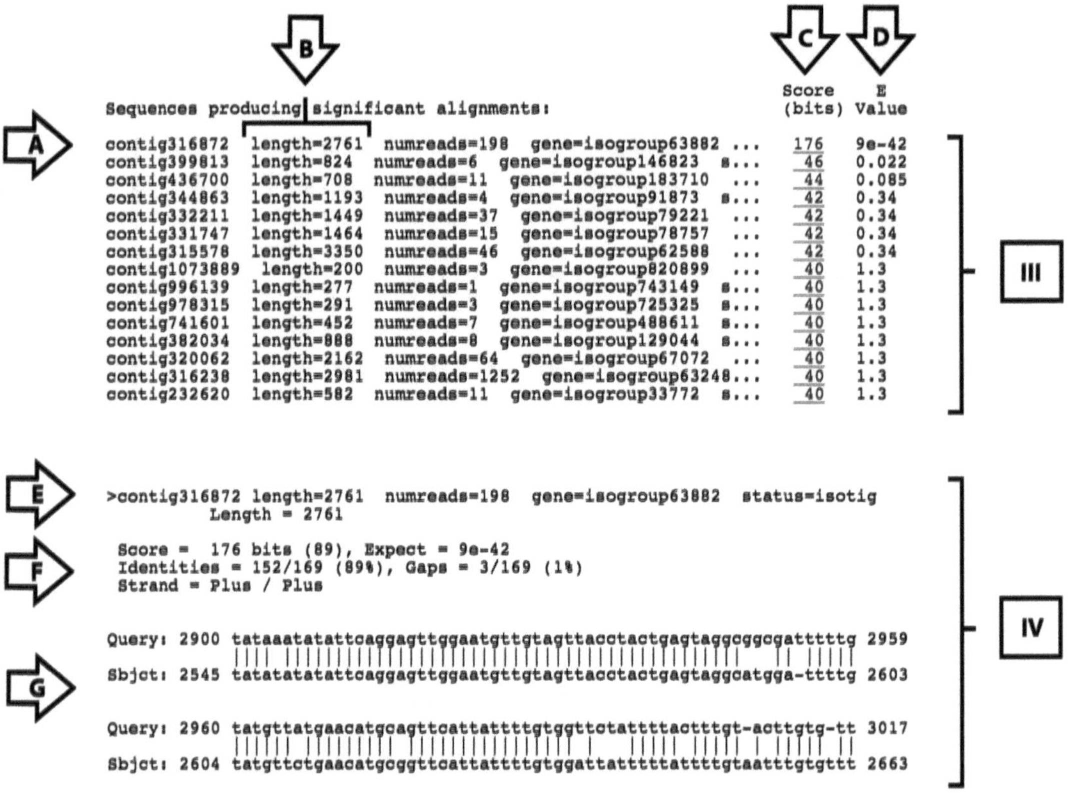

Fig. 11 Continuation of BLAST output from Fig. 10. (**III**) Summary of information for each hit (contig), including length, score, and ***E***-value (***A–D***). The score (***C***) measures similarity between the query and the hit sequence. The ***E***-value (***D***) is an index of the statistical significance of the hit reported. (**IV**) A detailed pairwise alignment of the query to each of the recovered hits. The header (***E***) is a repetition of the header information in the previous output section. This final output also provides the identities and the strand directionality of both sequences for which the alignment was performed (***F***). Identities refer to the number and percentage of identical nucleotides in the alignment. The pairwise alignment (***G***) provides the coordinates of the query and hit sequences that are aligned (i.e., nucleotides positions) in addition to a vertical line series that represent identical nucleotides shared between both sequences. As shown, the best hit recovered for human egr1 is contig316872

number—520748) [13] (Fig. 12). These microarrays provide a standard platform for conducting gene expression experiments. Data from published microarray experiments are available for download as CEL files. Additional files that are needed to analyze and annotate microarray data are available from Sal-Site, as are instructions for normalizing microarray data using Affymetrix® Expression Console™ or Partek® Genomics Suite™ to implement the Robust Multiarray Average (RMA) approach [14].

Given growing interest in axolotl regeneration, Sal-Site provides a searchable database (Fig. 13) of temporal gene expression profiles for the first 28 days of limb regeneration. This study is exemplary in scope and experimental design. The experiment used ten replicate axolotl GeneChips to estimate gene expression for two different amputation injuries and 20 different post-amputation time points (Fig. 13 I).

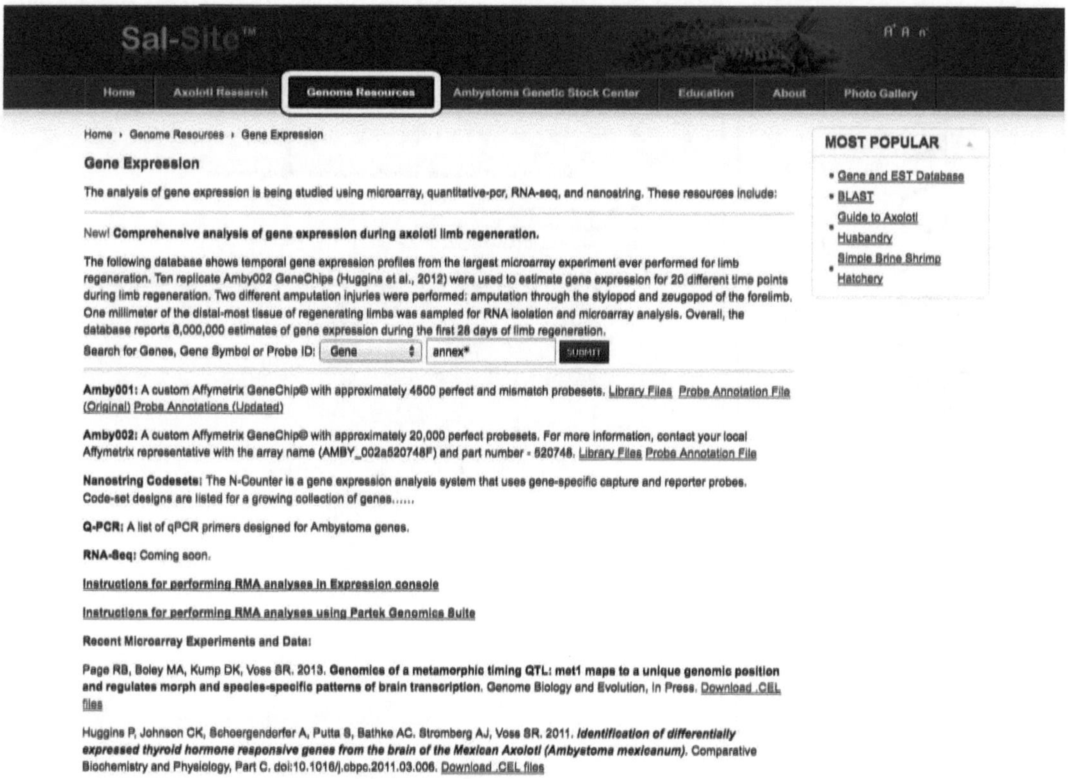

Fig. 12 Gene expression during axolotl limb regeneration. This webpage provides a searchable database of gene expression profiles during axolotl limb regeneration using gene symbol or Affymetrix probe ID. In addition, axolotl microarray library, probe annotation, and raw microarray data files are provided

Overall, the database reports 8,000,000 estimates of gene expression. These data can be queried using gene names, gene acronyms, or Affy probe IDs. For example, by typing the gene symbol *egr1* in the appropriate field and clicking the submit button, a gene expression profile is shown with log2 intensity on the *Y*-axis and post-amputation time points along the *X*-axis (Fig. 13 II). Additional experimental details about the microarray probe ID, contig sequence ID, and gene name are included in the output.

2.4 Ambystoma Genetic Stock Center

Approximately 150 years ago, Mexican axolotls (*Ambystoma mexicanum*) were collected from their ancestral aquatic habitats near present day Mexico City to establish laboratory populations in Europe and subsequently the US. The AGSC at University of Kentucky houses a historically significant population of axolotls that was founded by Rufus R. Humphrey in 1936, and successfully propagated by the University of Indiana through 2005. Over time, domestication created homogeneous wild type and mutant stocks that are amendable to laboratory culture and breeding. These stocks provide a sustainable resource for research and educational efforts around the world. The AGSC operates as a premier animal

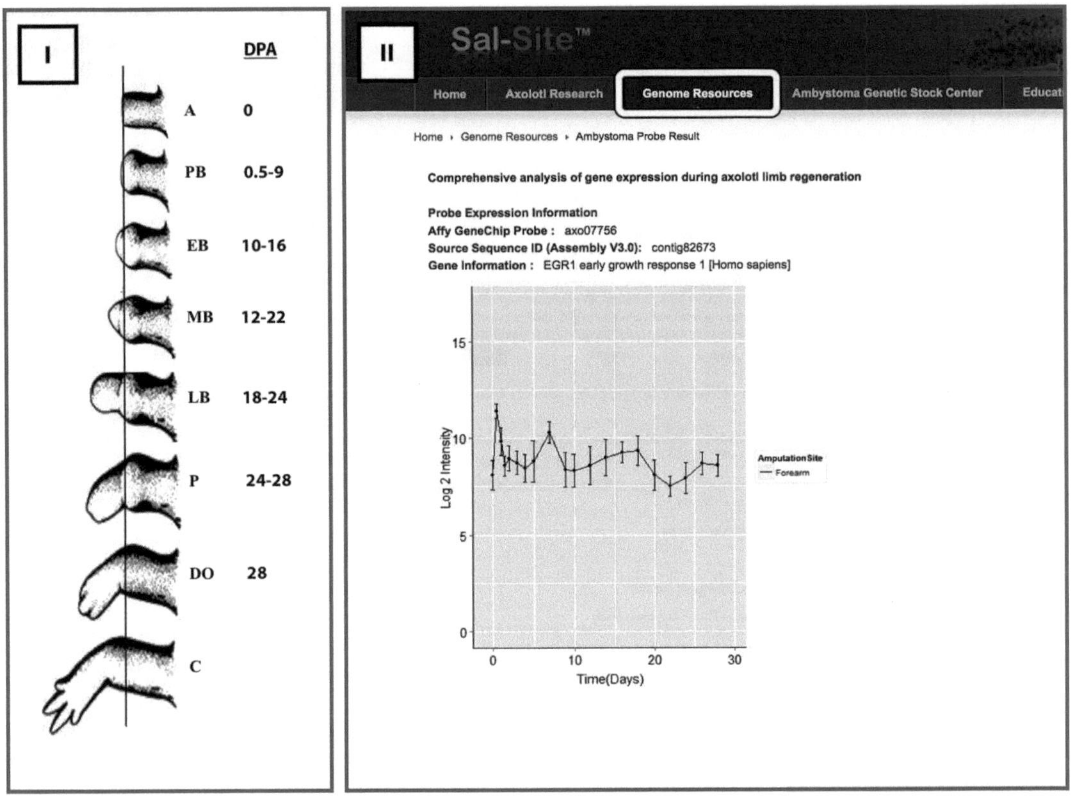

Fig. 13 Comprehensive microarray analysis of gene expression during axolotl limb regeneration is available on the Sal-Site genome resources page. (**I**) The *cartoon* shows regeneration stages after axolotl limb amputation (A). The stages include pre-bud (PB), early bud (EB), medium bud (MB), late bud (LB), palette (P), digital outgrowth (DO), and completed (C). The *vertical line* shows the plane of amputation. The DPA (days post-amputation) numbers shows when individuals were observed for each regeneration stage. (**II**) Temporal gene expression profile for *egr1* during the first 28 days of limb regeneration

resource center providing *A. mexicanum* embryos, larvae, juveniles, adults to laboratories and classrooms nationally and internationally. Essentially all laboratory axolotls in the USA, and the majority if not all axolotls in labs around the world, descend from the AGSC collection. The AGSC also supplies animals and display aquaria to museums, zoos, workshops, and aquariums. Sal-Site provides contact information for AGSC personnel, pricing for axolotl stocks (embryos, larvae, juveniles, adults, and transgenics), axolotl care supplies, information about animal care and use, ordering, and answers to frequently asked questions (Fig. 14).

2.5 Education

Sal-Site provides educational materials about axolotls that are useful to researchers and teachers (Fig. 15). Education kits are available to involve students in axolotl experiments. Further, Sal-Site provides descriptions and illustrations for axolotl embryo stages, limb bud stages, and limb regeneration stages. AGSC staffers provide advice in setting up exhibit aquaria. Locally, the AGSC provides tours for high school and college students.

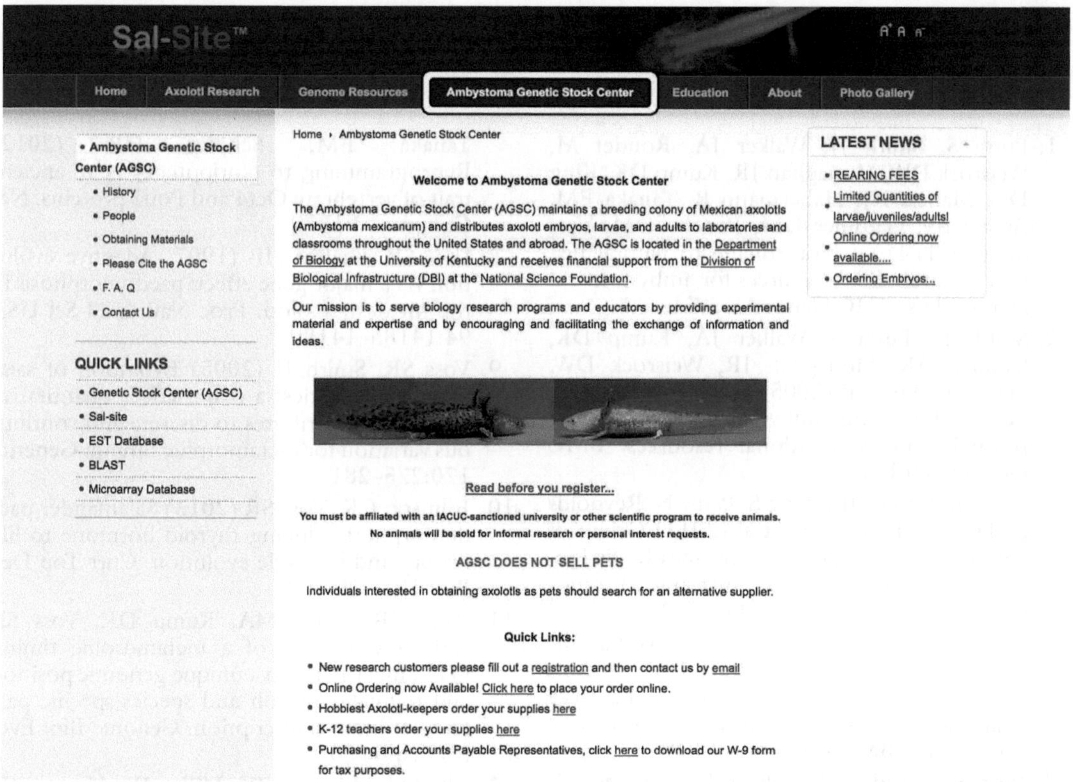

Fig. 14 The *Ambystoma* Genetic Stock Center (AGSC). Sal-Site provides a portal to the AGSC, where researchers and educators can purchase axolotls and supplies

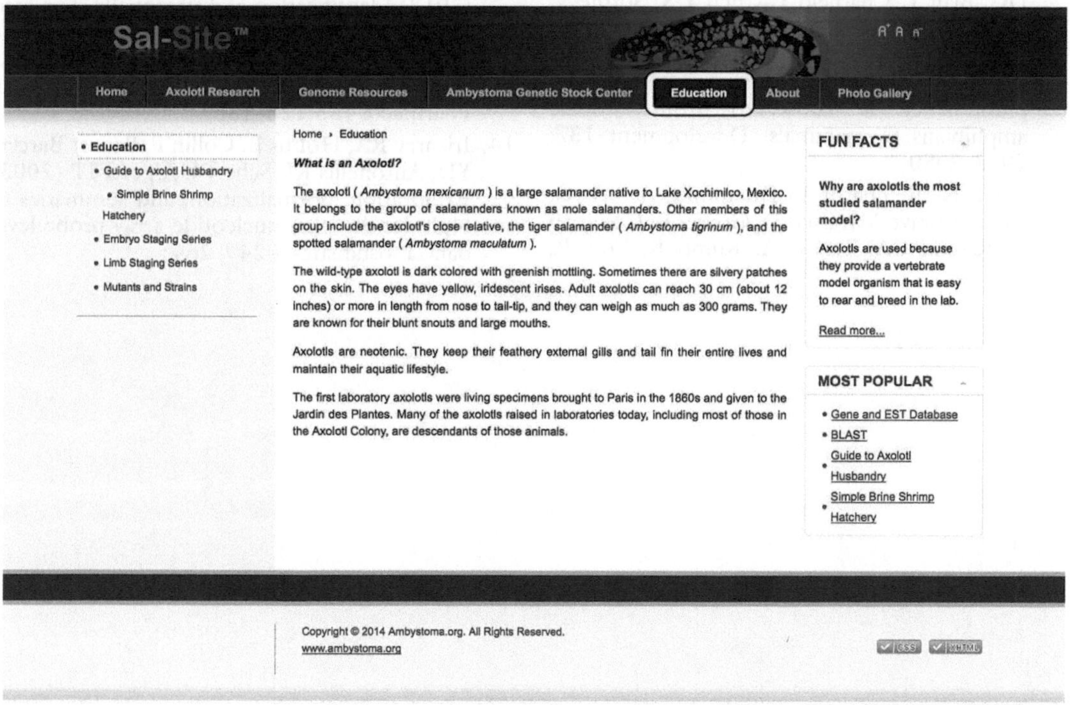

Fig. 15 The Education hyperlink. Sal-Site provides useful information for working with axolotls, including information about axolotl care, use, and development

References

1. Putta S, Smith JJ, Walker JA, Rondet M, Weisrock DW, Monaghan JR, Kump DK, King DC, Maness NJ, Habermann B, Tanaka EM, Bryant SV, Gardiner DM, Parichy DM, Voss SR (2004) From biomedicine to natural history research: EST resources for ambystomatid salamanders. BMC Genomics 5:54
2. Smith JJ, Putta S, Walker JA, Kump DK, Samuels AK, Monaghan JR, Weisrock DW, Staben C, Voss SR (2005) Sal-Site: integrating new and existing ambystomatid salamander research and informational resources. BMC Genomics 6:181
3. Voss SR, Kump DK, Putta S, Pauly N, Reynolds A, Henry RJ, Smith JJ et al (2011) Origin of amphibian and avian chromosomes by fission, fusion, and retention of ancestral chromosomes. Genome Res 21:1306–1312
4. Smith JJ, Putta S, Zhu W, Pao GM, Verma IM, Hunter T, Bryant SV, Gardiner DM, Harkins TT, Voss SR (2009) Genic regions of a large salamander genome contain long introns and novel genes. BMC Genomics 10:19
5. Voss SR, Prudic KL, Oliver JC, Shaffer HB (2003) Candidate gene analysis of metamorphic timing in ambystomatid salamanders. Mol Ecol 12:1217–1223
6. Dixon JE, Allegrucci C, Redwood C, Kump DC, Bian Y, Chatfield J, Chen Y-S, Sottile V, Voss SR, Alberio R, Johnson AD (2010) Axolotl Nanog activity in mouse embryonic stem cells demonstrates that ground state pluripotency is conserved from urodele amphibians to mammals. Development 137: 2973–2980
7. Tapia N, Reinhardt P, Duemmler A, Wu G, Araúzo-Bravo MJ, Esch D, Greber B, Cojocaru V, Rascon CA, Tazaki A, Kump K, Voss R, Tanaka EM, Schöler HR (2012) Reprogramming to pluripotency is an ancient trait of vertebrate Oct4 and Pou2 proteins. Nat Commun 3:1279
8. Voss SR, Shaffer HB (1997) Adaptive evolution via a major gene effect: paedomorphosis in the Mexican axolotl. Proc Natl Acad Sci USA 94:14185–14189
9. Voss SR, Smith JJ (2005) Evolution of salamander life cycles: a major-effect quantitative trait locus contributes to discrete and continuous variation for metamorphic timing. Genetics 170:275–281
10. Johnson CK, Voss SR (2013) Salamander paedomorphosis: linking thyroid hormone to life history and life cycle evolution. Curr Top Dev Biol 103:229–258
11. Page RB, Boley MA, Kump DK, Voss SR (2013) Genomics of a metamorphic timing QTL: met1 maps to a unique genomic position and regulates morph and species-specific patterns of brain transcription. Genome Biol Evol 5:1716–1730
12. Altschul SF, Gish W, Miller W, Myers EW, Lipman DJ (1990) Basic local alignment search tool. J Mol Biol 215:403–410
13. Huggins P, Johnson CK, Schoergendorfer A, Putta S, Bathke AC, Stromberg AJ, Voss SR (2012) Identification of differentially expressed thyroid hormone responsive genes from the brain of the Mexican Axolotl (*Ambystoma mexicanum*). Comp Biochem Physiol C Toxicol Pharmacol 155:128–135
14. Irizarry RA, Hobbs B, Collin F, Beazer Barclay YD, Antonellis KJ, Scherf U, Speed TP (2003) Exploration, normalization, and summaries of high-density oligonucleotide array probe level data. Biostatistics 4:249–264

Chapter 26

Data Mining in Newt-Omics, the Repository for Omics Data from the Newt

Mario Looso and Thomas Braun

Abstract

Salamanders are an excellent model organism to study regenerative processes due to their unique ability to regenerate lost appendages or organs. Straightforward bioinformatics tools to analyze and take advantage of the growing number of "omics" studies performed in salamanders were lacking so far. To overcome this limitation, we have generated a comprehensive data repository for the red-spotted newt *Notophthalmus viridescens*, named newt-omics, merging omics style datasets on the transcriptome and proteome level including expression values and annotations. The resource is freely available via a user-friendly Web-based graphical user interface (http://newt-omics.mpi-bn.mpg.de) that allows access and queries to the database without prior bioinformatical expertise. The repository is updated regularly, incorporating new published datasets from omics technologies.

Key words Newt, Salamander, Database, Transcriptomics, Proteomics

1 Introduction

1.1 The Newt-omics Repository

Currently, a small panel of model organisms such as the mouse, zebra fish, or *C. elegans* dominates biomedical research. Analysis of these standard model organisms is facilitated by large public repositories providing genome, transcriptome, and proteome information. Detailed information for single genes usually includes facts about intron and exon regions, splicing signals as well as the sequence itself and functional annotations. Available repositories allow easy querying of datasets, often without the need of specialized bioinformatical expertise. Access is usually granted by Web services. Salamanders do not belong to the group of well-investigated organisms rather than to the group of niche model organisms [1]. In the past, genome sequence data or comprehensive transcriptome or expression data were not available for such niche models. In newts, the first large-scale expression data were obtained using custom-made microarrays generated from lens [2] and other tissues [3], which yielded valuable but incomplete insights.

Anoop Kumar and András Simon (eds.), *Salamanders in Regeneration Research: Methods and Protocols*, Methods in Molecular Biology, vol. 1290, DOI 10.1007/978-1-4939-2495-0_26, © Springer Science+Business Media New York 2015

This situation dramatically changed within the past 5 years. New sequencing methods were developed for analysis of smaller genomes, transcriptomes, and proteomes. Analysis of niche model organisms has profited considerably from these new techniques, which are continuously evolving and becoming more and more affordable and easier to use. The most important techniques in this context are probably high-throughput mass spectrometry and next-generation sequencing, which enable de novo proteome, genome, or transcriptome analysis of niche models even in small labs. So far several studies on regenerative processes in niche models have been published [4–9], which in the newt have focused on the heart and the lens [10–14].

The vast amount of data generated by modern sequencing and mass spectrometry analysis creates new challenges. Datasets in raw format "as they are" are rather difficult to handle for nonspecialists and not very intuitive to use. Data formats are difficult to read by eye and the large volume of files sometimes makes even trivial tasks as, e.g., opening of a file cumbersome, further adding to problems with data storage and exchange. One potential solution for such problem is the exchange of processed and analyzed datasets (often excel files) but at the cost of transparency and comparability. To overcome such limitations and to give access to datasets generated on different platforms and in different labs, we established the newt-omics repository [15] for the red-spotted newt *Notophthalmus viridescens*. As the estimated genome size of the newt is up to ten times larger than that of humans, a genome project has not been initiated yet, despite the increasing speed and capacities of modern sequencing machines and assembly algorithms. Therefore, we decided to focus on the transcriptome instead of the genome as the central element for our database. All information stored in the database is linked directly to this central element. Newt-omics includes a de novo-assembled transcriptome based on a set of sequencing approaches on different sequencing platforms. In addition, expression data from multiple high-density microarrays as well as RNA-seq experiments were included. The transcriptome was translated to open reading frames and utilized for mass spectrometry-based protein identification. Newt-omics stores and interlinks all available datasets and provides a user-friendly gateway, which should help to make molecular data for *Notophthalmus viridescens* more accessible for researchers with limited experience in bioinformatics.

The Newt-omics hosts sequence data from transcriptomics and proteomics approaches, experimental expression data, primary sequence annotations, and functional annotations. In terms of the transcriptome, a de novo transcriptome was generated and annotated. Expression data from microarray and RNA-seq experiments are linked to the de novo transcriptome. Further, protein expression data derived from mass spectrometry experiments are included. The design of the Newt-omics is based on an entity–relationship model (ERM) connecting various datasets to avoid compartmentalization. Connection of datasets is realized by selection of biological entities such as transcripts or peptides. Experimental settings are used as attributes of biological entities. The core object of our model is the transcript (due to the lack of a genome) that links queries from distinct experiments to general data such as annotations and vice versa. Individual sections of the model can be updated separately (Fig. 1). This is crucial, as, e.g., annotations or functional annotations like ontologies change more often than associated peptide sequences. To facilitate access to the database, we implemented a Web-based user interface using PHP/HTML/JS scripts. The only prerequisite to browse and query the database is a Web browser like Firefox or IE. The database is freely available via http://newt-omics.mpi-bn.mpg.de and bioinformatics.mpi-bn.mpg.de.

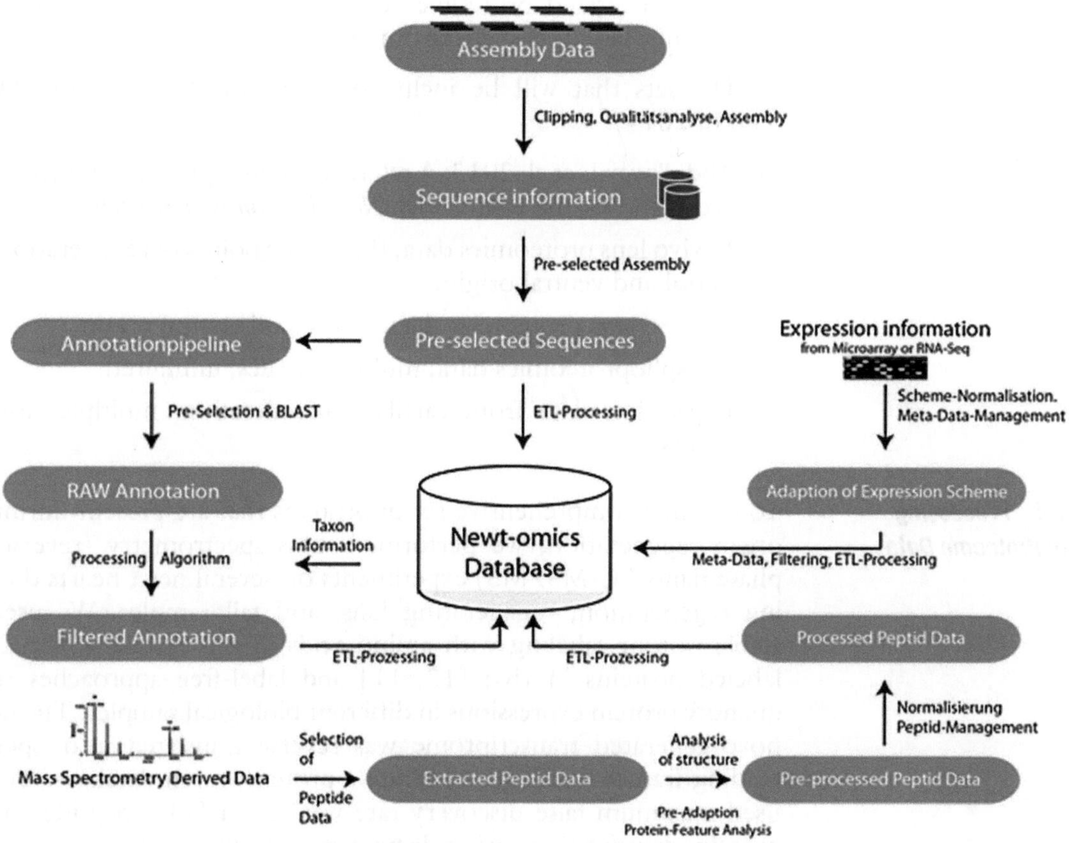

Fig. 1 Flow chart of information processing in the newt data repository. The newt repository is generated by automated pipelines for multiple raw data formats that perform an extract–transform–load (ETL) process to fill the database. Each part of the database is updated independently of other parts facilitating the update process to include new datasets

The database is updated once a year, incorporating further datasets from the red-spotted newt generated by the authors' group as well as other published transcriptome and proteome datasets.

Published datasets included in the current version 1.0:

1. Sousounis K. et al. 2013: Transcriptome analysis of newt lens regeneration reveals distinct gradients in gene expression patterns.
2. Looso M. et al. 2013: A de novo assembly of the newt transcriptome combined with proteomic validation identifies new protein families expressed during tissue regeneration.
3. Sousounis K et al. 2013: A microarray analysis of gene expression patterns during early phases of newt lens regeneration.
4. Looso et al. 2012: Spiked-in pulsed in vivo labeling identifies a new member of the CCN family in regenerating newt hearts.
5. Looso et al. 2010: Advanced identification of proteins in uncharacterized proteomes by pulsed in vivo stable isotope labeling-based mass spectrometry.

Datasets that will be included in the database version 2.0 (end of 2014):

1. Abdullayev I. et al. 2013: A reference transcriptome and inferred proteome for the salamander *Notophthalmus viridescens.*
2. In vivo lens proteomics data, three time points of regeneration, dorsal and ventral origin.
3. In vitro lens proteomics data, dorsal and ventral origin.
4. Phosphoproteomics data, multiple tissues, uninjured.
5. Expression data from cartilage regeneration, multiple time points.

1.2 Processing of Proteome Data

To obtain a comprehensive set of proteins that are present during organ regeneration, we performed mass spectrometry (reverse-phase nano-LC-MS/MS) experiments on several newt hearts during regeneration, regenerating lens, and tail samples. We used stable isotope labeling with amino acids in cell culture (SILAC) labeled proteins in vivo [12, 11] and label-free approaches to quantify protein expressions in different biological samples. The de novo-generated transcriptome was reverse translated into open reading frames (ORF) to generate a protein search database. The used maximum false discovery rate was set <1 % for peptide and protein identifications using decoy target databases [15].

1.3 Assignment of Annotations to Sequence Data

BLAST homology searches were performed using blastn, blastx, and tblastx tools on NCBI's NR (protein), NR (Nucleotide), EST, and UniProt databases. All hits were stored according to their taxa. For each single taxon (>90), a quality rating based on regular expressions was performed. Keywords providing a robust quality level were preferred. To achieve a maximum quality with a minimum number of annotations per taxon, we prioritize hits in the NCBI NR or UniProt database over an EST hit. Based on this assumption, we computed a quality rank based on a dictionary of keywords (e.g., cDNA, clone, mRNA, predicted) reflecting hits of limited information. Database entries containing one or more such keywords were marked as low-quality hits (*see* Ref. 16 for details). Remarkably a substantial number of sequenced and assembled transcripts did not share any reliable sequence similarity to known database entries.

1.4 Assignment of Functional Annotations

Subsequent to transcript annotation, we assigned transcripts to different functional groups. Since Gene Ontology (GO) annotations for amphibians are limited [17], we used the UniProt BLAST results from mouse, zebra fish, cow, human, and chicken. GO annotations for each transcript were stored in a taxon-dependent manner. We stored the best hit for each taxon having at least one GO annotation. Further, we used functional domains, interaction partners, protein families, and pathways from the UniProt entries.

2 Materials

To provide access to the freely accessible newtomics resource, we implemented a PhP Web-based graphical user interface (GUI). This can be used by any user by pointing any browser (GUI optimized for Firefox) to newtomics.mpi-bn.mpg.de.

3 Methods

3.1 General Considerations

For searching newt-related data, you will need a Web browser (database front end is optimized for Firefox) to connect to www.newtomics.mpi-bn.mpg.de. Based on the point of interest, the users need to choose one of the main sections of the database to generate a query. The database is grouped into the sections sequence data, annotation, functional annotation, expression data (including RNA-seq and microarrays), and proteome (Fig. 2). Since a genome is missing as central entity, you will find the transcript entity defined as core element that links additional entities. Any query that you formulate in the graphical user interface (GUI) to one of the mentioned sections will eventually utilize this central transcript element for presentation of results.

3.2 Guided Tour on the Graphical User Interface

The GUI of the newtomics database is accessible by pointing your browser to newtomics.mpi-bn.mpg.de. The Firefox browser is recommended for optimal display. On top of the GUI, you will find five main windows. Beginning from the left you find the **Home**

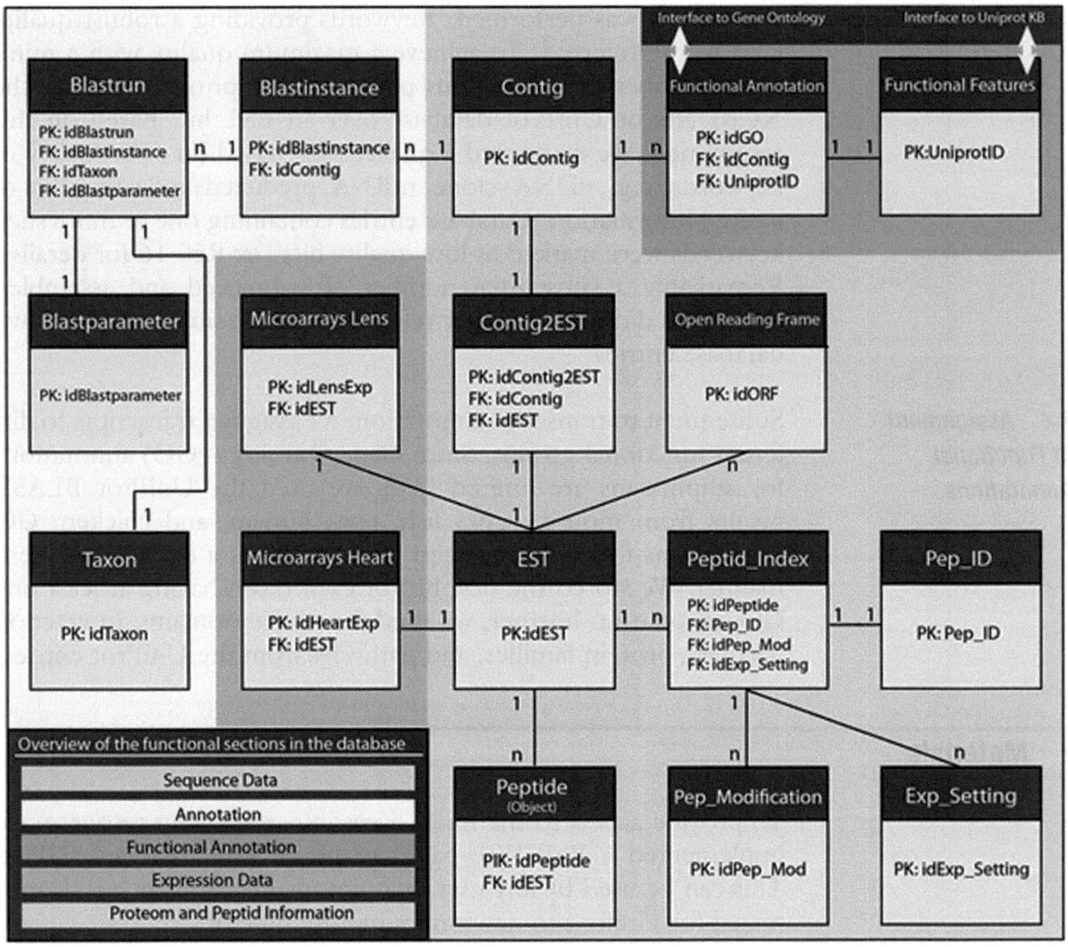

Fig. 2 Design of the database. The database is divided into five main parts, namely, the sequence part, the annotation part, the functional annotation part, the expression part, and the proteome part. The central element is the transcript that interlinks all other sections. Queries on multiple parts of the database utilize the central table. The scheme can get upgraded easily by adding new tables with transcript IDs as foreign key

window, which is the starting window giving general information about the amount of data stored in the database (details can be found above in the database section). To perform database searches in one of the data sections mentioned above, the users need to start from either of the four panels "Transcripts," "Blast," "Peptides," or "Expression" (Fig. 3). The Tutorial window will give some examples on the search forms; the **About** window on the right gives general information on the database development and linked publications.

Once you have chosen a single transcript of interest in any of the four main search windows mentioned above, you are directed to the *Single Transcript view* module. The transcript is the core object of the model due to the lack of a genome. The central view

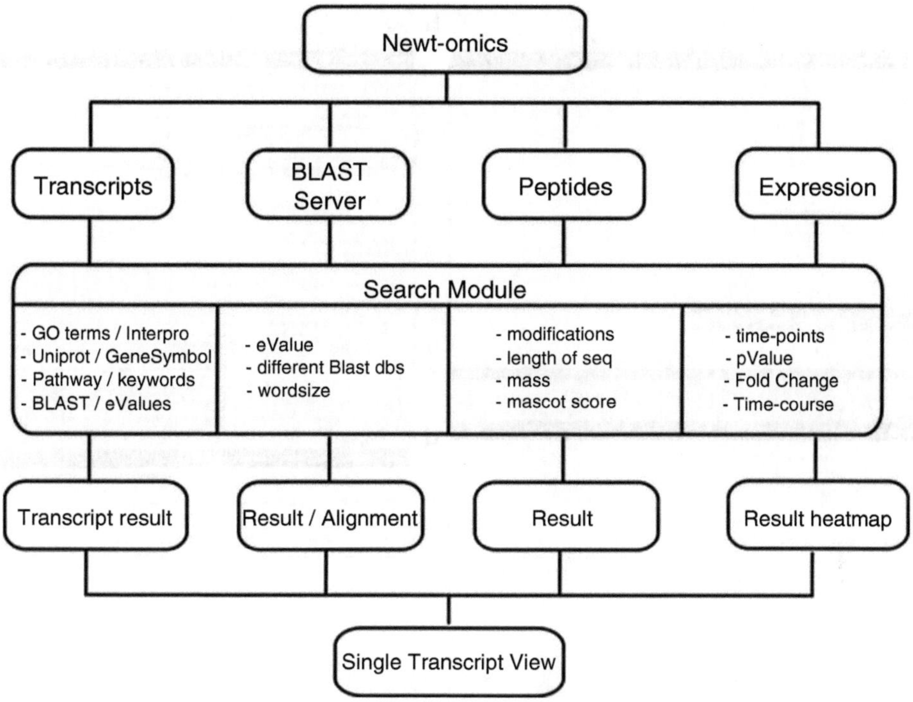

Fig. 3 Modes of database assessment. The database can be accessed by the graphical user interface (GUI) via the main search forms, focusing on the transcript itself, the sequence similarity (BLAST), the proteome (Peptides), or the expression values. Each search form offers specific search parameters in the corresponding search module. Each search module generates a specific result table or visualization. The single entities of the result structure are linked to the single transcript view, presenting all available information for a single database entry

enables you to access any information stored in relation to a single transcript in the database. The direct link http://newtomics.mpi-bn.mpg.de/contig5.php?line=Contig5389 provides an example. You will see that the view contains five main tabs on the top: **Transcript**, **Annotation**, **Functional Annotation**, **Expression,** and **Peptide** (from left to right).

1. The **Transcript** panel is activated by default and displays the transcript sequence of a selected transcript. Furthermore, you will get a list of the IDs and coordinates of individual reads that were assembled to generate the transcript (Fig. 4a). In the general overview you will find all IDs that are used in the database to reference the selected transcript. You will find these IDs useful when you want to reference a specific entry or want to access a specific entry by the ID search form explained below. To download the transcript assembly or the single ESTs that were used for the assembly in FASTA format, you can use the dedicated links (*see* **Note 1**).

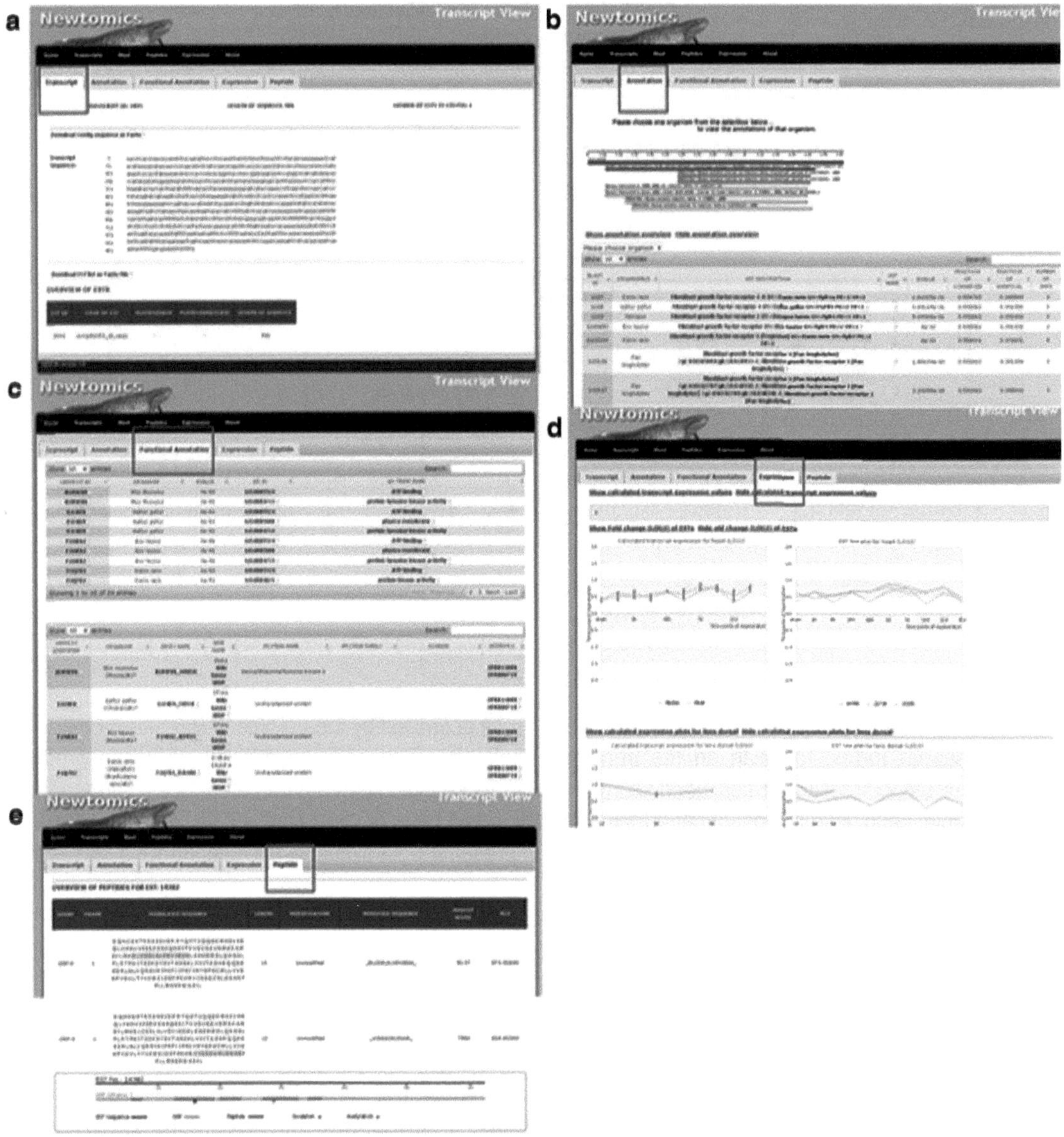

Fig. 4 Transcript view of the Web-based GUI of the newt-omics repository. Section (**a**) gives detailed information about the sequence, sequence download options, and information about the subsequences used in an alignment step. Section (**b**) gives an overview on detected sequence similarities. The view can be filtered for single taxa, and a mouse-sensitive graph gives positional information on single alignments. Section (**c**) represents identified functional annotations such as Gene Ontology terms or known pathways. Section (**d**) illustrates all quantitative measurements that were performed for the chosen transcript. The expression values are visualized by *heat maps* and *line plots*. Section (**e**) visualizes the coding potential of transcripts under investigation. The complete sequence as well as detected open reading frames are visualized; detected peptides are marked on the corresponding position

2. In the **Annotation** panel you will find all information derived from similarity searches for the selected transcript (Fig. 4b). It provides positional information via a graphical alignment overview as well as a tabular format. The view is further divided into two subsections:
   (a) In the upper part you will find a graphical overview on detected sequence similarities. You activate the graphic by selecting one single taxa of interest in the *Please choose organism* field (left). Once selected you will receive an overview about all sequences with a significant similarity to the selected transcript and the position of the alignments. Simultaneously you will get a filtered table (lower part of the view) to the hits that are visualized in the graphical overview. You can use the link *Show annotation overview* to estimate the general taxa distribution in respect to similarity (*see* **Note 2**).
   (b) In the lower part of the page, you get a tabular form that lists all detected significant similarities to other species. As indicated above, selection of a specific organism filters the list. You can estimate the quality of each hit by the attributes of the similarity search such as *e*-value and corresponding alignment parameters. By clicking on the entry hit description or hit name, you are redirected to the corresponding NCBI database entry. You can sort the complete tabular for each column, extremely useful in respect to *e*-value or organism (*see* **Note 3**).
3. By changing to the **Functional Annotation** panel, you get a list of Gene Ontology terms affiliated with a selected transcript (Fig. 4c, upper table) and a list of corresponding protein families and InterPro domains (lower table). The assignments are derived from UniProt similarity searches. To estimate the quality of the assignment, further attributes are listed, such as the origin of the organism, *e*-value, and UniProt ID. The upper table provides a versatile tool to check your candidate of interest for associated GO terms. The table can be sorted by clicking on the column header. Each entry is linked to the corresponding source database (*see* **Note 4**). In the lower table you can obtain further information about detected protein families and InterPro domains. Tables can also be sorted by columns and provide links to iHOP (25) and WikiGenes (24).
4. By highlighting the **Expression** panel, you will receive information about expression values that are available for your transcript of interest (including individual sequences that were used for the transcript assembly, Fig. 4d). A set of timeline graphs and heat maps visualizes expression experiments on the regenerating heart and dorsal/ventral lens. First, you will find the calculated mean and median fold changes with standard

deviation (upper left line plot), representing a time course of regenerating hearts with ten sampling points. Furthermore, you can inspect the expression changes for single on-array replicates comprising the transcript of interest on the second line plot on the right. By using the visualization, you can evaluate the uniformity and thereby the reliability of the replicate set. If you click on the *Show FC of ESTS* link, you will get a heat map on calculated log2-based changes for individual transcripts. Furthermore, all single reads are displayed in separate tables giving you detailed fold changes and *p* values calculated from biological replicates. On the lower heat map, you will find the expression values of regenerating dorsal and ventral lens tissue, based on the identical array platform (*see* **Note 5**).

5. In the **Peptide** panel, you will get an overview about the coding potential of your transcript of interest. The panel provides details about experimentally validated peptides for selected transcripts (Fig. 4e). Since transcripts are assembled from single ESTs, the coding potential of transcripts is based on individual ESTs, which are displayed separately. For each EST one or more open reading frames are listed (column 1). Identified peptides are highlighted in yellow within the complete translated EST sequence (column 2). Furthermore, you can go back to the peptide sequence as well as its score and *m/z* ratio. The EST and peptide information can be used to identify frame shifts in the sequence (*see* **Note 6**).

3.3 Protocols

The protocol section describes how to use the search modules. Search can be initiated at different points of view, starting with a keyword or ID, a sequence of interest, peptide attributes such as mass, or queries for specific expression patterns.

Protocol: Identification of your candidate of interest and retrieval of corresponding information by IDs or keywords. After opening the **Transcript** window in your browser, you will find three search options (Fig. 5a).

1. The first option is a transcript assembly based search form. You can directly access the database by internal IDs, the transcript length, or name. Use of this interface is recommended if you want to browse for general information available in the database for a limited number of already known candidates. This search form is also important to retrieve information about single known candidates after a database update. Search terms are selected from the drop-down list and search parameters are added to the appearing search field. You can further modify the queries including *larger/smaller than* operators. Multifactor searches are initiated by adding search fields, which might be linked using AND or OR operators.

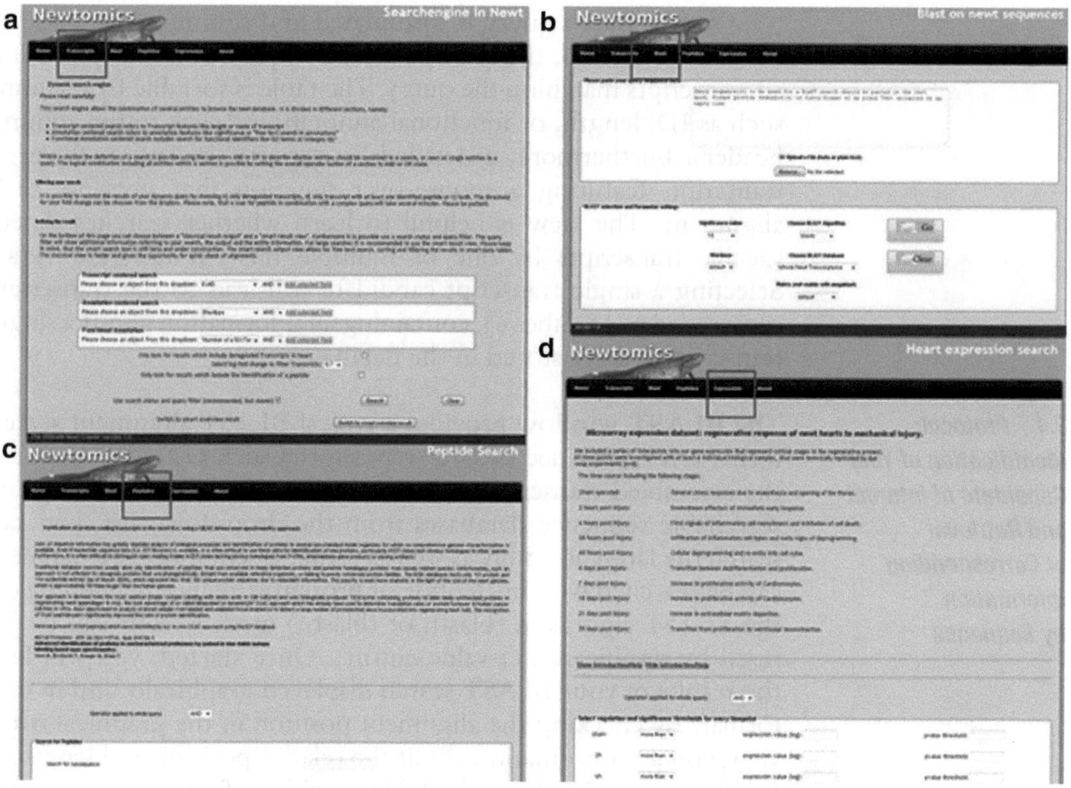

Fig. 5 Web-based GUI of the newt-omics repository. The GUI presents the main entrance panels available for queries of the repository. Screenshot (**a**) presents the search form based on single transcripts, screenshot (**b**) presents the BLAST interface for homology searches, screenshot (**c**) presents the expression based search form, and screenshot (**d**) illustrates the peptide search form. All search results will link to single transcripts matching the search query, visualized by the central transcript view (Fig. 4)

2. The second option is annotation-centered enabling queries to the annotation section of the database. It is required to specify the annotation algorithm, the database that will be used for similarity searches, keywords such as protein symbols or database identifiers, and finally a significance threshold. In general, this search option is used to query the database for single candidates/groups of candidates of interest for which a gene symbol or key words are available. Searches might be filtered by additional AND or OR operators.
3. The third option is annotation centered as well but focuses on functional annotations such as GO terms and GO identifiers, UniProt identifiers, protein family names, pathways, and domains. Furthermore, results might be filtered for expression changes on transcript level or coding potential. The search option should be employed if you are interested in a set of transcripts that are related to a specific term of interest, e.g., a specific molecular function.

Once a search query is formulated and submitted by clicking the submit button, the search result is displayed in a table, listing all transcripts matching the query. The table is sortable by columns such as ID, length, or functional annotations (click to the column header). Furthermore, a detailed list reports similarities for each transcript, featuring a mouse-over function that presents each alignment. The view is helpful to learn whether searches detect specific transcripts by one or multiple hits (in different taxa). Selecting a single transcript candidate will lead to the transcript-centered view (*see* above), containing all information about a single transcript that are stored in the database.

3.4 Protocol: Identification of Your Candidate of Interest and Retrieval of Corresponding Information by Sequence

The **BLAST window** provides a typical BLAST alignment search form (as it is provided by large repositories such as NCBI) to query the assembled transcriptome of the newt (Fig. 5b). Selecting the searchable sequence databases from the drop-down list, you can search the latest assembly as well as earlier assemblies. The database can be searched with nucleotide or protein sequence depending on the BLAST algorithm (blastn or tblastn) used. Results can be filtered by significance (*e*-value cutoff). Once started, you will find the results of your BLAST search displayed graphically and in text format. By checking the alignment position in the graphical overview, you get information about the relative position and length of the alignment. The lower hit table provides information about statistical significance linked directly to the corresponding transcripts (transcript entered view *see* above). Additionally, each sequence alignment is reported at the end of the page in text format.

3.5 Protocol: Search for Expression Patterns of Interest

Selection of the **Expression window** enables access to the dataset by general expression patterns. Expression data are based on microarray and RNA-seq experiments from the regenerating heart and lens (Fig. 5d). Initially, a tissue of interest is chosen. Next you specify the general conditions you are interested in. For instance, if you are interested in a set of transcripts that is significantly upregulated in the heart within the first day of regeneration:

1. Select the heart tissue.
2. Select greater than in the drop-down menu for 2, 6, and 24 h.
3. Type 1.5 (log2) into the expression value field.
4. Type 0.05 into the *p* value field.
5. Check AND als logical operator on top and start the search.

The result table is organized as a heat map. Each column represents an array spot, whereas a transcript (second column) comprises one or more array spots. Expression values are color coded. Red indicates higher expression compared to undamaged tissue, and green indicates lower expression compared to undamaged tissue.

Statistical significance can be assessed at the right part of the table, showing corrected *p* values for all expression changes. Direct links in column two provide access to the transcript-centered view (*see* above).

3.6 Protocol: Investigation of the Coding Potential of Transcripts

The **Peptides window** allows searches for experimentally validated transcript sequences (Fig. 5c). The database can be queried for peptide attributes such as length, score, or mass. Further, searches can be filtered for single experiments in distinct tissues. Multiple search forms can be combined by selecting the AND or OR operator from the drop-down menu as in the other main windows. After a result list is obtained (for instance search for peptide length *more than* 25), the provided links provide access to an overview list for the peptide. Following such a link (in this example with the ID 1000), a summary of annotations and functional annotations (has to be enlarged by clicking either on *Show contig results* or *Show functional annotation results*) is provided. Within the enlarged result tables, the transcript-centered view (*see* above) can be selected by clicking on one of the transcripts of interest (column 2).

4 Notes

1. The transcriptome assembly presented in the database is based on an EST sequencing approach. Each assembled transcript is derived from one or more EST sequences. The OVERVIEW OF ESTS table gives information about the number of ESTs that contributed to a transcript as wells as the number of spots/replicates that are represented in the microarray study, since all ESTs were used as spots on the custom-made arrays. You might use the number of spots representing an individual sequence but were not included in the array analysis (panel expression) to estimate the quality/expression level of these spots. If a large proportion of sequences is missing, low-level expression of the EST can be assumed.
2. The *Show annotation overview* view is useful to analyze similarity hits provided for a transcript of interest. First, you should check the taxa distribution for widely explored mammalians such as human and mouse. Next, you compare this percentage to the percentile for "close" relatives of the newt such as *Xenopus* and axolotl. If you see a large proportion of hits for mammalians, you will obtain sequences with high similarity and robust annotation. In case you receive a high percentage of "close" relatives, you will also see closely related sequences, but the quality of annotations will be low.
3. Multiple BLAST hits for multiple taxa are stored in the database. These hits are presented in the annotation panel of

the transcript-centered view. Sorting this table for organism and *e*-value will provide important insights in the robustness of sequence similarities. For instance, a key word search for BMPR2 is performed on all BLAST hit descriptions stored in the database. The result list will name all transcripts having at least one similar counterpart in another organism with the description line harboring BMPR2. To estimate the robustness of that assignment, the number of different organisms in the HSP group and the relative size of the corresponding *e*-value should be used.

4. Evaluating the impact of Gene Ontology assignments can be done by sorting the GO table by the GO ID column. There will be GO terms with the same protein assignment for multiple organisms as well as assignments only for single taxa. You can use this information to classify multiple taxa-based assignments as more reliable or highly conserved.
5. If you compare single EST IDs from heart and lens experiments, you will find an overlap but not congruence of the IDs. This is due to quality assessment during array analysis. A small overlap between heart and lens or within dorsal and ventral lens gives you a hint whether a transcript is highly or lowly expressed in the corresponding tissue.
6. In the peptide view, each corresponding EST is listed with the ORF ID and all peptides identified within an ORF. In case one ORF yields multiple peptide identifications, the probability for identification of a correct ORF is high. Browsing of the transcriptome might provide sequences that have two or even more ORFs with peptide identifications. Usually such a result indicates false-positive peptides due to sequencing errors. Such information might be used to correct the single EST sequence or even the assembled transcript.

Acknowledgment

This work was supported by the Max-Planck-Society, the Excellence Cluster Cardiopulmonary System (ECCPS), the University of Giessen-Marburg Lung Center (UGMLC), the Cell and Gene Therapy Center (CGT) of the University of Frankfurt, the DFG (SFB TRR81), and the German Center for Cardiovascular Research (DZHK). The authors would like to thank Marc Bruckskotten for his invaluable input for creating newt-omics and for preparation of the figures.

References

1. Looso M (2014) Opening the genetic toolbox of niche model organisms with high throughput techniques: novel proteins in regeneration as a case study. Bioessays 36(4):407–418
2. Sousounis K, Michel CS, Bruckskotten M, Maki N, Borchardt T, Braun T, Looso M, Tsonis PA (2013) A microarray analysis of gene expression patterns during early phases of newt lens regeneration. Mol Vis 19:135–145
3. Mercer SE, Cheng CH, Atkinson DL, Krcmery J, Guzman CE, Kent DT, Zukor K, Marx KA, Odelberg SJ, Simon HG (2012) Multi-tissue microarray analysis identifies a molecular signature of regeneration. PLoS One 7(12):e52375. doi:10.1371/journal.pone.0052375
4. Wu CH, Tsai MH, Ho CC, Chen CY, Lee HS (2013) De novo transcriptome sequencing of axolotl blastema for identification of differentially expressed genes during limb regeneration. BMC Genomics 14:434. doi:10.1186/1471-2164-14-434
5. Rao N, Jhamb D, Milner DJ, Li B, Song F, Wang M, Voss SR, Palakal M, King MW, Saranjami B, Nye HL, Cameron JA, Stocum DL (2009) Proteomic analysis of blastema formation in regenerating axolotl limbs. BMC Biol 7:83. doi:10.1186/1741-7007-7-83
6. Stewart R, Rascon CA, Tian S, Nie J, Barry C, Chu LF, Ardalani H, Wagner RJ, Probasco MD, Bolin JM, Leng N, Sengupta S, Volkmer M, Habermann B, Tanaka EM, Thomson JA, Dewey CN (2013) Comparative RNA-seq analysis in the unsequenced axolotl: the oncogene burst highlights early gene expression in the blastema. PLoS Comput Biol 9(3):e1002936. doi:10.1371/journal.pcbi.1002936
7. Boser A, Drexler HC, Reuter H, Schmitz H, Wu G, Scholer HR, Gentile L, Bartscherer K (2013) SILAC proteomics of planarians identifies Ncoa5 as a conserved component of pluripotent stem cells. Cell Rep 5(4): 1142–1155
8. Sandmann T, Vogg MC, Owlarn S, Boutros M, Bartscherer K (2011) The head-regeneration transcriptome of the planarian Schmidtea mediterranea. Genome Biol 12:R76. doi:10.1186/gb-2011-12-8-r76
9. Chapman JA et al (2010) The dynamic genome of Hydra. Nature 464:592–596
10. Borchardt T, Looso M, Bruckskotten M, Weis P, Kruse J, Braun T (2010) Analysis of newly established EST databases reveals similarities between heart regeneration in newt and fish. BMC Genomics 11:4. doi:10.1186/1471-2164-11-4
11. Looso M, Borchardt T, Kruger M, Braun T (2010) Advanced identification of proteins in uncharacterized proteomes by pulsed in vivo stable isotope labeling-based mass spectrometry. Mol Cell Proteomics 9:1157–1166
12. Looso M, Michel CS, Konzer A, Bruckskotten M, Borchardt T, Kruger M, Braun T (2012) Spiked-in pulsed in vivo labeling identifies a new member of the CCN family in regenerating newt hearts. J Proteome Res 11:4693–4704
13. Looso M et al (2013) A de novo assembly of the newt transcriptome combined with proteomic validation identifies new protein families expressed during tissue regeneration. Genome Biol 14:R16. doi:10.1186/gb-2013-14-2-r16
14. Sousounis K, Looso M, Maki N, Ivester CJ, Braun T, Tsonis PA (2013) Transcriptome analysis of newt lens regeneration reveals distinct gradients in gene expression patterns. PLoS One 8(4):e61445. doi:10.1371/journal.pone.0061445
15. Elias JE, Gygi SP (2007) Target-decoy search strategy for increased confidence in large-scale protein identifications by mass spectrometry. Nat Methods 4:207–214
16. Bruckskotten M, Looso M, Reinhardt R, Braun T, Borchardt T (2012) Newt-omics: a comprehensive repository for omics data from the newt Notophthalmus viridescens. Nucleic Acids Res 40:D895–900
17. Barrell D, Dimmer E, Huntley RP, Binns D, O'Donovan C, Apweiler R (2009) The GOA database in 2009–an integrated Gene Ontology Annotation resource. Nucleic Acids Res 37: D396–403

Index

A

Accessory limb model (ALM)....101–113
A1 cells....173, 182
Adaptive immune cells....205
AEC. *See* Apical epithelial cap (AEC)
Aeromonas....40
Affymetrix Expression Console....312, 314
Affymetrix GeneChip....310, 314, 315, 331
Affymetrix microarray....310–311
AGP. *See* Anterior gradient protein (AGP)
AGSC. *See Ambystoma* Genetic Stock Center (AGSC)
AL-1 cells....187, 188, 190–195
ALM. *See* Accessory limb model (ALM)
Alvegesic....175, 243, 247, 249
Ambystoma
  A. maculatum....149, 151, 152, 154
  A. mexicanum....27–43, 49, 50, 116, 141–145, 206, 209, 269–277, 310, 316, 321, 322, 324, 325, 327, 328, 333, 334
  A. tigrinum....28, 30, 33, 141, 324
Ambystoma Genetic Stock Center (AGSC)....28, 42, 103, 270, 310, 322, 333–335
Ambystomatidae....27, 71
AMEM. *See* Amphibian minimal essential medium (AMEM)
Amikacin....38, 40
3-Aminobenzoic acid ethyl ester (MS-222)....254, 257
Aminoglycosides....254, 255
Ammonium chloride....230, 235
Amniotes....7, 9, 12, 148
Amphibian minimal essential medium (AMEM)....176, 177, 179–181, 189–191, 193, 254, 255
Amphibian papilla....253, 264
Amphiumids....9
Amplexus....18, 24, 48, 64
Ampullae....253, 259, 260
Amquel....41, 142, 281
Aneurogenic....147–155
Animal pole....64, 275, 277, 288, 300, 301, 303
Anterior gradient protein (AGP)....148, 149
Aortic arches....31
Apateon....10, 11
Apical epithelial cap (AEC)....102, 103
Apomorphy....8, 11
Artemia....38, 39, 41, 51, 52, 56, 60, 66, 270
Ascorbic acid....95, 230, 235
Autopodial elements....148
Autotomy....71, 73, 76
Axolomics....291
Axolotl genome....159, 280, 291, 293, 322, 324, 327
Azide....230, 231, 235, 237, 239

B

Bacterial Artificial Chromosome (BAC)....269, 323, 324
Basale commune....7, 9, 12
Basic Local Alignment Tool (BLAST)....317–318, 324, 325, 327–332, 340, 341, 343, 347–350
Benzocaine....117, 119, 272, 276, 277
B1H1 cells....182
Bicistronic....160
Bioanalyzer....310, 313
Biofilters....52
Biotin-16-dUTP....85, 87
Blackworms....22, 24, 38, 39, 42, 142–145
BLAST. *See* Basic Local Alignment Tool (BLAST)
Blastema cells....101, 102, 171, 213
Bolitoglossa....71, 72, 76
Brachial....36, 102, 106, 148
Brachial nerves....102, 148
Brine shrimps....38, 39, 41, 154, 270, 299, 302

C

Cacodylate buffer....255, 256, 258, 261
Caecilians....10
Calretinin....257, 263
Capillary electrophoresis....280
Cardiac troponins....228
Cardiomyocytes....19, 227–239, 242, 243, 245, 248–249
Cas9 RNA-guided nuclease....279–294
Catecholaminergic neurons....92
cDNA sequencing....280
Centricon....130, 132
Chromatophores....33
Chytridiomycosis....74
Ciprofloxacin....244, 254
Clodronate liposome delivery....206
Clonal density....172
Clustered Regularly Interspersed Short Palindromic Repeats (CRISPR)....280, 291

Anoop Kumar and András Simon (eds.), *Salamanders in Regeneration Research: Methods and Protocols*, Methods in Molecular Biology, vol. 1290, DOI 10.1007/978-1-4939-2495-0, © Springer Science+Business Media New York 2015

CMV promoter....................93, 94, 97, 135, 192
Collagen.................... 33, 173, 176, 177, 179, 180, 183, 217, 227, 241, 242
Columellar bone....................258
Contig ID....................326, 330
Contigs....................316, 324–328, 330, 332, 333, 349
CRISPR. *See* Clustered Regularly Interspersed Short Palindromic Repeats (CRISPR)
Cryogenic....................177, 181
Cryopreservation....................181, 192
Cryptobranchoidea....................5
Cutsmart buffer....................283, 284, 286
Cynops pyrrhogaster....................29, 83
Cysteine....................53, 59, 298
Cytospin centrifuge....................207, 214

D

Daphnia....................22, 23
DAPI. *See* 4′,6-diamino-2-phenylindole (DAPI)....................93, 200, 202, 207, 213, 219, 220, 231, 243, 245, 246, 248–249, 257, 263
de Bruijn graph....................316
Demembranation....................256, 259, 261, 263, 264
Dextran-rhodamine....................207, 210, 211, 213, 220
4′,6-Diamino-2-phenylindole (DAPI).......93, 200, 202, 207, 213, 219, 220, 231, 243, 245, 246, 248–249, 257, 263
Dimethyl sulfoxide (DMSO).................... 177, 181, 190, 192
Diploid genome....................31
Direct development....................5, 8, 71
Dissorophoidea....................10
DMSO. *See* Dimethyl sulfoxide (DMSO).................... 177, 181, 190, 192
Dopaminergic neurons....................92, 93

E

EasySep™ magnet....................208, 209, 218–220
ECM. *See* Extracellular matrix (ECM)
Ectopic....................102, 105, 116, 148, 149
  transplantation....................149
edgeR package....................318
EDTA. *See* Ethylene diaminetetraacetic acid (EDTA)
EdU. *See* 5-Ethynyl-2-deoxyuridine (EdU)
Electrode.................... 95, 96, 118–121, 123, 150, 151, 154, 161, 162, 165
Electron microscope....................260, 262
Electroporation....... 92, 94–98, 115–124, 127, 128, 148, 149, 160, 162, 165, 178, 182, 184, 188, 190, 192, 194–196
Entity relationship model (ERM)....................339
Ependymoglial cells....................92–94, 97
Epimorphic....................19, 187
EST.................... 291, 324–328, 340, 341, 346, 349, 350
Ethylene diaminetetraacetic acid (EDTA).................... 177, 179, 189, 206–208, 211, 212, 217–219, 225, 256, 264, 272, 281
5-Ethynyl-2-deoxyuridine (EdU)....................230, 231, 235, 237–239, 244–246, 248–250
  labeling....................245, 248
Euthanasia....................62
Extracellular matrix (ECM).................... 220, 228–230, 232, 237, 239, 241, 258

F

False discovery rate (FDR)....................315, 340
Fast Green....................117, 119, 121, 123, 130, 134, 138, 139, 161, 162, 165
Fibrosis....................34, 206
Ficoll PM 400....................282
FIMO clay....................150, 154
Flow cytometer.................... 207, 208, 213, 220–222
Fossils....................3, 9–13

G

Gelatin.................... 103, 104, 108–112, 154, 173, 176, 177, 179–181, 183, 184
  bead....................103, 109–112
Gene Ontology (GO)....................315, 341, 344, 345, 350
Genome....................3–13, 31, 47, 83, 92, 94, 116, 128, 159, 172, 269, 270, 279, 280, 291, 293, 297, 302, 304, 309–311, 316, 317, 321–334, 337–339, 341, 342
  resources....................310, 321–334
Genotyping....................282, 289, 291–293, 298, 302, 304
Gentamicin sulfate....................149, 151, 175, 255, 298, 302
GentleMacs™ apparatus....................212
Germplasm....................37
Gerobatrachus....................10, 12
GFAP. *See* Glial fibrillary acidic protein (GFAP)
Giemsa Stain....................208, 214, 215
Gigantism....................6
Glial fibrillary acidic protein (GFAP).................... 93, 94, 199, 200, 202
Glutaraldehyde.......... 105, 108, 109, 112, 255, 256, 258, 261
GPI-anchored protein....................148
Green fluorescent protein (GFP).................... 32, 33, 93, 94, 116, 160–165, 174, 192, 287
Grindal worm.................... 52, 56, 60, 66

H

Hanks balanced salt solution (HBSS).................... 206–208, 211, 212, 217–219, 225
Heart cultures....................247
Hematopoiesis....................31
Hemocytometer.................... 191, 237, 282, 288
HiDI formamide....................290
Histoacryl....................243, 247, 249
Holtfreter's solution.................... 38, 53, 104, 117, 142, 144, 207, 209, 211, 218, 272, 281, 287, 288, 298
Human chorionic gonadotropin.................... 37, 53, 59, 298
Hyaluronic acid....................230, 239

6-Hydroxydopamine ... 92
Hynobiidae ... 5
Hypomethylation ... 173

I

Iberian ribbed newts ... 297–304
Illumina paired-end sequencing ... 311, 316
Image J ... 161, 163
Immuno-FISH ... 81–89
Immunoglobulin ... 32, 222
Immunogold labelling ... 256, 260, 264
Inflammatory response ... 205
Inflammatory signals ... 206, 222
Ingenio ... 178, 190
Inner ear ... 253–265
Insulin ... 176, 189
Insulin-transferrin-selenium (ITS) ... 189, 194
Intraventricular injection ... 93–97
I-SceI meganuclease ... 270, 272, 273
I-SceI transgenesis ... 275
Isogroup ... 325
Isolectin B4 ... 207, 212, 213, 219, 220, 222
Isopentane ... 230, 234

J

JMP Genomics ... 312, 314, 315

K

Keratins ... 173

L

Lagena ... 253
Lateral line system ... 31, 51
Lentectomy ... 81–89
Leucistic larvae ... 33
Leydig cells ... 34, 35
L-15 Leibovitz medium ... 244, 249
Lobectomy ... 36
Lymphocytes ... 33, 205
Lysine ... 85, 129, 136, 198, 200, 206, 214, 230, 235, 257

M

Macrophages ... 33, 205, 206, 209, 215, 220–223, 241
Macropinocytosis ... 220
Marc's modified ringers ... 282
Mass spectrometry ... 338, 340
May-Grunwald stain ... 208, 214
Membranous labyrinth ... 260
Mesencephalon ... 34
Messenger RNA (mRNA) synthesis ... 281, 286, 304
Microarray ... 310–315, 317, 318, 323, 324, 331, 332, 334, 337, 340, 341, 348, 349
Microcapillary needle ... 130, 134
Microdissection ... 94
Microinjection ... 116–124, 130, 160, 165, 282, 292, 300, 301, 303
Microloader tips ... 94, 95, 97, 117, 119, 123, 130, 161, 162, 282, 287
Micromanipulator ... 118–120, 123, 124, 161–163, 165, 271, 275, 288, 299, 301, 302
Micropipette puller ... 94, 95, 97, 119, 271, 274, 282
MicroRNA ... 159–165
Microtome ... 85, 248
Moloney murine leukemia virus (MMLV), 128
Monoclonal antibody ... 173, 219
Morpholinos ... 116, 280
Mutagenized animals ... 280, 290
Myeloid cells ... 206, 212–215, 220–223
Myofibrils ... 242
Myotubes ... 172, 173, 229
MySQL database ... 312

N

Nanodrop spectrophotometer ... 313
Neotenic ... 5, 10, 27, 34
Nerve dependence ... 147, 148
Neural plasticity ... 35
Neural stem cells ... 91
Neurogenesis ... 91, 92, 115, 197
Neuromast ... 31
Neurospheres ... 197–203
Neurotransmission ... 147
Neutrophils ... 33, 205, 215, 217, 221
Newt-Omics ... 337–350
Notophthalmus viridescens ... 17–24, 29, 49, 50, 91, 93, 171, 253, 254, 338, 340
Novaqua ... 41, 142, 281
Novel splice isoforms ... 316
Nucleofection ... 178, 181–184, 190, 193–196

O

Olfactory epithelium ... 31
Oligonucleotides ... 281, 283, 324
Ontogenic transition ... 71
Open reading frames ... 129, 326, 328, 338, 340, 344, 346
Orthotopic transplantation ... 149
Osmium tetroxide ... 255, 256, 263
Osmolarity ... 173, 175, 176, 189, 198, 202
Otoconia ... 258, 260
Ototoxic ... 253, 254
Ovomucoid ... 199

P

Pacific giant salamander ... 8
Paedomorphosis ... 10, 141
Paleozoic ... 4, 10–13
Parabiosis ... 147–155

Paraformaldehyde (PFA) 89, 224, 230, 245, 256, 257, 261, 262
Paralogous gene 316, 326
Paraplast 85
PBS. *See* Phosphate buffered saline (PBS)
pDR274 expression vector 281
Pearson's correlation 315
Perennibranchiate 5
Permian 10–12
Phagocytic cells 205
Phallacidin 256
Phosphate buffered saline (PBS) 84, 85, 87, 117, 118, 121, 123, 124, 161–163, 175, 189, 191, 192, 200, 202, 230, 232, 235, 236, 239, 245, 254, 256, 257, 261–263
Pigmented epithelial cells (PECs) 81–83, 87, 89
Placode 31, 49
Platinum Gate TALENs 302
Plethodontidae 5, 71, 72
Pleurodeles waltl 29, 47–67, 149, 297–304
Pluripotency 6
pMLM3613 Cas9 expression vector 281
Polyadenylation signal 291
Poly-D-lysine 129, 136, 198, 200
Polyethylenimine 129
Poly-L-Lysine 85, 206, 214
Povidone-iodine 229, 232, 236
Preaxial dominance 7, 8, 10, 12, 13
Proteids 9
Proteomes 337, 338, 340–343
Pseudogenes 7
Pseudotyped retrovirus 127–139, 188

R

Red efts 18, 19, 23
Red fluorescent protein 160, 161
Red spotted newt 17, 19, 338, 340
Retinoid 173
Retroelements 6
Retrotransposition 6
Retroviral 6, 128–131, 133, 136
infections 128
Retrovirus 6, 127–139, 188
Rhombencephalon 34
Robust Multiarray Average (RMA) algorithm 314, 315
RNA-guided nucleases (RGNs) 280, 283, 290, 291, 294
RNeasy Mini Kit 310, 313–314
RSEM software 317

S

Saccule 253, 259–261, 264
Salamander embryos 147–155
Salamander-specific novelties 12
Salamandrella keyserlingii 8
Sal-site 291, 310, 311, 316, 321–335
Saponin 85, 256, 261
Schwann cells 148
Semi-circular canals 253, 260
Sensory epithelia 253–265
Short guide RNAs (sgRNAs) 280, 281, 285–287, 291–294
Silicone elastomer 117
Silver paint 256
Sirenids 5, 9
Sodium thioglycolate 298
Sorenson's buffer 207, 208
Spanish ribbed newt 47–67
Spermatophore 5, 18, 24, 36, 37, 40, 42, 48, 287, 292
S-phase entry 229, 242, 245
Spinal cord 19, 92, 115–124, 162, 164, 205, 269, 297, 310
Stable isotope labeling with amino acids in cell culture (SILAC) 340
Steinberg 149, 151–155
Stratum corneum 34
Streptavidin 207, 212, 213, 219, 231, 264
Styrofoam 40, 52, 63, 192
Sulfamerazine 75, 175, 243, 247, 249
SYBR green 208, 216
Sytox Blue 244, 246, 248, 250

T

TALEN. *See* Transcription activator-like effector nucleases (TALEN)
293T cells 128–133, 136, 138
TCH. *See* Thiocarbohydrazide (TCH)
Telencephalon 34, 35
Temnospondyls 10
Tetrapods 7–12, 30, 31, 36
Tetrodotoxin 23
TH4B cells 173
Thiocarbohydrazide (TCH) 255, 260, 261
Thymidine 230, 235
Thyroid hormone 29, 141, 310
Thyroid stimulating hormone (TSH) 141
Thyroxine 141–145
Tissue primordia 32
T7 mMessage mMachine kit 281, 286
Tol2 transposase 94, 270, 272, 273, 276
Tol2 transposon 94, 116
Trabeculation 242
Transcript 6, 135, 279, 316, 317, 324–326, 331, 339, 341–350
Transcription activator-like effector nucleases (TALEN) 280, 299, 301, 302, 304
Transcriptome 6, 280, 309, 316, 317, 323–326, 337, 338, 340, 348–350
Transdifferentiation 81–89, 254

Transgenesis 47, 116, 127, 269–277, 297, 299
Triazol 310, 312, 313
Tricaine...........53, 62, 64, 104, 134, 139, 149, 175, 178, 207,
209–211, 217, 223, 229, 232, 243, 247, 249, 282, 288
Triiodothyronine (T_3) 141
Trinity platform...316
Tropomyosin...32
Trypsin 177, 179, 180, 189, 191, 232, 236
Tubifex.. 22, 23, 52, 56, 60, 66, 67
Tungsten............................. 94, 104, 110, 111, 149–153, 155
TurboDNase...286
Tweezer electrode 118, 121

U

Ultracentrifuge 130, 132, 137
UniProt.. 340, 341, 345, 347
3′ Untranslated region..159
Uranyl acetate... 257, 262, 264
Urodeles.. 12, 17, 29, 31–33,
47, 49, 67, 101, 141, 171, 187, 205, 254, 297, 304

V

Vimentin 243, 245–248
Viral vectors... 116, 128, 188

W

Winged infusion set 208, 209, 217, 218, 225

Z

Zeugopodial elements 11
Zinc finger nucleases (ZFNs)..280

MIX
Papier aus verantwortungsvollen Quellen
Paper from responsible sources
FSC® C105338

If you have any concerns about our products,
you can contact us on
ProductSafety@springernature.com

In case Publisher is established outside the EU,
the EU authorized representative is:
Springer Nature Customer Service Center GmbH
Europaplatz 3, 69115 Heidelberg, Germany

Printed by Libri Plureos GmbH
in Hamburg, Germany